Banach spaces for analysts

Already published

1   W.M.L. Holcombe *Algebraic automata theory*
2   K. Petersen *Ergodic theory*
3   P.T. Johnstone *Stone spaces*
4   W.H. Schikhof *Ultrametric calculus*
5   J.-P. Kahane *Some random series of functions, 2nd edition*
6   H. Cohn *Introduction to the construction of class fields*
7   J. Lambek & P.J. Scott *Introduction to higher-order categorical logic*
8   H. Matsumura *Commutative ring theory*
9   C.B. Thomas *Characteristic classes and the cohomology of finite groups*
10 M. Aschbacher *Finite group theory*
11 J.L. Alperin *Local representation theory*
12 P. Koosis *The logarithmic integral I*
13 A. Pietsch *Eigenvalues and s-numbers*
14 S.J. Patterson *An introduction to the theory of the Riemann zeta-function*
15 H.J. Baues *Algebraic homotopy*
16 V.S. Varadarajan *Introduction to harmonic analysis on semisimple Lie groups*
17 W. Dicks & M. Dunwoody *Groups acting on graphs*
18 L.J. Corwin & F.P. Greenleaf *Representations of nilpotent Lie groups and their applications*
19 R. Fritsch & R. Piccinini *Cellular structures in topology*
20 H Klingen *Introductory lectures on Siegel modular forms*
21 P. Koosis *The logarithmic integral II*
22 M.J. Collins *Representations and characters of finite groups*
24 H. Kunita *Stochastic flows and stochastic differential equations*
25 P. Wojtaszczyk *Banach spaces for analysts*
26 J.E. Gilbert & M.A.M. Murray *Clifford algebras and Dirac operators in harmonic analysis*
27 A. Frohlich & M.J. Taylor *Algebraic number theory*
28 K. Goebel & W.A. Kirk *Topics in metric fixed point theory*
29 J.F. Humphreys *Reflection groups and Coxeter groups*
30 D.J. Benson *Representations and cohomology I*
31 D.J. Benson *Representations and cohomology II*
32 C. Allday & V. Puppe *Cohomological methods in transformation groups*
33 C. Soulé et al *Lectures on Arakelov geometry*
34 A. Ambrosetti & G. Prodi *A primer of nonlinear analysis*
35 J. Palis & F. Takens *Hyperbolicity and sensitive chaotic dynamics at homoclinic bifurcations*
36 M. Auslander, I. Reiten & S. Smalo *Representation theory of Artin algebras*
37 Y. Meyer *Wavelets and operators*
38 C. Weibel *An introduction to homological algebra*
39 W. Bruns & J. Herzog *Cohen-Macaulay rings*
40 V. Snaith *Explicit Brauer induction*
41 G. Laumon *Cohomology of Drinfield modular varieties I*
42 E.B. Davies *Spectral theory and differential operators*
43 J. Diestel, H. Jarchow & A. Tonge *Absolutely summing operators*
44 P. Mattila *Geometry of sets and measures in Euclidean spaces*
45 R. Pinsky *Positive harmonic functions and diffusion*
46 G. Tenenbaum *Introduction to analytic and probabilistic number theory*
50 I. Porteous *Clifford algebras and the classical groups*

Banach Spaces
For Analysts

P. Wojtaszczyk

Institute of Mathematics, Polish Academy of Sciences

CAMBRIDGE
UNIVERSITY PRESS

CAMBRIDGE UNIVERSITY PRESS
Cambridge, New York, Melbourne, Madrid, Cape Town, Singapore, São Paulo, Delhi

Cambridge University Press
The Edinburgh Building, Cambridge CB2 8RU, UK

Published in the United States of America by Cambridge University Press, New York

www.cambridge.org
Information on this title: www.cambridge.org/9780521566759

First published 1991
Reprinted 1996
First paperback edition 1996

A catalogue record for this publication is available from the British Library

ISBN 978-0-521-35618-3 hardback
ISBN 978-0-521-56675-9 paperback

Transferred to digital printing 2009

Contents

Preface

Banach space theory became a recognised part of the mathematical scene with the appearance of Banach [1932]. From its birth it maintained close ties with the rest of analysis. It turned out that Banach space theory offered powerful tools to other branches of analysis. The most useful ones are duality theory for spaces and operators, results about infinite dimensional convexity and results connected with the Baire category theorem, most notably the closed graph theorem. These powerful general concepts are now well known among analysts and appear in almost every textbook on functional analysis or real variable theory. They were already well understood in the forties and fifties, and at that time it seemed to many that Banach space theory was dead or at least relegated to an obscure corner of science. However, this was not the case. The sixties, and especially the seventies and eighties, saw an enormous eruption of activity in the field. Old problems were solved and, more importantly, new problems and new ties with the rest of mathematics emerged. Also, new and powerful methods and directions of research appeared.

For those in the field progress seemed to be constantly accelerating, but to those outside it may have looked as if the theory was dying again. Probably one of the reasons for this perception was that this research activity (maybe because it was so dynamic) unfortunately did not produce many books on the subject, and those that did appear were usually devoted to a special topic. There are two notable exceptions to this statement, Beauzamy [1982] and Lindenstrauss-Tzafriri [1977, 1979]. There are old favourites, still beautiful and in good shape like Banach [1932], Dunford-Schwartz [1958] and Day [1958 and newer versions], but naturally they do not present the more recent results. So the newcomer to the field, after learning what was generally known thirty years ago, had a choice of either starting on more specialized books or turning to Beauzamy or Lindenstrauss-Tzafriri. Now I like both these books and the present one is not intended to replace either of them. Beauzamy is a nice, easily readable introduction and Lindenstrauss-Tzafriri, although difficult in places for the novice, has every mark of a classic, especially if the long-promised volumes III and IV are added to the first two. But both these books have one feature in common: they study Banach spaces for their own sake and their own beauty. In this way they are great for the future specialist who is already under the spell of Banach spaces. However, there are some mathematicians (even a majority) who are not convinced that Banach space theory is the most enchanting branch of mathematics, and although I am not one of them, I understand them.

In fact, Banach spaces are not only beautiful and interesting, but also useful. This point is not, in my opinion, made clear in Lindenstrauss-Tzafriri or Beauzamy. The methods created in Banach space theory since the late 60's can be applied in other areas of analysis. Hence somebody interested primarily in harmonic analysis, functions of a complex variable, orthonormal series, approximation theory or probability theory can find it useful to know some Banach space theory. This book is directed towards such a person. Ideally I think of it as a textbook for a graduate course for students who have already learned some functional analysis and are interested in analysis or in some area of it. I also hope that a mature analyst may read it, or some of it, as part of the ongoing education process which is an important part of the life of any active mathematician. I hope too that the Banach space specialist will find some portions of the book interesting because they present some applications of the subject he was not aware of.

Let me digress a bit and comment on the possible profit a classical analyst might derive from Banach space theory. I do not claim that Banach space theory can solve all his important problems. But it may help him to see the problem in a new light which makes it easier to isolate essential features. He can also use general theorems and techniques in the solution. The general framework can also suggest interesting new problems. To mention one example, the power of duality methods makes it a standard procedure to try to find the dual of any Banach space considered, a problem that could not even arise without this more general framework. Another example is the corona theorem. It is now a major part of the theory of analytic functions, but its origin lies in the attempt to get some understanding of the maximal ideal space H_∞, a question which is unthinkable without the general theory of Banach algebras.

Mathematics and each of its parts lives and grows on the exchange of ideas, between mathematicians, between various branches of mathematics and between mathematics and other areas of human activity. Banach space theory is no exception to this rule. It utilises ideas and techniques from other fields and in turn provides other branches of mathematics with some of its own. In this respect this book is one-sided. It concentrates heavily on ideas and results that have a clear potential to be useful in other branches of analysis.

Formally the book is divided into three parts, numbered by roman numerals. Each of these parts is divided into chapters, distinguished by capital letters. Each chapter is divided into sections with arabic numbers. Each such section contains at most one Theorem, Proposition or Lemma. The Theorem appearing in section II.B.7 is later referred to as Theorem II.B.7, or within the same chapter just as Theorem 7 (or Lemma or Proposition as the case may be). Part I is of introductory character. It contains well known and some less well known results (without proofs) that will be used later. Chapter I.A contains basic results from general functional analysis and Chapter I.B discusses the examples of Banach spaces that are considered later and quotes some analytical results about these spaces. The main function of Part I is to establish notation and conventions and to serve as a refresher and reference for the background material needed later on. Part II is essentially an introduction to the language and basic techniques of Banach space theory. The real heart of the matter is Part III, where a selection of topics is studied in depth. The reader can get an idea of the contents of each chapter from the Table of Contents, so I will not attempt a more detailed description here. Also, each chapter of Parts II and III starts with a short description of its contents, so the reader can find additional details there.

Each chapter concludes with Notes and Remarks containing bibliographical data and comments about generalizations, extensions or applications of the results presented in the chapter. I have tried really hard to find the correct reference and credits. On the other hand, I have not conducted a full scale historical investigation into the origin and development of each idea and result. *I would like to offer sincere apologies for any omissions and inaccuracies in this respect.* In the main text theorems are only given names if it is common usage. The absence of a name in the text or of a credit in the Notes and Remarks does not mean that the result is due to the author. It means either that the result is so well known that I judged it to be folklore or simply an omission on my part.

Each chapter of Parts I and II contains exercises. These exercises range from routine to very hard and I have not given any indication of their difficulty. There is a hint for each exercise located in the special chapter at the end of the book. These hints range from almost complete solutions to the reference only. I have tried to point out, if possible, where the solution of an exercise can be found in the literature and to give proper credit. There are also some repetitions in the exercises. I have simply put the same or similar problems into different chapters if

they fit well into the material. This should be useful for those read-
ers (the majority?) who read only selected parts of the book. It also
indicates different approaches to the same question.

The bibliography contains only the works actually referred to some-
where in the book. I have made no attempt to make it complete. Also, it
does not include any data about translations (this is particularly important with respect to publications in Russian) or reprints and republishing
(it happens sometimes that an East European book is published origi-
nally in English and later republished without any changes by a West
European or American company). If the work appeared in Russian, this
is indicated in the references and the author's English translation of
the title is given. This translation should be close to the one used in
Mathematical Reviews, but need not be identical.

As with most mathematics books, it will be unusual for a reader to
read this book from beginning to end. It is also unnecessary. The reader
interested in a particular theorem or chapter should start right there.

The choice of material in the book reflects my philosophy, taste and,
last but not least, knowledge and ability. Here I would like to mention
some subjects which really should have been included but which were
not because of limitations of space and time and (probably most important) my poor understanding of them. The first is the deep connections
between Banach spaces, descriptive set theory and the classical theory
of sets of uniqueness for trigonometric series (see Kechris-Louveau [1987]
and Lyons [1989]). The second is the applications of the study of finite
dimensional spaces to problems on convex bodies. This in turn has applications to harmonic analysis, number theory and other subjects. This
whole area is currently one of frantic activity and is undergoing constant
and fascinating changes. Probably anything I could write about it now
would be outdated by the time this book reaches the reader. As an introduction to this area I suggest Pisier [P], Milman-Schechtman [1986]
or Milman [1986]. The next subject which I regrettably had to omit
is the connection between Banach space theory and probability theory.
Actually probabilistic methods underly much of the current research in
Banach spaces. This shows even in this book despite my unfortunate
lack of knowledge of probability theory. The study of probability in Banach spaces is developing too: see Linde [1983]. The last subject I would
like to have included is the general area of vector valued functions and
operators acting on such functions. There is considerable activity in this
area as well. As an introduction to it I suggest Burkholder [1986] and
Pisier [P1].

While writing the book I received helpful advice from many mathematicians. I am grateful to all of them for their time, advice and help. First of all I would like to express my profound gratitude to my teacher and colleague, Prof. Aleksander Pełczyński. He convinced me that I should write the book in the first place and offered plenty of advice on what it should contain. Much of the time I did not follow his advice, but the effort needed to refute his arguments helped greatly to clarify my own ideas. I would like to thank Prof. Keith Ball and Prof. Ben Garling for reading large parts of the manuscript and providing numerous and invaluable pieces of advice on language and presentation. The following other mathematicians helped me greatly by generously offering their advice, knowledge and insight: Dan Amir, Sheldon Axler, Don Burkholder, Joe Cima, John Fournier, Ben Garling, Nassif Ghossoub, Yehoram Gordon, Paweł Hitczenko, Bill Johnson, Serguey Kislyakov, Boris Kashin, Stanisław Kwapień, Elton Lacey, Joram Lindenstrauss, Niels Nielsen, Gilles Pisier, Richard Rochberg and Walter Schachermayer.

Some work on the book was done during visits to St. John's College Cambridge; Monash University, Clayton; University of British Columbia, Vancouver; Odense University and Texas A&M University, College Station. I would like to thank Ben Garling, Ala Sterna-Karwat, Nassif Ghossoub, Niels Nielsen and Elton Lacey for arranging these visits and for being great hosts, each in his or her unique way. Clearly most of the actual work was done in Warsaw where my own Mathematical Institute of the Polish Academy of Sciences gave me the freedom to do the job and an excellent library to help me. The final version was written, and all typing done, at Texas A&M University. I would like sincerely to thank Robin Campbell for her unique ability to transform my scratchy and unreadable (sometimes even to myself) handwriting into the beautifully typeset text.

Finally I would like to express my deep gratitude to my wife Anna and my children Ola and Kuba for all the love, support and distractions they generously provided over the years. Without their presence (and at suitable times their absence!) writing this book would have been much more difficult.

Part I

Introduction

This introductory part contains background material. It is not intended to be a course in any subject. It is simply a collection of definitions and facts given without proof (with one exception). We provide references to works which contain detailed exposition, full proofs, examples and motivation. In a sense this whole part is a quick reference guide to results which will be used in the later parts. The introduction is divided into two chapters, Chapter I.A describing what we will need from general functional analysis and Chapter I.B which contains results about concrete spaces and operators. Since I.A is really a review of a standard course in functional analysis, references are given only at the end of the chapter. In I.B we give references after each paragraph.

The references given in this part are usually to standard textbooks and monographs, not to original works. If a particular result or subject cannot be easily located using the table of contents or index we try to provide more detailed information (sections or pages). Sometimes we formulate a result in a form which is more convenient to us but different from the one given in the reference. Usually in such a case it is easy to derive our formulation from the one given in the reference.

I.A. Functional analysis

1. A *linear topological space* X is a linear space over the real or complex numbers endowed with a topology τ such that the map $(x, y) \mapsto x + y$ is continuous from $(X, \tau) \times (X, \tau)$ into (X, τ) and the map $(t, x) \mapsto tx$ is continuous from $\mathbb{R} \times X$ (or $\mathbb{C} \times X$) into X. Such a topology is fully described by a basis of neighbourhoods of 0. A subset $V \subset X$ is called *convex* if whenever $x_1, x_2 \in V$ then the whole interval $\alpha x_1 + (1 - \alpha)x_2$ for $0 \le \alpha \le 1$ is in V. A linear topological space is called *locally convex* if it has a basis of convex neighbourhoods of 0. A *functional* on X is a continuous linear map from X into scalars. The set of all functionals on X will be denoted X^*, and called the *dual space*. A *linear operator* (or just *operator*) $T: X \to Y$ (where X and Y are linear topological spaces) is a continuous linear map. A *subspace* of X will always (unless explicitly stated otherwise) denote a closed linear subspace. Given a set $V \subset X$ by span$V$ we denote the closure of the set of all linear combinations of elements from V (i.e. the subspace of X spanned by V).

2. A linear topological space X is called an *F-space* if its topology is given by a metric ρ such that $\rho(x, y) = \rho(x - y, 0)$ and X is complete with respect to this metric. A *quasi-norm* on a linear space X is a function q from X into the nonnegative reals satisfying

(a) $q(x) = 0$ if and only if $x = 0$,

(b) $q(\lambda x) = |\lambda| q(x)$ for all scalars λ and all $x \in X$,

(c) there exists a constant C_X such that $q(x + y) \le C_X(q(x) + q(y))$ for all $x, y \in X$.

One easily checks that each quasi-norm defines a linear topology on X. The basis of neighbourhoods of the point x consists of 'balls' around x, i.e. sets $\{y \in X : q(x - y) < \varepsilon\}, \varepsilon > 0$. A linear space X with a quasi-norm will be called a *quasi-normed space*. A very important special case of quasi-norm is a *norm*. This is a quasi-norm on X for which the constant C_X in (c) above equals 1 (one gets from (b) that $C_X \ge 1$). The usual notation for the norm of x is $\|x\|$. Every norm $\|\cdot\|$ on X defines a metric $\rho(x, y) = \|x - y\|$. A *Banach space* X is a linear space equipped with a norm $\|\cdot\|$ and such that X is complete with respect to the metric ρ. Clearly every Banach space is an F-space. The symbol B_X will denote the closed unit ball of X, i.e. $B_X = \{x \in X : \|x\| \leq 1\}$. Its interior is an open convex set, so every Banach space is locally convex. This need not be true for general quasi-norms. Note also that a subspace of a Banach space (resp. quasi-normed space) is a Banach space (resp. quasi-normed space). We simply have to restrict the norm (resp. quasi-norm).

3.   Let X and Y be two Banach spaces and let $T : X \to Y$ be a linear map. Then T is continuous if and only if $\|T\| = \sup\{\|Tx\| : \|x\| \leq 1, x \in X\} < \infty$. The quantity $\|T\|$ is a norm on the linear space $L(X, Y)$ of all operators from X into Y. The space $L(X, Y)$ with the above defined norm is a Banach space. Unless otherwise indicated the convergence of operators will be understood in this norm.

4.   In particular, for a Banach space X the space X^* is also a Banach space with the norm of a functional $x^* \in X^*$ defined as $\|x^*\| = \sup\{|x^*(x)| : x \in X, \|x\| \leq 1\}$.

5. Open Mapping Theorem.   Let X and Y be Banach spaces and let $T : X \to Y$ be a linear operator such that $\overline{T(B_X)}$ contains some open ball in Y. Then $T(X) = Y$ and there exists a positive number r such that $T(B_X) \supset r \cdot B_Y = \{y \in Y : \|y\| < r\}$.

6. Closed Graph Theorem.   Suppose that $T : X \to Y$ is a linear map (not assumed to be continuous, but defined everywhere on X) from an F-space X into an F-space Y. Assume that $\{(x, Tx) : x \in X\} \subset X \times Y$ is closed in the product topology. Then T is continuous.

7. Banach-Steinhaus Theorem.   Suppose $(T_\gamma)_{\gamma \in \Gamma}$ is a family of linear operators from a Banach space X into a Banach space Y. Assume that for every $x \in X$ we have $\sup\{\|T_\gamma x\| : \gamma \in \Gamma\} < \infty$. Then $\sup\{\|T_\gamma\| : \gamma \in \Gamma\} < \infty$.

In particular we get that the pointwise limit of a sequence of linear operators (if it exists everywhere) is a linear operator.

8. Hahn-Banach Theorem.   Let X be a linear space over the real numbers (without any topology) and let $Y \subset X$ be a linear subspace. Assume also that we have a function $p : X \to \mathbb{R}$ such that $p(x + y) \leq p(x) + p(y)$ for all $x, y \in X$ and $p(tx) = tp(x)$ for all $x \in X$ and $t \in \mathbb{R}, t \geq 0$. Assume moreover that we have a linear map $f : Y \to \mathbb{R}$ such that $f(y) \leq p(y)$ for all $y \in Y$. Then there exists a linear map $F : X \to \mathbb{R}$ such that $F \mid Y = f$ and $-p(-x) \leq F(x) \leq p(x)$ for all $x \in X$.

This general algebraic theorem has many very important special cases. Some of them are stated below in **9-11**. Note that despite the fact that **8** is true only for real spaces, the consequences listed below are true also for spaces over the complex scalars.

9. Taking in **8** $p(y) = \|y\|$ we obtain:
If X is a Banach space and $Y \subset X$ is a subspace and if $y^* \in Y^*$, then there exists $x^* \in X^*$ such that $\|x^*\| = \|y^*\|$ and $x^* \mid Y = y^*$. In particular we get

$$\|x\| = \sup\{|x^*(x)|: x^* \in X^*, \|x^*\| \leq 1\}.$$

10. A judicious choice of p yields also the following:
If X is a locally convex space and $A, B \subset X$ are disjoint closed convex sets with A being compact, then there exists a continuous linear functional f on X and a real number α such that $Re\ f(A) < \alpha$ and $Re\ f(B) > \alpha$.
In particular we see that X^* separates the points of X.

11. We also have the following version of **10**.
If X is a locally convex space and $A, B \subset X$ are disjoint convex sets with A open, then there exists a continuous linear functional f on X and a real number α such that $Re\ f(A) < \alpha$ and $Re\ f(B) \geq \alpha$.

12. If $T: X \to Y$ is a linear operator then it induces an operator $T^*: Y^* \to X^*$, called the *adjoint* (or *dual*) operator and defined as $T^*(y^*)(x) = y^*(Tx)$. One easily checks that $\|T\| = \|T^*\|$.

13. If $T: X \to Y$ is onto then T^* is 1-1 and there exists a constant $c > 0$ such that $\|T^*y^*\| \geq c\|y^*\|$ for all $y^* \in Y^*$.

14. If $T: X \to Y$ is such that there exists $c > 0$ such that $\|Tx\| \geq c\|x\|$ for all $x \in X$ then T^* maps Y^* onto X^*.

15. A linear map $T: X \to Y$ (X, Y Banach spaces) is *compact* if the set $\overline{T(B_X)}$ is compact in the norm topology of Y. One easily checks that each compact map is automatically continuous. The set of all compact operators from X into Y, denoted $K(X,Y)$, is a subspace (i.e. closed and linear) of $L(X,Y)$. Also an operator $T: X \to Y$ is compact if and only if $T^*: Y^* \to X^*$ is compact. It is easy to see that if $T \in K(X,Y)$ and $S_1 \in L(Z_1, X)$ and $S_2 \in L(Y, Z_2)$ then $S_2TS_1 \in K(Z_1, Z_2)$.

16. An operator $T: X \to X$ is called *power-compact* if T^n is compact for some $n \in \mathbb{N}$.

17. An operator $T: X \to Y$ is called *invertible* if there exists an operator $S: Y \to X$ (usually we write T^{-1} instead of S) such that $ST = id_X$ and $TS = id_Y$. By id_X (or id_Y) we mean the identity operator on X (or on Y), i.e. $id_X(x) = x$ for all $x \in X$. It is important that we need both conditions $ST = id_X$ and $TS = id_Y$; one of them is not enough.

18. If X is a complex Banach space (i.e. it is a linear space over the complex numbers) and $T: X \to X$ is a linear operator, then the *spectrum* of T, denoted $\sigma(T)$, is the set of all $\lambda \in \mathbb{C}$ such that $(\lambda id_X - T)$ is not an invertible operator. The set $\sigma(T)$ is a non-empty, compact subset of $\mathbb{C}$. A number $\lambda \in \mathbb{C}$ is called an *eigenvalue* of T if there exists a vector $x \in X, x \neq 0$, called an *eigenvector* associated with the eigenvalue λ, such that $Tx = \lambda x$. Clearly each eigenvalue of T belongs to $\sigma(T)$. With each eigenvalue λ we associate its *spectral manifold* $E_\lambda = E_\lambda(T) = \bigcup_{n \geq 1} \ker(\lambda id_X - T)^n$.

Clearly E_λ is an increasing union of subspaces of X. The number $\dim E_\lambda$ (possibly ∞) is called the *multiplicity* of the eigenvalue λ.

19. If an operator $T: X \to Y$ is power-compact then $\sigma(T)$ is finite or consists of a sequence of points tending to zero together with zero itself. Every point $\lambda \in \sigma(T)$, except possibly zero, is an eigenvalue of T of finite multiplicity.

20. An operator $P: X \to X$ is called a *projection* if $P^2 = P$. Then $P(X)$ is a closed subspace of X and $Px = x$ for $x \in P(X)$. A subspace $Y \subset X$ which equals $P(X)$ for some projection $P: X \to X$ is called *complemented*.

21. Let V be a convex set in a locally convex topological vector space X. A point $v \in V$ is called an *extreme point* if it is not in the interior of any closed interval contained in V, i.e. if $v_1, v_2 \in V$ and $v = \alpha v_1 + (1-\alpha)v_2$ with $0 < \alpha < 1$ then $v = v_1 = v_2$. If A is any subset of X then the convex hull of A, denoted conv A, equals

$$\left\{ x \in X: x = \sum_{j=1}^{n} \alpha_j a_j \text{ with } \sum_{j=1}^{n} \alpha_j = \sum_{j=1}^{n} |\alpha_j| = 1 \right.$$

$$\left. \text{and } a_j \in A \text{ for } j = 1, 2, \ldots, n, \ n \in \mathbb{N} \right\}.$$

22. Krein-Milman Theorem.   If V is a compact, convex subset of a locally convex, topological vector space, then V equals the closure of the convex hull of its extreme points.

In particular this theorem implies that each convex, compact subset of a locally convex, topological vector space has an extreme point.

References.   Everything said above can be found in most textbooks on functional analysis. In particular everything can be found in Dunford-Schwartz [1958] or Edwards [1965] and everything except power-compact operators can be found in Rudin [1973].

I.B. Examples of spaces and operators

1. Whenever we consider a measure space (Ω, Σ, μ) we assume that the measure μ is complete and that there are no atoms of infinite measure. We say that the measure space (Ω, Σ, μ) is *separable* if there exists a countable family of sets $(A_j)_{j=1}^\infty \subset \Sigma$ such that the smallest complete σ-field containing $(A_j)_{j=1}^\infty$ is Σ. The following characterizes separable measure spaces.

Suppose that (Ω, Σ, μ) is a separable non-atomic measure space, with μ a positive measure and $\mu(\Omega) = 1$ (i.e. μ is a probability measure). Then (Ω, Σ, μ) is isomorphic to the unit interval $[0,1]$ with the Lebesgue measure.

This can be found in Halmos [1950] §41. A general classification theorem for arbitrary measure spaces is due to Maharam [1942] and can also be found in Lacey [1974] §14.

2. For any measure space (Ω, Σ, μ) with μ positive we define $L_p(\Omega, \Sigma, \mu)$, $0 < p \leq \infty$, to be the space of Σ-measurable functions (more precisely of classes of functions where we identify functions which are equal μ-a.e.) such that $\int_\Omega |f(\omega)|^p d\mu(\omega) < \infty$. For $p = \infty$ we mean supess $|f(\omega)| < \infty$. Usually the σ-field Σ is clear so we will use the notation $L_p(\Omega, \mu)$ and when either the measure or the set are clear from the context we will suppress them also. We will use the notation $\|f\|_p = (\int_\Omega |f(\omega)|^p d\mu(\omega))^{\frac{1}{p}}$ for $0 < p < \infty$ and $\|f\|_\infty = \text{supess}|f(\omega)|$. If $1 \leq p \leq \infty$ then $\|\cdot\|_p$ is a norm and for $0 < p < 1$ it is a quasi-norm and it satisfies the inequality

$$\|f + g\|_p \leq (\|f\|_p^p + \|g\|_p^p)^{\frac{1}{p}}.$$

Thus $L_p(\Omega, \mu)$, $0 < p \leq \infty$, are linear spaces. For $1 \leq p \leq \infty$ they are Banach spaces. For $0 < p < 1$ we can introduce the natural metric on $L_p(\Omega, \mu)$ by $\rho(f,g) = \|f - g\|_p^p$. $L_p(\Omega, \mu)$ is complete with respect to this metric and the topology induced by this metric makes $L_p(\Omega, \mu)$ an F-space. It is easy to see that $L_p(\Omega, \mu)$ is not locally convex if $0 < p < 1$. By $L_0(\Omega, \mu)$ we mean the space of all (classes of) measurable, almost everywhere finite functions on Ω with the topology of convergence in measure. If $\mu(\Omega)$ is finite, this topology can be given by the metric $\rho(f,g) = \int_\Omega (\frac{|f-g|}{(1+|f-g|)})d\mu$. When (Ω, μ) is σ-finite we

write $\Omega = \bigcup_{j=1}^{\infty} E_j$ with the E_j's disjoint and $\mu(E_j) < \infty$ and define
$\rho(f,g) = \sum_{j=1}^{\infty} 2^{-j} \mu(E_j)^{-1} \int_{E_j} (\frac{|f-g|}{(1+|f-g|)}) d\mu$. Clearly $L_0(\Omega, \mu)$ (at least
for a σ-finite measure space) is an F-space.

All this can be found in Dunford-Schwartz [1958], Edwards [1965],
Banach [1932] and in most textbooks on functional analysis.

3.   For a number p, $1 \le p \le \infty$ we will denote by p' the *conjugate
exponent*, i.e. p' satisfies $\frac{1}{p} + \frac{1}{p'} = 1$ (so clearly $p' = \frac{p}{(p-1)}$) with the convention $1' = \infty$ and $\infty' = 1$. The following inequality, called **Hölder's
inequality**, is of fundamental importance:

For any p, $1 \le p \le \infty$ and any two functions f, g we have

$$\int_\Omega |f(\omega) g(\omega)| d\mu(\omega) \le \left(\int_\Omega |f(\omega)|^p d\mu(\omega) \right)^{\frac{1}{p}} \left(\int_\Omega |g(\omega)|^{p'} d\mu(\omega) \right)^{\frac{1}{p'}}$$

or in short $\|f \cdot g\|_1 \le \|f\|_p \|g\|_{p'}$.

If this inequality is actually an equality then $g(\omega) = c \cdot \overline{\mathrm{sgn}(f(\omega))}$
$|f(\omega)|^{p-1}$ μ-a.e. where c is a constant and $\mathrm{sgn}(\alpha)$ is the complex signum
of α, i.e. $\mathrm{sgn}\,\alpha = \alpha/|\alpha|$.

The proof can be found in almost every textbook of functional analysis or measure theory, e.g. Dunford-Schwartz [1958], Edwards [1965].

4.   The duals of $L_p(\Omega, \mu)$ spaces for $1 \le p < \infty$ are described as follows (this is really a restatement of Hölder's inequality): $L_p(\Omega, \mu)^* =
L_{p'}(\Omega, \mu)$ and the duality is given as $g(f) = \int_\Omega g(\omega) f(\omega) d\mu(\omega)$ for
$g \in L_{p'}(\Omega, \mu)$ and $f \in L_p(\Omega, \mu)$. The space $L_\infty(\Omega, \mu)^*$ is (unless (Ω, μ)
consists of a finite number of atoms) much bigger than $L_1(\Omega, \mu)$ and is
impossible to describe explicitly. For $p < 1$ the dual depends on the
measure space. If (Ω, μ) has no atoms then 0 is the only continuous
linear functional on $L_p(\Omega, \mu)$. If (Ω, μ) is purely atomic and each atom
has measure 1 then $L_p(\Omega, \mu)^* = L_\infty(\Omega, \mu)$ for $0 < p < 1$.

All this can be found in Dunford-Schwartz [1958] or Edwards [1965]
or in most textbooks on functional analysis.

5.   The following notational conventions will be used throughout this
book. When the measure on the set Ω and the σ-field are clear from the
context, we will use the notation $L_p(\Omega)$, in particular $L_p(\mathbb{R})$ and $L_p[0, 1]$
will mean that we consider the usual Lebesgue measure and $L_p(\mathbb{T})$ will
be with respect to Lebesgue measure on $\mathbb{T}$, but normalized to make
the measure of the whole circle 1. The Lebesgue measure of a set A
is denoted by $|A|$. If the measure space consists of a certain number

of atoms each having measure 1, then $L_p(\Omega)$ will be denoted $\ell_p(\Omega)$. If those atoms are naturally identified with the integers $\mathbb{Z}$ or the natural numbers $\mathbb{N}$ we will use the notation ℓ_p. If the measure space Ω consists of n atoms each of measure 1, then $\ell_p(\Omega)$ will be denoted by ℓ_p^n. All this applies to $0 < p \le \infty$.

6.   The spaces $L_p(\Omega, \mu)$ for various values of p are clearly related. One of the expressions of this relationship are interpolation theorems. In this book we will use only the simplest cases of the best known results.

Riesz-Thorin Theorem.   Let (Ω, μ) be a measure space and let $B \subset L_{p_1}(\Omega, \mu) \cap L_{p_2}(\Omega, \mu)$ be a linear set which is dense in both $L_{p_i}(\Omega, \mu)$, $i = 1, 2$, with $1 \le p_1 < p_2 \le \infty$. Assume that we have a linear map T defined on B with values in the measurable functions on a measure space (Ω_1, μ_1). Assume also that for every $f \in B$ we have $\|Tf\|_{p_i} \le C\|f\|_{p_i}$ for $i = 1, 2$. Then for every p, $p_1 \le p \le p_2$ and $f \in B$ we have

$$\|Tf\|_p \le C\|f\|_p$$

so T extends to a continuous linear operator from $L_p(\Omega, \mu)$ to $L_p(\Omega_1, \mu_1)$.

The proof of this theorem can be found in Katznelson [1968], Zygmund [1968], Dunford-Schwartz [1958], or in books devoted to interpolation of operators like Bennet-Sharpley [1988] or Krein-Petunin-Semenow [1978].

7.   We say that a function f on a measure space (Ω, μ) is of *weak type* p, $0 < p < \infty$ and write $f \in L_{p,\infty}(\Omega, \mu)$ if there exists a constant K such that for every $\lambda \in \mathbb{R}_+$ we have

$$\mu\{\omega \in \Omega \colon |f(\omega)| \ge \lambda\} \le \left(\frac{K}{\lambda}\right)^p.$$

We denote the space of all functions of weak type p by $L_{p,\infty}$ and the inf of all constants K which can appear in the above inequality for a given f by $\|f\|_{p,\infty}$. This quantity is not a norm but it is a convenient notation.

If T is a linear map defined on a dense, linear subset $B \subset L_p(\Omega, \mu)$ with values in the measurable functions on (Ω_1, μ_1), then we say that T is of *weak type (p-p)* if there exists a constant K such that

$$\mu_1\{\omega_1 \in \Omega_1 \colon |Tf(\omega_1)| \ge \lambda\} \le K\left(\frac{\|f\|_p}{\lambda}\right)^p$$

for every $\lambda \in \mathbb{R}_+$ and $f \in B$. If $p = \infty$ we interpret this condition to mean $\|Tf\|_\infty \leq K\|f\|_\infty$.

Marcinkiewicz Theorem. Suppose that B is a linear set dense in both $L_{p_1}(\Omega, \mu)$ and $L_{p_2}(\Omega, \mu)$, $1 \leq p_1 < p_2 \leq \infty$ and suppose that T is a linear map defined on B with values in the measurable functions on a measure space (Ω_1, μ_1). If T is of weak type $(p_1 - p_1)$ and of weak type $(p_2 - p_2)$ then for every $p, p_1 < p < p_2$ there exists a constant C_p such that $\|Tf\|_p \leq C_p\|f\|_p$ for all $f \in B$.

So we get that T extends to a continuous operator from $L_p(\Omega, \mu)$ into $L_p(\Omega_1, \mu_1)$ for $p_1 < p < p_2$. For the constant C_p we have the estimate $C_p \leq K \max(\frac{1}{(p-p_1)}, \frac{1}{(p_2-p)})$. The proof can be found in Stein [1970], Zygmund [1968] or in any of the books on interpolation of operators mentioned in **6**.

8. The *Rademacher functions* $(r_n(t))_{n=1}^\infty$ are defined on [0,1] as $r_n(t) = \text{sgn} \, \sin 2^n t\pi$. The alternative description is that $r_1(t) = 1$ if $0 \leq t < \frac{1}{2}$ and $r_1(t) = -1$ if $\frac{1}{2} < t \leq 1$ and $r_{n+1}(t)$ takes value 1 on the left hand half of each interval where $r_n(t)$ is constant and takes value -1 on the right hand half of each such interval. Using probabilistic language we can say that $(r_n(t))_{n=1}^\infty$ is a sequence of independent random variables, each taking value 1 with probability $\frac{1}{2}$ and value -1 with probability $\frac{1}{2}$.

Whatever the description, we easily see that $(r_n(t))_{n=1}^\infty$ is an orthonormal system in $L_2[0, 1]$. Clearly this system is not complete. The main fact about Rademacher functions, which will be used repeatedly in this book is

Khintchine's inequality. There exist constants A_p, B_p, $0 < p < \infty$ such that for all (finite) sequences of scalars $(a_n)_{n=1}^\infty$ and every p, $0 < p < \infty$ we have

$$A_p \left\| \sum_{n\geq 1} a_n r_n \right\|_p \leq \left\| \sum_{n\geq 1} a_n r_n \right\|_2 = \left(\sum_{n\geq 1} |a_n|^2 \right)^{\frac{1}{2}} \leq B_p \left\| \sum_{n\geq 1} a_n r_n \right\|_p.$$

There is also a similar inequality for lacunary sequences of exponents. More precisely we have the following: If $(n_k)_{k=1}^\infty$ is a sequence of natural numbers such that $\inf_k (n_{k+1}/n_k) = \lambda > 1$ then there exist constants $A_p = A(p, \lambda)$ and $B_p = B(p, \lambda)$ for $0 < p < \infty$ such that for every sequence of scalars $(a_k)_{k=1}^\infty$ and every p, $0 < p < \infty$ we have

$$A_p \left\| \sum_{k=1}^\infty a_k e^{in_k\theta} \right\|_p \leq \left\| \sum_{k=1}^\infty a_k e^{in_k\theta} \right\|_2 \leq B_p \left\| \sum_{k=1}^\infty a_k e^{in_k\theta} \right\|_p.$$

This can be found in many places, e.g. Zygmund [1968] or Stein [1970].

9.   If K is a topological space, then by $C(K)$ we mean the space of all scalar valued (i.e. real or complex valued), bounded, continuous functions on K. This is a Banach space with the norm $\|f\|_\infty = \sup\{|f(k)| : k \in K\}$. Clearly if K is a compact space then $C(K)$ consists of all continuous, scalar valued functions. If the space K is locally compact then by $C_0(K)$ we denote the subspace of $C(K)$ of functions 'vanishing at infinity', i.e. such that for every $\varepsilon > 0$ there exists a compact set $V_\varepsilon \subset K$ such that $\sup\{|f(k)| : k \in K \backslash V_\varepsilon\} < \varepsilon$. If K has the discrete topology then $C(K)$ is denoted by $\ell_\infty(K)$ and if $K = \mathbb{N}$ then $\ell_\infty(\mathbb{N})$ is abbreviated to ℓ_∞. The symbol c_0 denotes $C_0(\mathbb{N})$, i.e. the space of all sequences converging to zero. If the set K happens to be of finite cardinality n then $C(K) = \ell_\infty(K)$ is also denoted ℓ_∞^n. The symbol c denotes the subspace of ℓ_∞ consisting of convergent sequences.

10.   If on a Banach space X we have also defined a multiplication (not necessary commutative) $x \cdot y$ in such a way that $x \cdot (z + y) = x \cdot z + x \cdot y, (z + y) \cdot x = z \cdot x + y \cdot x$ and $(\lambda x) \cdot y = x \cdot (\lambda y) = \lambda(x \cdot y)$ (λ is a scalar) and $\|x \cdot y\| \le \|x\|\|y\|$ then we have a *Banach algebra*. A *linear multiplicative functional* on such an algebra is a non-zero linear functional which also preserves the multiplication. The set of all linear multiplicative functionals on X is denoted $\mathcal{M}(X)$. If X is a commutative algebra with unit then $\mathcal{M}(X) \subset X^*$ is an ω^*-compact set (see II.A.9). There is a natural, norm-1 map $i \colon X \to C(\mathcal{M}(X))$ defined as $i(x)(\varphi) = \varphi(x)$ (see II.A.10). If this map is an isometry the algebra is called a *uniform algebra*. Thus a uniform algebra is a norm-closed subalgebra of the algebra $C(S)$. The following gives a criterion for this map to be onto.

Gelfand-Naimark Theorem.   Suppose that on a commutative, complex Banach algebra X with unit we have an involution $*$ such that $(x + y)^* = x^* + y^*, (x \cdot y)^* = y^* \cdot x^*, (\lambda x)^* = \bar{\lambda} x^*, x^{**} = x$ and $\|xx^*\| = \|x\|^2$. Then the map i is a linear multiplicative isometry from X onto $C(\mathcal{M}(X))$.

This theorem implies in particular that $L_\infty(\Omega, \mu)$ is isometric to a $C(K)$ for some compact space K. Also any space $C(K)$ as defined in **9** is isometric to a space $C(\tilde{K})$ for some compact space $\tilde{K}$.

This can be found in any book on Banach algebras, e.g. Żelazko [1973], and in many general texts on functional analysis, e.g. Rudin [1973].

11. Riesz Theorem. The dual of the space $C(K)$, K compact, equals the space of all regular, Borel measures (scalar valued, but obviously not necessarily positive) on K. The duality is defined as $\mu(f) = \int_K f d\mu$ and the norm of the measure as a functional on $C(K)$ equals its total variation.

The space of all Borel, regular measures with finite total variation on a space K will be denoted $M(K)$. So for compact K we have $C(K)^* = M(K)$. This is a classical theorem and can be found in most texts on measure theory and functional analysis, e.g. Dunford-Schwartz [1958] or Edwards [1965].

12. Stone-Weierstrass Theorem. Suppose that $A \subset C(K)$, K compact, is an algebra (i.e. sums, products and scalar multiples of functions from A are in A). Suppose also that if $f \in A$ then $\bar{f} \in A$ (this condition is empty if we consider real scalars) and that $1 \in A$. If A separates points of K (i.e. for every $k_1, k_2 \in K$, $k_1 \neq k_2$ there is an $f \in A$ such that $f(k_1) \neq f(k_2)$), then A is dense in $C(K)$.

If $1 \notin A$ but all the other assumptions are satisfied then $\bigcap_{f \in A} f^{-1}(0) = \{a\}$ for some $a \in K$ and the closure of A equals $\{f \in C(K): f(a) = 0\}$. If A does not separate points, then we can pass to the quotient space $K/\sim$ where $\sim$ is the equivalence relation on K defined by $k_1 \sim k_2$ if and only if $f(k_1) = f(k_2)$ for all $f \in A$. The space $K/\sim$ is compact and the algebra in $C(K/\sim)$ corresponding to A separates points of $K/\sim$. This can be found in many texts on functional analysis, topology or approximation theory. A detailed discussion can be found in Edwards [1965].

13. A *locally compact abelian group* G is an abelian group which at the same time is a locally compact space and such that all group operations are jointly continuous. On each such group there exists a unique (up to multiplication by a scalar) invariant, Borel measure, i.e. a Borel measure m such that $m(A) = m(A + g)$ for every Borel set A and every $g \in G$. This measure is called the *Haar measure*. Note that $A + g = \{h \in G: h = a + g$ with $a \in A\}$. If μ is any measure in $M(G)$ and f is a Borel function on G, then the *convolution* $f * \mu$ is defined to be the function $f * \mu(s) = \int_G f(s - t) d\mu(t)$, provided that this integral exists. If $f \in L_p(G, m)$, $1 \leq p \leq \infty$ then $f * \mu \in L_p(G, m)$ and $\|f * \mu\|_p \leq \|\mu\| \|f\|_p$. Also if $f \in C(G)$ then $f * \mu \in C(G)$. We can define the convolution of two functions f and g by the formula $f * g(s) = \int_G f(s - t) g(t) dm(t)$ if this integral exists.

If $g \in L_1(G, m)$ this definition clearly coincides with the previous one after we identify g with the measure gdm. If $f \in L_1(G, m)$ and $g \in L_\infty(G)$ then $f * g \in C(G)$. We can also define the convolution of two measures $\mu, \nu \in M(G)$ as the measure defined for $A \subset G$ as $\mu * \nu(A) = \int_G \mu(A - t)d\nu(t)$. One checks that (in all the above cases) the convolution is commutative and distributive, i.e. $f * (\lambda_1 g_1 + \lambda_2 g_2) = \lambda_1 f * g_1 + \lambda_2 f * g_2$ where λ_1, λ_2 are scalars. This means that every measure μ defines an operator $T_\mu : L_p(G) \to L_p(G)$, $1 \leq p \leq \infty$ and $T_\mu : C(G) \to C(G)$ defined by $T_\mu(f) = f * \mu$. We have $\|T_\mu\| \leq \|\mu\|$ in all cases. When we consider $T_\mu : L_1(G) \to L_1(G)$ or $T_\mu : C(G) \to C(G)$ then we actually have $\|T_\mu\| = \|\mu\|$. All this is basic harmonic analysis and can be found (in various degrees of generality) in any textbook on harmonic analysis, e.g. Rudin [1962a].

14.   If G is a compact abelian group, then a *character* of G is a continuous group homeomorphism from G into $\mathbb{T}$. The set of all characters of G is usually denoted by Γ. If we treat characters as complex valued functions on G then they form a complete orthonormal set in $L_2(G, m)$ where m is normalized so that $m(G) = 1$ (this is possible because the group is assumed to be compact). The characters are also linearly dense in $C(G)$. Given a measure $\mu \in M(G)$ (or a function $f \in L_1(G)$) we define its *Fourier coefficients*

$$\hat{\mu}(\gamma) = \int_G \bar{\gamma}d\mu \qquad \text{for} \qquad \gamma \in \Gamma.$$

The map $\hat{\ } : M(G) \to \ell_\infty(\Gamma)$ defined above is called the *Fourier transform*. Since characters are linearly dense in $C(G)$ this map is one-to-one, i.e. the Fourier coefficients determine the measure. One easily checks that $(\mu * \nu)\hat{\ }(\gamma) = \hat{\mu}(\gamma)\hat{\nu}(\gamma)$. If $\mu \in M(G)$ has Fourier coefficients $\hat{\mu}(\gamma)$ then we will write $\mu = \sum_{\gamma \in \Gamma} \hat{\mu}(\gamma) \cdot \gamma$ and call this series the Fourier series of μ even when we know nothing about the convergence.

On the circle group $\mathbb{T}$ the characters can be identified with the functions $(e^{in\theta})_{n \in \mathbb{Z}}$. We write $\hat{\mu}(n)$ for $\int_\mathbb{T} e^{-in\theta}d\mu(\theta)$.

Let us also recall here that a subset $H \subset \Gamma$ is called a $\Lambda(p)$ set, $p > 1$, if there exists a constant K such that

$$\left\| \sum_{\gamma \in H} a_\gamma \gamma \right\|_p \leq K \left\| \sum_{\gamma \in H} a_\gamma \gamma \right\|_1$$

for every sequence of scalars $(a_\gamma)_{\gamma \in H}$.

Let us also recall that a subset $H \subset \Gamma$ is called a Sidon set if there exists a constant $K > 0$ such that

$$\left\| \sum_{\gamma \in H} a_\gamma \gamma \right\|_\infty \geq K \sum_{\gamma \in H} |a_\gamma|$$

for every sequence of scalars $(a_\gamma)_{\gamma \in H}$.

The details can be found e.g. in Rudin [1962a].

15. The *Dirichlet kernel* is defined as

$$\mathcal{D}_n(\theta) = \sum_{k=-n}^{n} e^{ik\theta}, \qquad n = 0, 1, 2, \ldots .$$

If we have $\mu \in M(\mathbb{T})$ and $\mu = \sum_{-\infty}^{+\infty} \hat{\mu}(k)e^{ik\theta}$ then $\mathcal{D}_n * \mu = \sum_{-n}^{n} \hat{\mu}(k)e^{ik\theta}$. This means that the convolution with the Dirichlet kernel realizes the partial sum projection with respect to the Fourier series. We have

$$\int_{\mathbb{T}} |\mathcal{D}_n(\theta)| dm(\theta) = \frac{4}{\pi^2} \log(n+1) + o(1).$$

Note that this equals the norm of the operator $f \mapsto f * \mathcal{D}_n$ in the spaces $L_1(\mathbb{T})$ and $C(\mathbb{T})$. This operator will sometimes be called the *Dirichlet projection*.

For details see Zygmund [1968], Katznelson [1968] p. 50, Kashin-Saakian [1984] or any book on trigonometric series.

16. The *Fejér kernel* is defined as

$$\mathcal{F}_n(\theta) = \frac{1}{n+1} \sum_{k=0}^{n} \mathcal{D}_k(\theta) = \sum_{k=-n}^{n} \left(1 - \frac{|k|}{n+1}\right) e^{ik\theta}.$$

One checks that $\mathcal{F}_n(\theta) \geq 0$ for $n = 0, 1, 2, \ldots$ and so $\int_{\mathbb{T}} |\mathcal{F}_n(\theta)| dm(\theta) = \int_{\mathbb{T}} \mathcal{F}_n(\theta) dm(\theta) = 1$. If $f \in L_p(\mathbb{T})$, $1 \leq p < \infty$, or $f \in C(\mathbb{T})$ in the case $p = \infty$, then $\|f - f * \mathcal{F}_n\|_p \to 0$ as $n \to \infty$. The norm of the Fejér operator, $f \mapsto f * \mathcal{F}_n$, is 1 in spaces $L_p(\mathbb{T})$, $1 \leq p \leq \infty$. The details can be found in Zygmund [1968], Katznelson [1968] p. 12, Kashin-Saakian [1984] or in any book on trigonometric series.

17. The *de la Vallée-Poussin kernel* is defined as $\mathcal{V}_n = 2\mathcal{F}_{2n-1} - \mathcal{F}_{n-1}$. Clearly $\|\mathcal{V}_n\|_1 \leq 3$ and $\|f - \mathcal{V}_n * f\|_p \to 0$ as $n \to \infty$ for $f \in L_p(\mathbb{T})$, $1 \leq p < \infty$, or $f \in C(\mathbb{T})$. This follows immediately from **16**. The nice property of the de la Vallée-Poussin kernel is that for $f = \sum_{-n}^{n} a_j e^{ij\theta}$ we have $\mathcal{V}_n * f = f$. All this follows easily from properties of the Fejér kernels. The details are in Zygmund [1968] and Katznelson [1968].

18. The *Poisson kernel* is defined as

$$\mathcal{P}_r(\theta) = \sum_{-\infty}^{+\infty} r^{|j|} e^{ij\theta} = \frac{(1-r^2)}{(1-2r\cos\theta+r^2)} \quad \text{for } 0 < r < 1.$$

Clearly $\mathcal{P}_r(\theta) \geq 0$ and $\int_{\mathbb{T}} \mathcal{P}_r(\theta) dm(\theta) = 1$ for $0 < r < 1$. As for the Fejér kernel we have $\|f - f * \mathcal{P}_r\|_p \to 0$ as $r \to 1$ for $f \in L_p(\mathbb{T})$, $1 \leq p < \infty$, or $f \in C(\mathbb{T})$. The important feature of the Poisson kernel is that for $\mu \in M(\mathbb{T})$, $P_r * \mu$ can be treated as a function on $\mathbb{D}$ by $h(re^{i\theta}) = \mathcal{P}_r * \mu(\theta)$. Such a function is always harmonic in $\mathbb{D}$.

Basic properties of the Poisson kernel can be found in Zygmund [1968], Katznelson [1968], Hoffmann [1962], Duren [1970], Koosis [1980] and in many other places.

19. The *Hardy spaces* $H_p(\mathbb{D})$, $0 < p \leq \infty$, are defined as

$$H_p(\mathbb{D}) = \left\{ f(z): f(z) \text{ is analytic in } \mathbb{D} \text{ and} \right.$$

$$\left. \|f\|_{H_p} = \sup_{r<1} \left(\frac{1}{2\pi} \int_0^{2\pi} |f(re^{i\theta})|^p d\theta \right)^{\frac{1}{p}} < \infty \right\}.$$

For $0 < p < 1$ the Hardy space $H_p(\mathbb{D})$ is a complete, linear metric space and for $1 \leq p \leq \infty$ it is a Banach space. The limit $\lim_{r\to 1} f(re^{i\theta}) = f(e^{i\theta})$, called the boundary value of f, exists for almost all $\theta \in \mathbb{T}$, and $\|f\|_{H_p} = \|f\|_{L_p(\mathbb{T})}$. This shows that we can identify $H_p(\mathbb{D})$ with $\overline{\text{span}}\{e^{in\theta}\}_{n\geq 0} \subset L_p(\mathbb{T})$. This space of functions on $\mathbb{T}$ will be denoted by $H_p(\mathbb{T})$. If $f \in H_p(\mathbb{T}), p \geq 1$ then we can recover the original analytic function as $f(re^{i\theta}) = f * \mathcal{P}_r(\theta)$ where $\mathcal{P}_r$ denotes the Poisson kernel. This correspondence is also natural in terms of Fourier series. If $f = \sum_{n=0}^{\infty} a_n z^n \in H_p(\mathbb{D})$, $p \geq 1$ then the boundary value $f \in L_p(\mathbb{T})$ has Fourier series $\sum_{n=0}^{\infty} a_n e^{in\theta}$.

We will use the notation $H_p^0(\mathbb{D})$ or $H_p^0(\mathbb{T})$ for the subspace of H_p consisting of functions vanishing at $0 \in \mathbb{D}$, or equivalently such that $\int_{\mathbb{T}} f = 0$.

The details can be found in Hoffmann [1962], Zygmund [1968], Katznelson [1968], Duren [1970] or Koosis [1980] and in many other places.

20.   There is an orthogonal projection from $L_2(\mathbb{T})$ onto $H_2(\mathbb{T})$. In terms of the Fourier coefficients this projection can be written as $\mathcal{R}(\sum_{-\infty}^{+\infty} a_n e^{in\theta}) = \sum_{n=0}^{\infty} a_n e^{in\theta}$. This projection is called the *Riesz projection*. It is important to know how the Riesz projection acts on other $L_p(\mathbb{T})$ spaces. The basic fact is that $\mathcal{R}$ is of weak type (1-1) (this is called Kolmogorov's theorem), so it is a continuous projection from $L_p(\mathbb{T})$ onto $H_p(\mathbb{T})$ for $1 < p < \infty$ (see **7**). We have $\|\mathcal{R}\|_{L_p \to L_p} \le Cp^2(p-1)^{-1}$. The Riesz projection $\mathcal{R}$ is actually not continuous in $L_1(\mathbb{T})$ and in $L_\infty(\mathbb{T})$.

This projection is clearly closely related to the Cauchy formula. If $\mu \in M(\mathbb{T})$ then the *Cauchy formula*

$$f(z) = \frac{1}{2\pi i} \int_0^{2\pi} \frac{d\mu(e^{it})}{e^{it} - z}$$

defines an analytic function in $\mathbb{D}$. One easily checks that $f(z) = \sum_{n=0}^{\infty} \hat{\mu}(n) z^n$. Because $\mathcal{R}$ is of weak type (1-1) we obtain that $f(z) \in H_p(\mathbb{D})$ for $p < 1$. In a sense the Riesz projection is the Cauchy formula acting on $\mathbb{T}$.

As in **19** the details can be found in Hoffmann [1962], Zygmund [1968], Katznelson [1968], Duren [1970], Koosis [1980] and in many other places, e.g. a nice presentation of a proof that $\mathcal{R}$ is bounded on $L_p(\mathbb{T})$ for $1 < p < \infty$ can be found in Lindenstrauss-Tzafriri [1979].

21.   Most of what has been said in **19** and **20** can be extended to polydiscs $\mathbb{D}^n$ and balls $\mathbb{B}_n$ in $\mathbb{C}^n$. For $0 < p \le \infty$ we define

$$H_p(\mathbb{D}^n) = \left\{ f : f \text{ is analytic in } \mathbb{D} \text{ and} \right.$$

$$\left. \|f\|_{H_p} = \sup_{r<1} \left(\int_{\mathbb{T}^n} |f(rz)|^p dm(z) \right)^{\frac{1}{p}} < \infty \right\}$$

where m is the probability Haar measure on $\mathbb{T}^n$,

$$H_p(\mathbb{B}_n) = \left\{ f : f \text{ is analytic in } \mathbb{D} \text{ and} \right.$$

$$\left. \|f\|_{H_p} = \sup_{r<1} \left(\int_{\mathbb{S}_n} |f(rz)|^p d\sigma(z) \right)^{\frac{1}{p}} < \infty \right\}$$

where σ is the normalized rotation invariant measure on $\mathbb{S}_n$.

The boundary values $\lim_{r \to 1} f(rz) = f(z)$ exist almost everywhere on $\mathbb{T}^n$ or $\mathbb{S}_n$, and we have $\|f\|_{H_p} = \|f\|_p$. Thus we can identity $H_p(\mathbb{D}^n)$

with a subspace of $L_p(\mathbf{T}^n)$ denoted by $H_p(\mathbf{T}^n)$ and we can identify $H_p(\mathbf{B}_n)$ with a subspace of $L_p(\mathbf{S}_n)$ denoted by $H_p(\mathbf{S}_n)$. The natural orthogonal projection from $L_2(\mathbf{T}^n)$ onto $H_2(\mathbf{T}^n)$ extends to a bounded projection from $L_p(\mathbf{T}^n)$ onto $H_p(\mathbf{T}^n)$ for $1 < p < \infty$ (apply the Riesz projection $\mathcal{R}$ in each variable separately) but is not of weak type (1-1). On the ball this natural orthogonal projection from $L_2(\mathbf{S}_n)$ onto $H_2(\mathbf{S}_n)$, called the *Cauchy projection* and denoted by $\mathcal{C}$, is of weak type (1-1), and so it is a continuous projection from $L_p(\mathbf{S}_n)$ onto $H_p(\mathbf{S}_n)$. Using the Cauchy kernel for the ball $\mathbf{B}_n$ we can express this projection as

$$\mathcal{C}(f)(z) = \int_{\mathbf{S}_n} \frac{f(\zeta)}{(1 - \langle z, \zeta \rangle)^n} d\sigma(\zeta) \qquad \text{for} \qquad z \in \mathbf{B}_n.$$

and $f \in L_p(\mathbf{S}_n)$. This is simply the Cauchy formula for the ball $\mathbf{B}_n$.

The standard references for all this are Rudin [1969] and Rudin [1980].

22.   If $f \in L_1(\mathbf{T})$ then we know (see **18**) that $f(re^{i\theta}) = (f * \mathcal{P}_r)(\theta)$ is harmonic in $\mathbf{D}$, so there exists a unique harmonic function $\tilde{f}(re^{i\theta})$ in $\mathbf{D}$ such that $\tilde{f}(0) = 0$ and $f(re^{i\theta}) + i\tilde{f}(re^{i\theta})$ is an analytic function. We know that $f(re^{i\theta}) = \sum_{-\infty}^{+\infty} r^{|n|} \hat{f}(n) e^{in\theta}$ so

$$\tilde{f}(re^{i\theta}) = -i \sum_{-\infty}^{+\infty} \operatorname{sgn} n \; r^{|n|} \hat{f}(n) e^{in\theta}$$

with the convention that sgn $0 = 0$. One can show that $\lim_{r \to 1} \tilde{f}(re^{i\theta}) = \tilde{f}(e^{i\theta})$ exists almost everywhere on $\mathbf{T}$. There is also a very useful integral representation of $\tilde{f}(e^{i\theta})$, namely

$$\tilde{f}(e^{i\theta}) = P.V. \frac{1}{2\pi} \int_0^{2\pi} f(t - \tau) \cot \frac{\tau}{2} d\tau.$$

The function $\tilde{f}$ on $\mathbf{T}$ is called the *trigonometric* (or *harmonic*) *conjugate* of f. There is a close connection between the trigonometric conjugation and the Riesz projection. Looking at Fourier coefficients we see that $\mathcal{R}f = \frac{1}{2}\hat{f}(0) + (f + i\tilde{f})$. This shows that the map $f \mapsto \tilde{f}$ is continuous on $L_p(\mathbf{T})$ for $1 < p < \infty$ and of weak type (1-1). This is called Kolmogorov's theorem.

This is a standard result in the theory of Fourier series. The details can be found in Zygmund [1968], Katznelson [1968], Koosis [1980] and many other books.

23.   One of the basic tools in the study of spaces $H_p(\mathbb{D})$, $0 < p \le \infty$ is the so-called canonical factorization. For $f \in H_p(\mathbb{D})$ let $(z_n)_{n=1}^\infty$ denote the zeros of f each counted according to its multiplicity. The sequence $(z_n)_{n=1}^\infty$ satisfies the *Blaschke condition* $\sum_{n=1}^\infty (1 - |z_n|) < \infty$. Conversely, if we have a sequence of points (with repetitions) $(z_n)_{n=1}^\infty \subset \mathbb{D}$ such that $\sum_{n=1}^\infty (1 - |z_n|) < \infty$, then the following product (called the *Blaschke product*)

$$B(z) = \prod_{n=1}^\infty \frac{\bar{z}_n}{|z_n|} \frac{z_n - z}{1 - z\bar{z}_n} \qquad \left(\text{we take } \frac{0}{0} = 1 \right)$$

converges almost uniformly in $\mathbb{D}$ and the function $B(z)$ (also called the Blaschke product) has zeros exactly at the points $(z_n)_{n=1}^\infty$. Moreover $\|B\|_\infty = 1$ and $|B(e^{i\theta})| = 1$ a.e. on $\mathbb{T}$. From this we see that each $f \in H_p(\mathbb{D})$ can be written as $f = B \cdot f_1$ where f_1 is zero-free and $\|f\|_p = \|f_1\|_p$ and $|f(e^{i\theta})| = |f_1(e^{i\theta})|$ a.e. on $\mathbb{T}$.

In order to factor f_1 we need to present the construction of an *outer function*. Let $h \in L_1(\mathbb{T})$, $h \ge 0$ be such that $\int_\mathbb{T} \log h(t) dt > -\infty$.

It is known that for $f \in H_p(\mathbb{T})$, $0 < p \le \infty$, $|f|$ satisfies these assumptions. We take the harmonic conjugate of $\log h(t)$ and form the analytic function

$$H(z) = \exp(\log h + i\widetilde{\log h}).$$

One checks that $\lim_{r \to 1} H(re^{i\theta})$ exists almost everywhere and has modulus h on $\mathbb{T}$. Applying this to $|f(e^{i\theta})|$, for $f \in H_p(\mathbb{D})$ we get the outer function $F(z)$. This function is in $H_p(\mathbb{D})$ and on $\mathbb{T}$ we have $|f(e^{i\theta})| = |F(e^{i\theta})|$ a.e. Thus we can write every $f \in H_p(\mathbb{D})$ as $f = B \cdot F \cdot I$ where $B(z)$ is the Blaschke product formed from the zeros of f (counting multiplicity) and F is an outer function with $|F(e^{i\theta})| = |f(e^{i\theta})|$ a.e. on $\mathbb{T}$. The remaining factor I can be defined simply as $f \cdot (B \cdot F)^{-1}$. It is an analytic function with $\|I\|_\infty = 1$ and $|I(e^{i\theta})| = 1$ a.e. on $\mathbb{T}$. This factor is called a *singular inner function* (if it is not a constant). In general an *inner function* is a function g in $H_\infty(\mathbb{T})$ such that $|g(e^{i\theta})| = 1$ a.e. on $\mathbb{T}$. Thus the Blaschke product is also an inner function, and a singular inner function is a zero-free inner function (non-constant). The singular inner functions have the representation

$$I(z) = \exp -\frac{1}{2\pi} \int_{-\pi}^{\pi} \frac{e^{i\theta} + z}{e^{i\theta} - z} d\mu(\theta)$$

where μ is a positive, singular measure on $[-\pi, \pi]$. The above described factorization $f = B \cdot F \cdot I$ is called the *canonical factorization*. The outer functions provide the tool to build analytic functions in $H_p(\mathbb{D})$ with given modulus on $\mathbb{T}$. This will be used extensively in III.I.

Detailed proofs of the canonical factorization can be found in Hoffmann [1962], Duren [1970], Koosis [1980] and Garnett [1981].

24. The following inequality due to R.E.A.C. Paley will be very useful. Suppose that $f = \sum_{n=0}^{\infty} a_n z^n \in H_1(\mathbb{D})$. Then

$$\left(\sum_{k=1}^{\infty} |a_{2^k}|^2 \right)^{\frac{1}{2}} \leq C\|f\|_1.$$

Since this inequality will be used later to prove Grothendieck's Theorem III.F.7. which is the basis for many results presented in chapters III.F-I, I have decided to present sketches of the proofs here.

Proof A. (Paley [1933]) It follows from **23** that one can write $f(z) = \left(\sum_{n=0}^{\infty} b_n z^n \right) \left(\sum_{n=0}^{\infty} c_n z^n \right)$ where $\left(\sum_{n=0}^{\infty} |b_n|^2 \right)^{1/2} \left(\sum_{n=0}^{\infty} |c_n|^2 \right)^{1/2} = \|f\|_1$. From this we infer that $a_{2^k} = \sum_{s,r:s+r=2^k} c_s b_r$, so

$$\sum_{k=1}^{\infty} |a_{2^k}|^2 = \sum_{k=1}^{\infty} \left(\sum_{s=0}^{2^k} c_s b_{2^k-s} \right)^2$$

$$\leq 2 \left(\sum_{k=1}^{\infty} \left(\sum_{s=0}^{2^{k-1}-1} c_s b_{2^k-s} \right)^2 + \sum_{k=1}^{\infty} \left(\sum_{s=0}^{2^{k-1}-1} b_s c_{2^k-s} \right)^2 \right)$$

$$\leq 2 \left(\sum_{k=0}^{\infty} \left(\sum_{s=0}^{2^{k-1}-1} |c_s|^2 \right) \left(\sum_{s=2^{k-1}+1}^{2^k} |b_s|^2 \right) \right.$$
$$\left. + \sum_{k=0}^{\infty} \left(\sum_{s=0}^{2^{k-1}-1} |b_s|^2 \right) \left(\sum_{2^{k-1}+1}^{2^k} |c_s|^2 \right) \right)$$

$$\leq 4 \sum_{s=0}^{\infty} |c_s|^2 \sum_{s=0}^{\infty} |b_s|^2 = 4\|f\|_1^2$$

Proof B. (Smith [1983]) Let $\alpha_k = 10^{-1} a_{2^k} \left(\sum_{s=1}^{\infty} |a_{2^s}|^2 \right)^{-1/2}$ and define inductively

$$g_1 = \alpha_1 e^{2i\theta}$$
$$g_k = \frac{1}{2}\alpha_k e^{2^k i\theta} + (1 - |\alpha_k|^2)g_{k-1} - \frac{1}{2}\bar{\alpha}_k e^{-2^k i\theta} g_{k-1}^2.$$

Since for $a, b \in \mathbb{D}$ we have $\left| \frac{b}{2} + (1 - |b|^2)a - \frac{a^2\bar{b}}{2} \right| \leq 1$ we see that $\|g_k\| \leq$ 1. One checks that for $l \leq k$ we have $\hat{g}_k(2^l) = (1-|\alpha_1|^2)(1-|\alpha_2|^2)\cdots(1-|\alpha_{l-1}|^2)\alpha_l$ and $\hat{g}_k(s) = 0$ for $s \geq 0$ and $s \neq 2^l$. Since $\sum_{k=1}^{\infty} |\alpha_k|^2 \leq 100^{-1}$ this product is bounded away from 0, so

$$\|f\|_1 \geq \left| \frac{1}{2\pi} \int_0^{2\pi} \bar{g}_k f \right|$$

$$= \sum_{l=1}^{k} (1 - |\alpha_1|^2) \cdots (1 - |\alpha_{l-1}|^2) \bar{\alpha}_l a_{2^l}$$

$$\geq c \left(\sum_{l=1}^{k} |a_{2^l}|^2 \right)^{1/2}.$$

Since this holds for all k we have the claim.

25.   We will also use the following inequality due to Hardy. If $f = \sum_{n=0}^{\infty} a_n z^n \in H_1(\mathbb{D})$ then $\sum_{n=0}^{\infty} \frac{|a_n|}{(n+1)} \leq \pi \|f\|_1$.

The proof can be found in Duren [1970] p. 48, Hoffmann [1962] p. 70, Katznelson [1968] p. 91.

The inequalities of Paley and Hardy fail for $L_1(\mathbb{T})$. This is one possible way to show that the Riesz projection is not continuous on $L_1(\mathbb{T})$.

26.   The *disc algebra* is the space of all functions analytic in $\mathbb{D}$ which admit a continuous extension to $\overline{\mathbb{D}}$. We can identify $A(\mathbb{D})$ with the subspace $\text{span}\{e^{in\theta}\}_{n \geq 0} \subset C(\mathbb{T})$. This is the space of all functions $f \in C(\mathbb{T})$ such that $\hat{f}(n) = 0$ for $n = -1, -2, \ldots$. It will be denoted by $A(\mathbb{T})$. The symbols $A_0(\mathbb{D})$ and $A_0(\mathbb{T})$ will denote the subspace of $A(\mathbb{D})$ consisting of functions vanishing at $0 \in \mathbb{D}$. The annihilator of $A(\mathbb{T}) \subset C(\mathbb{T})$ is described as follows.

F.-M. Riesz Theorem.   If $\mu \in M(\mathbb{T})$ and $\int_{\mathbb{T}} f d\mu = 0$ for all $f \in A(\mathbb{T})$ then μ is absolutely continuous with respect to the Lebesgue measure. Consequently $d\mu = hdm$ with $\bar{h} \in H_1^0(\mathbb{T})$.

The proof can be found in Hoffmann [1962], Duren [1970], Koosis [1980] and Katznelson [1968] p. 89.

27.   If $f \in H_p(\mathbb{D})$ and $z \in \mathbb{D}$, then the map $f \mapsto f(z)$ defines a linear functional on $H_p(\mathbb{D})$, $0 < p \leq \infty$. The norm of this functional is estimated in the following simple inequality

$$|f(z)| \leq 2^{\frac{1}{p}} (1 - |z|)^{-\frac{1}{p}} \|f\|_p.$$

The proof is in Duren [1970] p. 36.

28.   If $V \in \mathbb{C}^n$ is an open set, then we can define the *Bergman space* $B_p(V)$, $0 < p \leq \infty$ as the space of all holomorphic functions in $L_p(V)$, where on V we have $2n$-dimensional Lebesgue measure. We will be mostly interested in the case $V = \mathbb{D}$ or sometimes $V = \mathbb{B}_n$ and $V = \mathbb{D}^n$. In all those cases $B_p(V)$ is complete. Also the point evaluations are continuous and we have inequalities of the form: for every compact set $K \subset V$ there exists a constant $C = C_K$ such that $|f(z)| \leq C\|f\|_p$ for $z \in K$. We will also discuss the ball algebra $A(\mathbb{B}_d)$ which is the space of all functions analytic in $\mathbb{B}_d \subset \mathbb{C}^d$ which have a continuous extension to $\overline{\mathbb{B}_d}$. The polydisc algebra $A(\mathbb{D}^d)$ is the space of functions analytic in $\mathbb{D}^d$ which admit a continuous extension to $\overline{\mathbb{D}^d}$. The space $A(\mathbb{D}^d)$ can be naturally identified with a subspace of $C(\mathbb{T}^d)$.

   A good reference is Rudin [1980] where the case $\mathbb{B}_n$ (so in particular $\mathbb{D}$) is discussed in detail. Also Axler [1988] discusses $B_p(\mathbb{D})$ in detail. For the polydisc the standard reference is Rudin [1969].

29.   If f is a scalar valued function on a metric space (K, ρ), we say that f satisfies the *Hölder condition* of order α, $0 < \alpha \leq 1$ (for $\alpha = 1$ the term Lipschitz condition is also used) if

$$\varphi_\alpha(f) = \sup\{|f(k_1) - f(k_2)| \cdot \rho(k_1, k_2)^{-\alpha} : k_1, k_2 \in K, k_1 \neq k_2\} < \infty.$$

Clearly for every $k \in K$ the quantity $|f(k)| + \varphi_\alpha(f)$ is a norm and the corresponding Banach space of all functions satisfying the Hölder condition of order α is denoted by $Lip_\alpha(K, \rho)$. We will consider only K being $\mathbb{T}, \mathbb{R}$ or $[0,1]$ with the natural metric. One can also replace the function $t^{-\alpha}$ in the definition by a more general function.

30.   Now we want to discuss the *Sobolev norms* which measure the smoothness of functions of several variables. The theory is usually done in a much more general setting, but we restrict our attention to functions on $\mathbb{T}^s$, $s = 1, 2, \ldots$. Let $D(k)$, $k = 0, 1, 2, \ldots$ be the set of all multiindices $A = (i_1, i_2, \ldots, i_s)$ such that the i_j's are nonnegative integers and $\rho(A) = \sum_{j=1}^s i_j \leq k$. Each $A \in D(k)$ defines a partial derivative denoted ∂_A.

   We define the Sobolev space $C^k(\mathbb{T}^s)$ for $k \geq 0, s \geq 1$, to be the space of all functions on $\mathbb{T}^s$ such that $\partial_A f$ is continuous for all $A \in D(k)$. The norm which makes $C^k(\mathbb{T}^s)$ into a Banach space can be defined as

$$\|f\|_\infty^k = \sup\{\|\partial_A f\|_\infty : A \in D(k)\}.$$

If we want to extend this definition for $p < \infty$ we encounter some technical problems, so to avoid them, we first define the norm on C^∞ functions as

$$\|f\|_p^k = \left(\sum_{A \in D(k)} \|\partial_A f\|_p^p \right)^{\frac{1}{p}}.$$

By the Sobolev space $W_p^k(\mathbf{T}^s)$ we understand the completion of the C^∞ functions under this norm. Quite often we will treat $C^k(\mathbf{T}^s)$ as a subspace of $C(V)$ where V is the disjoint union of $|D(k)|$ copies of $\mathbf{T}^s$. (Here the modulus signs represent cardinality.) The natural isometric injection $j : C^k(\mathbf{T}^s) \to C(V)$ is defined by the rule that $j(f)$ on the torus $\mathbf{T}^s$ which is a subset of V with the index $A \in D(k)$ is simply $\partial_A f$. We can also treat $C^k(\mathbf{T}^s)$ as a subspace of $C(\mathbf{T}^s; E)$, the space of continuous, E-valued functions on $\mathbf{T}^s$ where E is the Banach space $\ell_\infty^{|D(k)|}$. The natural injection maps $f \mapsto (\partial_A f(t))_{A \in D(k)}$.

The same operations work also for $W_p^k(\mathbf{T}^s)$ and we can treat $W_p^k(\mathbf{T}^s)$ as a subspace of $L_p(V)$ (V the same as above) or a subspace of $L_p(\mathbf{T}^s; \ell_p^{|D(k)|})$.

The standard reference for the theory of Sobolev spaces is Adams [1975]. It is also presented in Stein [1970] and Stein-Weiss [1971].

31.   We have defined $W_p^k(\mathbf{T}^s)$ as the completion of the C^∞ functions under a certain norm. Nevertheless it is interesting to know up to what degree the elements of $W_p^k(\mathbf{T}^s)$ are really functions. This is answered by the so-called embedding theorems. We will use the simplest one.

Sobolev Embedding Theorem.   The identity on C^∞ functions extends to a continuous, linear operator from $W_p^k(\mathbf{T}^s)$ into $W_q^{k-m}(\mathbf{T}^s)$ where $0 \leq m \leq k$ and $mp < s$ and $p \leq q \leq \frac{sp}{(s-mp)}$. In particular $W_p^k(\mathbf{T}^s)$ embeds into $L_q(\mathbf{T}^s)$ if $p \leq q \leq \frac{sp}{(s-kp)}$.

The proof can be found in Adams [1975] or Stein-Weiss [1971].

32.   We will also use the general multiplier theorem on $\mathbf{T}^s$. Let us recall that the characters of the group $\mathbf{T}^s$ are $e^{in_1\theta_1} \cdot e^{in_2\theta_2} \cdots e^{in_s\theta_s}$ when $(n_1, \ldots, n_s) \in \mathbf{Z}^s$. A bounded function $m(\xi)$, for $\xi \in \mathbf{Z}^s$, defines an operator on $L_2(\mathbf{T}^s)$, called a multiplier and defined by

$$e^{in_1\theta_1} \cdot e^{in_2\theta_2} \cdots e^{in_s\theta_s} \mapsto m(n_1, \ldots, n_s) e^{in_1\theta_1} e^{in_2\theta_2} \cdots e^{in_s\theta_s}.$$

The following theorem describes sufficient conditions for a multiplier to act on $L_p(\mathbf{T}^s), 1 < p < \infty$ and to be of weak type (1-1).

Multiplier Theorem.   Suppose that φ is a function on $\mathbf{R}^s$ such that

(1) φ is bounded,

(2) for every multiindex $A \in D(k)$ (the description of $D(k)$ is in **30**) with $k > \frac{s}{2}$ we have

$$\sup_{0 < R < \infty} R^{2\rho(A)-n} \int_{R < |x| < 2R} |\partial_A m(x)| dx < \infty.$$

Then $\varphi(\xi)$ for $\xi \in \mathbf{Z}^s \subset \mathbf{R}^s$ is a multiplier which is bounded on $L_p(\mathbf{T}^s)$ for $1 < p < \infty$ and of weak type (1-1).

This theorem is a well known Hörmander-Mihlin type multiplier theorem. It is usually formulated on $\mathbf{R}^s$ and in even greater generality, so I was unable to locate the exact reference. It follows e.g. from Theorem 4.4 of Chapter XIV of Torchinsky [1986] together with an easy observation that a map bounded on some $H_p, p < 1$ and some $L_q, q > 1$ is of weak type (1-1) (see Folland-Stein [1982] Th. 3.37). Also the proof in Garcia-Cuerva–Rubio de Francia [1985] pp. 210-14 gives it.

Part II

Basic concepts of Banach space Theory

II.A. Weak Topologies

In this chapter we discuss two topologies which are weaker than the norm topology. They play an important role, both in the general theory and in applications. These topologies allow easy use of compactness arguments, and so they are helpful in existence proofs. This is particularly true of the ω^*-topology which can be defined on any dual space. We also introduce and study the class of reflexive spaces.

1. On each Banach space X there exists a weak topology (or $\sigma(X, X^*)$-topology or ω-topology). For each point $x_0 \in X$ its basis of neighbourhoods is defined as

$$U(x_0; \varepsilon, x_1^*, \ldots, x_n^*) = \{x \in X \colon |x_j^*(x) - x_j^*(x_0)| < \varepsilon \quad \text{for} \quad j = 1, \ldots, n\}$$

where $x_1^*, \ldots, x_n^*$ is an arbitrary finite set in X^* and ε is an arbitrary positive number. Obviously this defines a locally convex topology on X. A sequence $(x_n)_{n=1}^\infty \subset X$ such that for some $x \in X$, $x_n \longrightarrow x$ in the $\sigma(X, X^*)$-topology is said to be weakly convergent. We write this convergence as $x_n \overset{\omega}{\longrightarrow} x$. This is clearly equivalent to the condition that for every $x^* \in X^*$ the sequence $x^*(x_n) \to x^*(x)$ as $n \to \infty$. A sequence $(x_n)_{n=1}^\infty \subset X$ such that for every $x^* \in X^*$ the scalar sequence $x^*(x_n)$ is convergent is said to be weakly Cauchy.

2 Examples. (a) Let $X = c_0$ and $x_n = (\underbrace{1, \ldots, 1}_{n \text{ times}}, 0, 0, \ldots)$. Then x_n is weakly Cauchy because for every $x^* = (\xi_j)_{j=1}^\infty \in c_0^* = \ell_1$ we have $x^*(x_n) = \sum_{j=1}^n \xi_j \to \sum_{j=1}^\infty \xi_j$ as $n \to \infty$. On the other hand $(x_n)_{n=1}^\infty$ is not weakly convergent, because $(1, 1, 1, \ldots) \notin c_0$.

Usually it is quite difficult to describe the weak topology on the whole space. Quite often it is easier to describe this topology when restricted to the unit ball of X.

(b) On the unit ball B of c_0 the $\sigma(c_0, \ell_1)$-topology coincides with the topology of pointwise (i.e. coordinatewise) convergence (i.e. the Tychonoff topology on $\prod_{n=1}^{\infty}[-1, 1]$ restricted to the unit ball of c_0).

Clearly every open set in the topology of pointwise convergence is open in the $\sigma(c_0, \ell_1)$-topology. Conversely, given a neighbourhood $B \cap U(x_0; \varepsilon, x_1^*, \ldots, x_n^*)$ with $\|x_0\| \leq 1$ we fix N such that $\sum_{k=N}^{\infty} |x_j(k)| < \frac{\varepsilon}{10}$ for $j = 1, \ldots, n$ and put $M = \max_{j=1,\ldots,n} \|x_j^*\|$. One checks that $\{x \in c_0 \colon \|x\| \leq 1$ and $|x(k) - x_0(k)| < \frac{\varepsilon}{10M}$ for $k = 1, \ldots, N\} \subset B \cap U(x_0; \varepsilon, x_1^*, \ldots, x_n^*)$. On the other hand the $\sigma(c_0, \ell_1)$-topology and the topology of pointwise convergence do not coincide on the whole space c_0. To see this let us consider $x_n = 2^n e_n$ for $n = 1, 2, \ldots$. Clearly x_n converges pointwise to zero, but for $x^* = \left(n^{-2}\right)_{n=1}^{\infty} \in \ell_1$ we have $x_n \notin U(0; 1, x^*)$.

For more examples see Exercises 1,2,3.

3 Lemma. *A weakly Cauchy sequence is norm-bounded.*

Proof: For a given weakly Cauchy sequence $(x_n)_{n=1}^{\infty} \subset X$ we define an operator $T \colon X^* \longrightarrow c$ by $T(x^*) = (x^*(x_n))_{n=1}^{\infty}$. By the closed graph theorem I.A.6, the operator T is continuous so $\|T\| = \sup_n \|x_n\| < \infty$. ∎

4 Theorem. (Mazur) *If A is a convex set in X then the norm closure $\overline{A}$ of A equals its $\sigma(X, X^*)$ closure-$\overline{A}^{\omega}$.*

Proof: Clearly $\overline{A} \subset \overline{A}^{\omega}$. On the other hand if $x \in \overline{A}^{\omega} \backslash \overline{A}$ then by the Hahn-Banach theorem I.A.10 there exists $x^* \in X$ such that $x^*(x) > \sup\{x^*(a) \colon a \in \overline{A}\} = \sup\{x^*(a) \colon a \in A\}$. This implies that $x \notin \overline{A}^{\omega}$.   ∎

5. Suppose $x_n \overset{\omega}{\longrightarrow} x$. Then, by Theorem 4 for each j we have $x \in \overline{\text{conv}}\{x_n\}_{n=j}^{\infty}$. Thus we have the following useful

Corollary. *If $x_n \overset{\omega}{\longrightarrow} x$ then there exists a sequence of convex combinations $y_j = \sum_{k=j}^{n(j)} \lambda_k x_k$ such that $\|y_j - x\| \to 0$.*   ∎

6. On a dual space X^* we can introduce the ω^*-topology or $\sigma(X^*, X)$-topology. The basis of neighbourhoods of a point $x_0^* \in X^*$ is given by

$$U(x_0; \varepsilon, x_1, \ldots, x_n) = \{x^* \in X^* \colon |x^*(x_j) - x_0^*(x_j)| < \varepsilon \text{ for } j = 1, \ldots, n\}$$

where $x_1, \ldots, x_n$ is an arbitrary finite subset of X and ε is an arbitrary positive number. Clearly this defines a locally convex topology on X^*.

7 Examples. (a) On $\ell_\infty = \ell_1^*$ the ω^*-topology restricted to the unit ball coincides with the topology of pointwise convergence. The argument is the same as in Example 2b).

(b) On ℓ_p and $L_p[0,1]$ for $1 < p < \infty$ the ω-topology and the ω^*-topology coincide.

8 Remark. Let $T\colon X \to Y$ be a continuous linear operator. Then T is continuous from $(X, \sigma(X, X^*))$ into $(Y, \sigma(Y, Y^*))$ and T^* is continuous from $(Y^*, \sigma(Y^*, Y))$ into $(X^*, \sigma(X^*, X))$. ∎

9. The great usefulness of the ω^*-topology stems from the fact that it allows us to use compactness. We have

Theorem. (Alaoglu) *The closed unit ball B_{X^*} of X^* is $\sigma(X^*, X)$-compact.*

Proof: Let K_t for $t \geq 0$ denote the set of scalars of absolute value $\leq t$. We define a map $\varphi\colon B_{X^*} \longrightarrow P = \prod_{x \in X} K_{\|x\|}$ by $\varphi(x^*) = \{x^*(x)\}_{x \in X}$. By the Tychonoff theorem P is a compact space when equipped with the product topology. The very definitions of ω^*-topology and product topology give that φ is a homeomorphic embedding. The proof is finished once we show that $\varphi(B_{X^*})$ is closed in P. If $p_0 = p_0(x) \in P \backslash \varphi(B_{X^*})$ then either

(a) for some x_1, x_2 such that $x_1 = \lambda x_2$ we have $p_0(x_1) \neq \lambda p_0(x_2)$

or

(b) for some x_1, x_2, x_3 such that $x_1 + x_2 = x_3$ we have $p_0(x_1) + p_0(x_2) \neq p_0(x_3)$.

In case (a) we define $U = \{p \in P\colon |p(x_1) - p_0(x_1)| < \delta$ and $|p(x_2) - p_0(x_2)| < \delta\}$ where $\delta = \frac{1}{2}|p_0(x_1) - \lambda p_0(x_2)|$. Case (b) is treated analogously. In both cases we get an open set $U \subset P$ with $p_0 \in U$ and $U \cap \varphi(B_{X^*}) = \emptyset$. Thus $\varphi(B_{X^*})$ is closed. ∎

10. Let X be a Banach space. There exists a canonical embedding $i\colon X \to X^{**}$ given by $i(x)(x^*) = x^*(x)$. When $X = c_0$ one easily sees that $i\colon c_0 \to \ell_\infty$ is the identity map. For $X = \ell_p$, $1 < p < \infty$ we have $X^{**} = X$ and i is the identity. Thus the following is not very surprising.

Proposition. *The map $i\colon X \to X^{**}$ is a linear isometry.*

Proof: We have

$$i(\alpha x_1 + \beta x_2)(x^*) = x^*(\alpha x_1 + \beta x_2) = \alpha x^*(x_1) + \beta x^*(x_2)$$
$$= [\alpha i(x_1) + \beta i(x_2)](x^*).$$

Since this holds for arbitrary scalars α, β and arbitrary $x_1, x_2 \in X$ and $x^* \in X^*$ we see that i is a linear map. From the Hahn-Banach theorem I.A.9 we get, for $x \in X$,

$$\|i(x)\| = \sup_{\|x^*\| \le 1} |i(x)(x^*)| = \sup_{\|x^*\| \le 1} |x^*(x)| = \|x\|;$$

thus i is an isometry.   ∎

Quite often we identify X with $i(X)$ and treat X simply as a subspace of X^{**}. Note that this identification identifies the $\sigma(X, X^*)$-topology with the $\sigma(X^{**}, X^*)$-topology restricted to $i(X)$. In other words i is a homeomorphism of $(X, \sigma(X, X^*))$ onto $i(X)$ with the $\sigma(X^{**}, X^*)$-topology.

11.   One can view the $\sigma(X, X^*)$-topology on X as the weakest topology such that all norm-continuous linear functionals are still continuous. Analogously the $\sigma(X^*, X)$-topology is the weakest topology which makes all functionals in $i(X) \subset X^{**}$ continuous. It is a useful fact that these are all the functionals which are continuous in those topologies. We have the following

Proposition.   (a) *The space of all continuous linear functionals on $(X, \sigma(X, X^*))$ equals X^*.*

(b) *The space of all continuous linear functionals on $(X^*, \sigma(X^*, X))$ equals X.*

Proof: Part (a) is easy since the $\sigma(X, X^*)$-topology is weaker than the norm topology. The proof of (b) is as follows: Let φ be any linear functional on X^* continuous in $\sigma(X^*, X)$. Then $\{x^* \in X^*\colon |\varphi(x^*)| < 1\}$ $\supset \{x^* \in X^*\colon |x_j(x^*)| < \varepsilon \quad j = 1, 2, \ldots, n\}$ for some $\varepsilon > 0$ and some $x_1, \ldots, x_j \in X^*$. The application of the following algebraic lemma completes the proof.

12 Lemma.   *Let $\varphi_0, \varphi_1, \ldots, \varphi_n$ be linear forms on a linear space X (without any topology). Then the following are equivalent:*

(a) $\varphi_0 \in \text{span}\{\varphi_j\}_{j=1}^n$;

(b) $\ker \varphi_0 \supset \bigcap\limits_{j=1}^n \ker \varphi_j$.

Proof: (a)$\Rightarrow$(b) is obvious. To prove (b)$\Rightarrow$(a) let K denote the scalar field and consider a map $\pi: X \to K^n$ defined by $\pi(x) = (\varphi_j(x))_{j=1}^n$. Since $\ker \pi = \bigcap_{j=1}^n \ker \varphi_j$ the condition (b) implies that φ_0 induces a linear form Λ on K^n, i.e. $\varphi_0(x) = \Lambda\pi(x)$. Thus there exists a sequence of scalars $(\alpha_j)_{j=1}^n$ such that $\varphi_0(x) = \sum_{j=1}^n \alpha_j \varphi_j(x)$.   ∎

13.   We know that c_0 is pointwise dense in $c_0^{**} = \ell_\infty$. This suggests the following

Theorem. (Goldstine)   *The closed unit ball of X is $\sigma(X^{**}, X^*)$-dense in the closed unit ball of X^{**}.*

Note that we identify X with its canonical embedding into X^{**} (cf. **10**).

Proof: Let V denote the $\sigma(X^{**}, X^*)$-closure of B_X and assume that there exists $x^{**} \in X^{**}$ with $\|x^{**}\| \le 1$ and $x^{**} \notin V$. From Theorem 9 we infer that V is $\sigma(X^{**}, X^*)$-compact. Thus (use the Hahn-Banach theorem I.A.10) there exists a continuous linear functional φ on $(X^{**}, \sigma(X^{**}, X^*))$ such that $\varphi(x^{**}) > \sup\{\varphi(v): v \in V\}$. Proposition 11 shows that $\varphi(x^{**}) = x^{**}(x_0^*)$ for some $x_0^* \in X^*$. We have

$$\|x_0^*\| = \sup\{|x_0^*(x)|:\ x \in X, \|x\| \le 1\} \le \sup\{|v(x_0^*)|:\ v \in V\}$$
$$= \sup\{|\varphi(v)|:\ v \in V\} < \varphi(x^{**}) = x^{**}(x_0^*) \le \|x^{**}\| \cdot \|x_0^*\| \le \|x_0^*\|.$$

This contradiction completes the proof.   ∎

14.   We now turn to the study of a class of spaces for which the duality theory is particularly simple. The space X is said to be reflexive if $i(X) = X^{**}$ where i is the canonical embedding. The spaces ℓ_p and $L_p, 1 < p < \infty$, are reflexive while $\ell_1, L_1, c_0, C(K), L_\infty$ are not, unless they are finite dimensional.

Theorem.   *The following conditions on the Banach space X are equivalent*

(a)   *X is reflexive;*

(b) X^* *is reflexive;*

(c) B_X *is* $\sigma(X, X^*)$ *compact;*

(d) *every subspace of X is reflexive;*

(e) *every quotient space of X is reflexive.*

Proof: (a)$\Rightarrow$(c). B_X with the $\sigma(X, X^*)$-topology is homeomorphic to $B_{X^{**}}$ with the $\sigma(X^{**}, X^*)$-topology which is compact by Theorem 9.

(c)$\Rightarrow$(a). We get that $i(B_X)$ is a compact subset of X^{**} with the $\sigma(X^{**}, X^*)$-topology. Thus by the above Theorem 13 we have $i(B_X) = B_{X^{**}}$, so also $i(X) = X^{**}$.

(d)$\Rightarrow$(a). Obvious.

(a)$\Rightarrow$(d). If Y is a norm-closed subspace of X then the Hahn-Banach theorem yields that Y is $\sigma(X, X^*)$-closed. Thus B_Y is $\sigma(X, X^*)$-compact. But on Y the $\sigma(X, X^*)$-topology coincides with $\sigma(Y, Y^*)$ so the implication ((c)$\Rightarrow$(a)) gives that Y is reflexive.

(a)$\Rightarrow$(b). If X is reflexive then the $\sigma(X^*, X)$ and $\sigma(X^*, X^{**})$-topologies coincide on X^*, so Theorem 9 gives that B_{X^*} is $\sigma(X^*, X^{**})$-compact. Thus (c)$\Rightarrow$(a) gives that X^* is reflexive.

(b)$\Rightarrow$(a). If X^* is reflexive then using the implication (a)$\Rightarrow$(b) we get that X^{**} is reflexive so by (a)$\Rightarrow$(d) X is reflexive.

(e)$\Leftrightarrow$(a). We have to note that the dual of a quotient space of X is a subspace of X^* and use a previous equivalences.   ■

15.   We would like to conclude this chapter with the following useful observation.

Proposition.   (a) *If X is separable then $(B_{X^*}, \sigma(X^*, X))$ is metrizable.*

(b) *If X^* is separable then $(B_X, \sigma(X, X^*))$ is metrizable.*

Proof: (a) Let $(x_n)_{n=1}^{\infty}$ be a dense set in B_X. We define a metric on B_{X^*} by

$$\rho(x_1^*, x_2^*) = \sum_{n=1}^{\infty} 2^{-n} |x_1^*(x_n) - x_2^*(x_n)|.$$

Clearly ρ is a $\sigma(X^*, X)$-continuous metric. Let $U = \{x^* \in B_{X^*} : |x^*(x) - x_0^*(x)| < \varepsilon\}$ for a fixed $x \in X$ and $x_0^* \in X^*$. Changing ε we can assume $x \in B_X$. If we take n such that x_n is close enough to x then

$$U \supset \{x^* \in B_{X^*} : |x^*(x_n) - x_0^*(x_n)| < \frac{\varepsilon}{2}\} \supset \{x^* \in B_{X^*} : \rho(x_0^*, x^*) < \frac{\varepsilon}{2}\}.$$

Taking finite intersections we infer that every $\sigma(X^*, X)$-open set in B_{X^*} contains a ball in metric ρ, so ρ defines the $\sigma(X^*, X)$-topology on B_{X^*}.

(b) is an immediate consequence of (a) and the remarks made after Proposition 10.   ∎

Notes and remarks.

Weak convergence of sequences in some special instances was used at the beginning of the century by D. Hilbert and M. Riesz. The general theory is presented in Banach [1932]. Banach, however, did not use the notion of weak topology, he relied exclusively on sequential arguments. This resulted in unnecessary separability assumptions in many theorems. The weak topology defined by weak neighbourhoods appears (for algebras of operators on Hilbert spaces) already in von Neumann [1930]. In the thirties it became clear that general topological notions are needed in order to establish non-separable theorems. *Theorem 9* for separable X is in Banach [1932] and the general form was announced by several authors with the proof first published by Alaoglu [1940]. *Theorem 4* and *Corollary 5* are from Mazur [1933] and *Theorem 13* was first proved in Goldstine [1938]. The notion of reflexive space is implicit in Banach [1932]. Reflexive spaces, under the name 'regular spaces', were an object of intensive study in the late thirties and early forties. *Theorem 14* summarizes some of this work. The implication $(c) \Rightarrow (a)$ for separable X is in Banach [1932] chapter XI Th. 13. It is interesting to note that he did not realize the converse, despite having a sequential version of *Theorem 9*. The equivalence $(a) \Leftrightarrow (b)$ was known to Plessner before 1936 (see Lusternik [1936]). The equivalence $(a) \Leftrightarrow (c)$ was proved in Šmulian[1939], Bourbaki [1938] and Kakutani [1939].

   As is clear from the above the content of this section is classical. The reader will find much more in Dunford-Schwartz [1958]. This material is also covered in most textbooks on functional analysis.

Exercises

1.   Show that, if X is an infinite dimensional Banach space then the $\sigma(X, X^*)$-topology on X is not metrizable.

2.   Let Γ be an uncountable set. Show that the weak topology on the unit ball of $\ell_2(\Gamma)$ is not metrizable.

3.   Let $f_n \in C(K)$, $\|f_n\| \le 1$ for $n = 1, 2, \dots$. Show that $f_n \xrightarrow{\omega} 0$ if and only if $f_n(k) \to 0$ for every $k \in K$.

4. Find a Banach space X and a sequence $(x_n^*)_{n=1}^\infty \subset X^*$, such that $\|x_n^*\| = 1$ for $n = 1, 2, \ldots$ and x_n^* tends w^* to 0 but every convex combination of x_n^*'s has norm 1.

5. Show that on the unit ball of $H_\infty(\mathbb{D})$ the topology of uniform convergence on compact subsets of $\mathbb{D}$ coincides with the $\sigma(H_\infty, (L_1/H_1))$-topology.

6. Let $T: H_\infty(\mathbb{D}) \to A(\mathbb{D})$ be defined by $Tf(z) = f(\frac{z}{2})$. Show that T maps w^*-convergent sequences onto norm-convergent sequences but is not continuous from the $\sigma(H_\infty, (L_1/H_1))$-topology into the norm topology.

7. (Riesz products) Let $(a_n)_{n=1}^\infty$ be a sequence of real numbers such that $|a_n| \le 1$ for $n = 1, 2, \ldots$. Let $(q_n)_{n=1}^\infty$ be a sequence of natural numbers such that $q_{n+1} \ge 3q_n$. Show that for every sequence $(\varphi_n)_{n=1}^\infty$ the products $\prod_{n=1}^N (1 + a_n \cos(q_n t + \varphi_n))$ converge in the $\sigma(M(\mathbb{T}), C(\mathbb{T}))$-topology as $N \to \infty$ to a positive measure μ. Show also that $\hat\mu(q_n) = e^{i\varphi_n} a_n$.

8. Suppose that $(a_n)_{n=-\infty}^{+\infty}$ is a sequence of numbers such that $\liminf_{N\to\infty} \left\| \sum_{-N}^N a_n e^{in\theta} \right\|_p < \infty$ for some $p, 1 \le p < \infty$. Show that if $1 < p < \infty$, then there exists a function $f \in L_p(\mathbb{T})$ such that $\hat{f}(n) = a_n$ for $n = 0, \pm 1, \pm 2, \ldots$, and if $p = 1$ then there exists a measure $\mu \in M(\mathbb{T})$ such that $\hat\mu(n) = a_n$ for $n = 0, \pm 1, \pm 2, \ldots$.

9. Let G be a compact abelian group with dual group Γ. For $g \in G$ let $I_g(f)(h) = f(h - g)$. Suppose that $T: L_1(G) \to L_1(G)$ is a linear operator such that $I_g T = T I_g$ for all $g \in G$. Show that there exists a measure μ on G such that $Tf = f * \mu$

10. Show that every positive, harmonic function in $\mathbb{D}$ is a Poisson integral of a positive measure on $\mathbb{T}$. This means (show this) that every analytic function in $\mathbb{D}$ with values in a half plane is in $H_p(\mathbb{D})$ for all $p < 1$.

II.B. Isomorphisms, Bases, Projections

In this chapter we discuss a diverse set of topics all centred on the concept of isomorphism, the basic equivalence relation between Banach spaces. We discuss projections and closely related direct sum decompositions of a Banach space. We present the 'decomposition method' which allows easy proofs of the existence of isomorphisms between Banach spaces. The concept of Schauder basis is introduced and some fundamental results about block-basic sequences are proved.

1. We will say that two Banach spaces X and Y are isomorphic (in symbols $X \sim Y$) if there exists an invertible operator I (called an isomorphism) from X onto Y. An operator $T\colon X \to Y$ such that $\|Tx\| \geq c\|x\|$ for some $c > 0$ and all $x \in X$ is called an isomorphic embedding (or simply embedding). Clearly then $T(X)$ is a closed subspace of Y and X is isomorphic to $T(X)$. An isometry is an operator $I\colon X \to Y$ such that $\|I(x)\| = \|x\|$ for all $x \in X$. Two Banach spaces X and Y are isometric (in symbols $X \cong Y$) if there exists an isometry from X onto Y.

2 Examples. (a) The identity embeds c_0 into c and into l_∞. The operator $I\colon c_0 \to c$ defined by $I((\xi_j)_{j=1}^\infty) = (\xi_1 + \xi_{j+1})_{j=1}^\infty$ is an isomorphism between c_0 and c. Obviously c_0 is not isomorphic to l_∞ since c_0 is separable while l_∞ is not.

(b) If (Ω, μ) is a measure space which contains an infinite sequence of disjoint sets $(A_j)_{j=1}^\infty$ with $0 < \mu(A_j) < \infty$, then l_p embeds into $L_p(\Omega, \mu), 1 \leq p \leq \infty$. The embedding is given by $T(\xi_j) = \sum_j \xi_j \mu(A_j)^{-\frac{1}{p}} \chi_{A_j}$

(c) The space $L_p(\mathbb{R})$ is isometric to $L_p[0, 1]$ for every $p, 1 \leq p \leq \infty$. The isometry $I\colon L_p[0, 1] \to L_p(\mathbb{R})$ can be defined as

$$I(f)(x) = f\left(\tan \pi\left(x - \frac{1}{2}\right)\right)\left(\pi\left(1 + \tan^2\left(\pi\left(x - \frac{1}{2}\right)\right)\right)\right)^{\frac{1}{p}}.$$

More generally if (Ω, μ) is a σ-finite measure space then $L_p(\Omega, \mu), 0 \leq p \leq \infty$, is isometric to L_p on a probability measure space. In order to see this let us take a sequence of disjoint subsets $(A_j)_{j=1}^\infty$ of Ω with $0 <$

$\mu(A_j) < \infty$ for $j = 1, 2, \ldots$ and $\Omega = \bigcup_{j=1}^{\infty} A_j$. Let us define a measure μ_1 on Ω by $\mu_1(A) = \sum_{j=1}^{\infty} 2^{-j} \mu(A_j)^{-1} \mu(A \cap A_j)$. One checks that μ_1 is a probability measure on Ω and that the map $I \colon L_p(\Omega, \mu) \to L_p(\Omega, \mu_1)$ defined by

$$I(f) = \sum_{j=1}^{\infty} f \cdot \chi_{A_j} \cdot 2^{j/p} \mu(A_j)^{\frac{1}{p}}$$

is an isometry onto.

3 Proposition. *Let X_1 and X_2 be two closed subspaces of a Banach space X both of codimension 1. Then $X_1 \sim X_2$.*

Proof: If $X_1 = X_2$ there is nothing to prove. Otherwise, the space $X_0 = X_1 \cap X_2$ has codimension 2 in X so there are $x_1 \in X_1$ and $x_2 \in X_2$ such that $\|x_1\| = \|x_2\| = 1$ and $X_1 = X_0 + \lambda x_1$ and $X_2 = X_0 + \lambda x_2$. We define $T \colon X_1 \to X_2$ by the following rule: $T|X_0 = id$ and $T(x_1) = x_2$.∎

Now we will give examples of rather non-obvious isomorphic (even isometric) embeddings.

4 Theorem. (Banach-Mazur) *Every separable Banach space X is isometric to a subspace of $C(\Delta)$, where Δ is the Cantor set.*

Proof: Let B^* be the closed unit ball of X^* with ω^*-topology. By II.A.9 and II.A.15 it is a compact metric space so by the well known Alexandroff-Hausdorff theorem (see Kuratowski [1968] 4§41.VI or Lacey [1974] §6.3) there exists a continuous map $\varphi \colon \Delta \xrightarrow{\text{onto}} B^*$. We define $T \colon X \to C(\Delta)$ by

$$T(x)(\delta) = \varphi(\delta)(x) \text{ for } \delta \in \Delta.$$

It is easy to check that it is an isometric embedding.   ∎

We can replace $C(\Delta)$ by $C[0,1]$ in Theorem 4. To see this, it is enough to find an isometric embedding of $C(\Delta)$ into $C[0,1]$. Such an embedding can be realized as an extension operator; to each $f \in C(\Delta)$ we assign $F \in C[0,1]$ such that $F|\Delta = f$ and F is linear on each interval of $[0,1] \setminus \Delta$.

5.   The sequence of elements $(x_n)_{n=1}^{\infty}$ of a Banach space X is called a Schauder basis (or simply a basis) if, for every $x \in X$, there exists a unique sequence of scalars $(a_n)_{n=1}^{\infty}$ such that $x = \sum_{n=1}^{\infty} a_n x_n$. Let us emphasize that the sign '$=$' means here that the series $\sum_{n=1}^{\infty} a_n x_n$ converges to x in the norm of X. Sometimes we will consider bases indexed by countable sets other than the natural numbers, but then the order of summation has to be specified. Let us recall (I.A.20) that an idempotent operator $P \colon X \to X$, i.e. an operator such that $P^2 = P$, is called a projection. Obviously P acts as the identity on $P(X)$, so $P(X)$ is a closed subspace of X. A subspace $Y \subset X$ such that there exists a projection $P \colon X \to X$ with $P(X) = Y$ is called complemented. Bases and projections are connected as follows:

6 Proposition. *Let $(x_n)_{n=1}^{\infty}$ be a basis in X. For each N we define a partial sum projection $P_N \colon X \to X$ by $P_N \left(\sum_{n=1}^{\infty} a_n x_n \right) = \sum_{n=1}^{N} a_n x_n$. Then $\sup_N \|P_N\| < \infty$.*

The number $\sup_N \|P_N\|$ is called the basis constant of the basis $(x_n)_{n=1}^{\infty}$ and is denoted by $bc(x_n)$.

Proof: The linearity of P_N and the identity $P_N^2 = P_N$ are obvious. The real difficulty is that we do not know that the P_N's are continuous. We define a new norm on X by $|||x||| = \sup_N \|P_N x\|$.

We want to show that $(X, |||\cdot|||)$ is complete. Once this is established the closed graph theorem I.A.6 will give a C such that $|||x||| \leq C\|x\|$ for $x \in X$, i.e. $\sup_N \|P_N\| \leq C$.

Let us take $(y^k) \subset X$ which is Cauchy in $||| \cdot |||$. This implies that there are $z_N \in X$ such that $\|P_N(y^k) - z_N\| \to 0$ as $k \to \infty$, uniformly in N. The sequence $(z_N)_{N=1}^{\infty}$ is Cauchy in $\| \cdot \|$ because, given $\varepsilon > 0$, if k is fixed so that $\|P_N(y^k) - z_N\| < \frac{\varepsilon}{3}$ for all N we have

$$\|z_N - z_M\| \leq \frac{2}{3}\varepsilon + \|P_N(y^k) - P_M(y^k)\| < \varepsilon$$

for N and M big enough. Let $z \in X$ be such that $\|z - z_N\| \to 0$ as $N \to \infty$. Since every linear operator on a finite dimensional space is continuous and $\dim P_M(X) = M$ we have

$$P_N(z_M) = P_N(\lim_k P_M(y^k)) = \lim_k P_N P_M(y^k)$$
$$= \lim_k P_{\min(N,M)}(y^k) = z_{\min(N,M)}.$$

This shows that there exists a sequence of scalars $(a_n)_{n=1}^{\infty}$ such that $z_N = \sum_{n=1}^{N} a_n x_n$, so that $z = \sum_{n=1}^{\infty} a_n x_n$. This gives $P_N(z) = z_N$ so

$$|||y^k - z||| = \sup_N \|P_N(y^k) - P_N(z)\| = \sup_N \|P_N(y^k) - z_N\| \to 0, \text{ as } k \to \infty.$$

∎

7 Corollary. If $(x_n)_{n=1}^{\infty}$ is a basis in the space X and for $x = \sum_{n=1}^{\infty} a_n x_n$ we define $x_n^*(x) = a_n$, then $(x_n^*)_{n=1}^{\infty}$ are continuous linear functionals and $\|x_n\| \cdot \|x_n^*\| \le 2bc(x_n)$. The functionals x_n^*, called biorthogonal or coefficient functionals, are uniquely determined by the conditions $x_n^*(x_m) = \delta_{n,m}$. ∎

Thus if $(x_n)_{n=1}^{\infty}$ is a basis in X, then for each $x \in X$ we have $x = \sum_{n=1}^{\infty} x_n^*(x)x_n$, where $(x_n^*)_{n=1}^{\infty}$ are biorthogonal functionals and the series converges in norm. The partial sum projection P_N admits the representation $P_N(x) = \sum_{n=1}^{N} x_n^*(x)x_n$.

8. The following routine but useful proposition gives an equivalence between bases and some systems of projections.

Proposition. If $(x_n)_{n=1}^{\infty}$ is a basis in X then its partial sum projections satisfy

(a) for all $x \in X$ we have $P_N(x) \to x$ as $N \to \infty$,

(b) $\dim P_N(X) = N$,

(c) $P_N P_M = P_{\min(N,M)}$.

Conversely, if we are given a sequence of projections $(P_N)_{N=1}^{\infty}$ satisfying (a), (b), (c) above then any sequence of non-zero vectors $(x_n)_{n=1}^{\infty}$ such that $x_1 \in P_1(X)$ and $x_n \in P_n(X) \cap \ker P_{n-1}$ for $n = 2, 3, \ldots$ is a basis in X. ∎

9. Any complete orthogonal system in a separable Hilbert space is a basis. Also the standard unit vectors in the spaces $l_p, 1 \le p < \infty$ and in c_0 form a basis. Now we will discuss more interesting examples of bases. We start with the Haar system, which is one of the most important orthonormal systems. Each number $n = 1, 2, 3, \ldots$ we represent as $n =$

$2^j + k$ with $j = 0, 1, 2, \ldots$ and $k = 0, 1, \ldots, 2^j - 1$. We define the Haar functions on $[0, 1]$, $(h_n)_{n=0}^\infty$ as follows:

$$h_0(t) = 1;$$

$$h_n(t) = \begin{cases} -2^{\frac{j}{2}} & \text{for} \quad k2^{-j} \le t < (2k+1)2^{-j-1}, \\ 2^{\frac{j}{2}} & \text{for} \quad (2k+1)2^{-j-1} \le t < (k+1)2^{-j}, \\ 0 & \text{otherwise.} \end{cases}$$

It is imperative that the reader draws the picture and understands how the supports of consecutive Haar functions are located.

10 Proposition. (a) *The Haar system is an orthonormal system in* $L_2[0, 1]$.

(b) *The Haar system is a basis in* $L_p[0, 1]$, $1 \le p < \infty$.

Proof. (a) This is obvious once one realizes that $\int_0^1 h_n(t)dt = 0$ for $n = 1, 2, \ldots$ and that for $m > n, h_n(t) \cdot h_m(t)$ equals either zero or a constant times h_m.

(b) Let $F_N, N = 2^j + k$, $j = 0, 1, 2, \ldots$, $k = 0, 1, \ldots, 2^j - 1$, be the space of functions which are constant on intervals of the family

$$S_N = \{(s2^{-j-1}, (s+1)2^{-j-1}): s = 0, 1 \ldots, 2k + 1, \text{ and}$$
$$(s2^{-j}, (s+1)2^{-j}) \ s = k + 1, \ldots, 2^j - 1\}.$$

Since $h_n \in F_N$ for $n = 0, 1, \ldots, N$ and $\dim F_N = N + 1$ we infer that $\mathrm{span}(h_n)_{n=0}^N = F_N$. This description implies that $\mathrm{span}(h_n)_{n=0}^\infty$ is dense in $L_p[0, 1]$, $1 \le p < \infty$. One also easily sees that the projection

$$P_N(f) = \sum_{I \in S_N} |I|^{-1} \int_I f(t)dt \chi_I$$

is a partial sum projection. Applying the Hölder inequality we have

$$\|P_N f\|_p^p = \sum_{I \in S_N} |I|^{-p} \left| \int_I f(t)dt \right|^p \int_0^1 \chi_I$$

$$\le \sum_{I \in S_N} |I|^{1-p} |I|^{p/q} \int_I |f|^p = \sum_{I \in S_N} \int_I |f|^p = \|f\|_p^p.$$

So $\|P_N\|_p = 1$ for $N = 0, 1, 2, \ldots$ and Proposition 8 completes the proof.

■

11.   Let us consider the trigonometric system $(e^{in\theta})_{n=-\infty}^{+\infty}$. We know (see I.B.20) that the Riesz projection $\mathcal{R}(f) = \sum_{n\geq 0} \hat{f}(n)e^{in\theta}$ is bounded in $L_p(\mathbb{T}), 1 < p < \infty$. Since the operator of multiplication by $e^{in\theta}$, $I_n(f)(\theta) = e^{in\theta}f(\theta)$, is an isometry from $L_p(\mathbb{T})$ into $L_p(\mathbb{T})$ we infer that for $-\infty < N < M < \infty$, the map $S_{N,M} = I_M(I - \mathcal{R})I_{N-M}\mathcal{R}I_{-N}$ is a bounded operator on $L_p(\mathbb{T}), 1 < p < \infty$. One checks that $S_{N,M}(f) = \sum_{n=N}^{M} \hat{f}(n)e^{in\theta}$.

Trigonometric polynomials are dense in $L_p(\mathbb{T})$ so Proposition 8 gives that the characters $(e^{in\theta})_{n=-\infty}^{+\infty}$ ordered $0, 1, -1, 2, -2, \ldots$ form a basis in $L_p(\mathbb{T}), 1 < p < \infty$.

12.   The Faber-Schauder basis in $C[0,1]$. Actually we will describe a whole class of bases in terms of partial sum projections. Let $(t_j)_{j=1}^{\infty}$ be a sequence of distinct points of $[0,1]$ with $t_1 = 0, t_2 = 1$ and such that $\overline{\{t_j\}_{j=1}^{\infty}} = [0,1]$. We define projections in $C[0,1]$ as

$$P_1(f)(t) = f(0),$$

$P_N(f)$ is a piecewise linear function with nodes

at $t_j, \; j = 1, 2, \ldots, N$ and $P_N(f)(t_j) = f(t_j), \; j = 1, 2, \ldots, N$.

Clearly the projections $P_N, N = 1, 2, \ldots$ satisfy conditions (a), (b), (c) of Proposition 8. The classical Faber-Schauder system is defined as $\varphi_1(t) = 1, \varphi_n(t) = \|h_{n-2}\|_1^{-1} \int_0^t h_{n-2}(s)ds$ for $n = 2, 3, \ldots$, where $(h_j)_{j=0}^{\infty}$ are the Haar functions. The Faber-Schauder system corresponds to the sequence $(t_j)_{j=1}^{\infty}$ being naturally ordered dyadic points, i.e. $0, 1, 1/2, 1/4, 3/4, 1/8, 3/8, \ldots$.

13.   A system $(x_n)_{n=1}^{\infty} \subset X$ is called a basic sequence if it is a basis in its closed linear span. Obviously every subsequence of a basis is a basic sequence. Also if (x_n) is a basic sequence in X and $i\colon X \to Y$ is an isomorphic embedding then $(i(x_n))$ is a basic sequence in Y. We can also note that the Haar system is a basic sequence in $L_\infty[0,1]$. Two basic sequences $(x_n)_{n=1}^{\infty}$ in X and $(y_n)_{n=1}^{\infty}$ in Y are called equivalent if the series $\sum_{n=1}^{\infty} a_n x_n$ converges if and only if $\sum_{n=1}^{\infty} a_n y_n$ converges.

This is equivalent to saying that there exists an isomorphism between $\overline{\text{span}}(x_n)_{n=1}^{\infty}$ and $\overline{\text{span}}(y_n)_{n=1}^{\infty}$ which sends x_n onto y_n for $n = 1, 2, 3, \ldots$.

14.   Now we will present two useful ways to find basic sequences. One is a small-perturbation argument and the other involves the so-called block-basic sequences. They are of fundamental importance in

Banach space theory and in many applications. Let us start with the perturbation arguments. They are based on the following well known

Lemma. Let $U\colon X \to X$ and $\|id_x - U\| < 1$. Then U is invertible.

Proof: The series $\sum_{n=0}^{\infty}(id_x - U)^n$ is absolutely convergent and defines U^{-1}. ∎

The following proposition sums up the most useful perturbation arguments.

15 Proposition. Let $(x_n)_{n=1}^{\infty}$ be a basic sequence in X and let $(x_n^*)_{n=1}^{\infty}$ be its biorthogonal functionals, i.e. $x_n^*(x_m) = \delta_{n,m}$.
(a) If $(y_n)_{n=1}^{\infty} \subset X$ is a sequence of vectors such that $\sum_{n=1}^{\infty} \|x_n - y_n\| \, \|x_n^*\| = \delta < 1$, then $(y_n)_{n=1}^{\infty}$ is a basic sequence equivalent to $(x_n)_{n=1}^{\infty}$.
(b) If, moreover, $\mathrm{span}(x_n)_{n=1}^{\infty}$ is complemented by a projection P with $\|P\| < \delta^{-1}$ then $\mathrm{span}(y_n)_{n=1}^{\infty}$ is complemented in X.

Proof: (a) Let us define $T\colon \mathrm{span}(x_n)_{n=1}^{\infty} \to X$ by $T(x) = \sum_{n=1}^{\infty} x_n^*(x)y_n$. Obviously $T(x_n) = y_n$. Since

$$\|x - T(x)\| \le \sum_{n=1}^{\infty} |x_n^*(x)| \, \|y_n - x_n\| \le \delta\|x\| \tag{1}$$

we infer that $(1 - \delta)\|x\| \le \|T(x)\| \le (1 + \delta)\|x\|$ so T is an isomorphic embedding. This shows that $(y_n)_{n=1}^{\infty}$ is a basic sequence equivalent to $(x_n)_{n=1}^{\infty}$.
(b) Let us define $A\colon X \to X$ by $A = id_x - P + TP$ where T is defined above. By (1) $\|I - A\| = \|P - TP\| \le \|P\|\delta < 1$, so Lemma 14 yields that A is invertible. One easily checks that APA^{-1} is a projection onto $\overline{\mathrm{span}}(y_n)_{n=1}^{\infty}$. ∎

16. If $(x_n)_{n=1}^{\infty}$ is a basic sequence in X then a sequence of non-zero vectors of the form $u_n = \sum_{n_k+1}^{n_{k+1}} a_n x_n$ for (n_k) a strictly increasing sequence of integers and (a_n) a sequence of scalars is called a block-basic sequence (block basis for short). Clearly $(u_k)_{k=1}^{\infty}$ is a basic sequence and $bc(u_k) \le bc(x_n)$.

17 Proposition. Let $(x_n)_{n=1}^{\infty}$ be a basis in X with $\|x_n\| = 1$ and biorthogonal functionals $(x_n^*)_{n=1}^{\infty}$. Let $(z_k)_{k=1}^{\infty}$ be a sequence in X such that

$$\|z_k\| \geq \delta > 0 \quad \text{for} \quad k = 1, 2, 3, \ldots \qquad (2)$$

$$\lim_{k \to \infty} x_n^*(z_k) = 0 \quad \text{for} \quad n = 1, 2, 3, \ldots \qquad (3)$$

Then there exists a subsequence (z_{n_s}) which is equivalent to a block basic sequence of (x_n).

Proof: Let us put $K = bc(x_n)$ and let P_n denote the partial sum projections of the basis (x_n). Fix $\eta < 3/4 \; \delta(K+1)^{-1}$. Inductively we choose two increasing sequences of integers n_s and N_s in such a way that

$$\|P_{N_{s-1}}(z_{n_s})\| < \eta 4^{-s}, \qquad (4)$$

$$\|z_{n_s} - P_{N_s}(z_{n_s})\| < \eta 4^{-s}. \qquad (5)$$

We start with $n_1 = 1$ and N_1 chosen to satisfy (5). This is possible since $P_n(x) \to x$ as $n \to \infty$ for $x \in X$. Having n_s and N_s for $s = 1, 2, 3, \ldots, r$ we pick n_{r+1} so big that (3) ensures that (4) holds. Next we find N_{r+1} so big that (5) holds for $z_{n_{r+1}}$. The sequence $u_s = P_{N_s} z_{n_s} - P_{N_{s-1}} z_{n_s}$ is a block-basic sequence. We can find biorthogonal functionals u_s^* such that $\|u_s^*\| \leq 2K\|u_s\|^{-1} \leq 2K(\delta - \eta/4)^{-1}$ (Corollary 7). Since $\|z_{n_s} - u_s\| \leq 2\eta 4^{-s}$ we infer from Proposition 15 (a) that (z_{n_s}) is the desired subsequence.   ∎

Corollary.   Every weakly null sequence $(x_n)_{n=1}^{\infty}$ in X with $\|x_n\| = 1$ has a basic subsequence.

Proof: Since span(x_n) is separable we can assume that X is separable as well. Using Theorem 4 we can treat X as a subspace of $C[0,1]$ which has a basis (**12**). Now the corollary follows from Proposition 17.   ∎

19.   Let $X_1, \ldots X_n$ be subspaces of a Banach space X. We say that X is a direct sum of X_j's and write $X = \sum_{j=1}^{n} X_j$ or $X = X_1 \oplus \cdots \oplus X_n$ if every $x \in X$ has a unique representation $x = \sum_{j=1}^{n} x_j$ with $x_j \in X_j$. If we are given Banach spaces $X_1, \ldots, X_n$ we can form their direct sum $X = \sum_{j=1}^{n} X_j$ as follows:
    X is the set of n-tuples of elements $(x_j)_{j=1}^{n}$ with $x_j \in X_j$, with coordinatewise algebraic operations and product topology.
We can introduce a norm in X in many equivalent ways. The most common are l_p-sums, $1 \leq p \leq \infty$, where $\|(x_j)\|_p = \left(\sum_{j=1}^{n} \|x_j\|^p \right)^{\frac{1}{p}}$. The direct sum with such a norm will be denoted $\left(\sum_{j=1}^{n} X_j \right)_p$.

20. The following proposition states the easy but fundamental properties of these constructions:

Proposition. (a) *The space X is a direct sum of its subspaces X_1 and X_2 if and only if there exists a projection $P\colon X \to X$ such that $X_1 = P(X)$ and $X_2 = \ker P$.*
 (b) *If $T_j\colon X_j \to Y_j$ for $j = 1, \ldots, n$, then we can define an operator*

$$T\colon \left(\sum_{j=1}^{n} X_j\right) \to \left(\sum_{j=1}^{n} Y_j\right)$$

by the formula $T((x_j)_{j=1}^{n}) = (T_j(x_j))_{j=1}^{n}$.
 (c) *If in (b) every T_j is an isomorphism then T is.*

Proof: Parts (b) and (c) are obvious. For (a) note that for $x = x_1 + x_2$ with $x_1 \in X_1$ and $x_2 \in X_2$ we define $P(x) = x_1$. P is continuous by the closed graph theorem. Conversely given a projection P we write $x = Px + (x - Px)$, and $Px \in P(X)$ and $(x - Px) \in \ker P$. The decomposition is unique. ∎

21. Quite often we will need infinite direct sums. Since a basis provides an infinite direct sum decomposition of a space into one-dimensional subspaces, we see that, unlike the finite case, the form of the norm on the whole sum is crucial. In practice we will consider only l_p-sums. They are formed as follows:
 Given a sequence of Banach spaces $(X_j)_{j=1}^{\infty}$ we put for $1 \le p < \infty$

$$\left(\sum_{j=1}^{\infty} X_j\right)_p = \left\{(x_j)_{j=1}^{\infty}\colon x_j \in X_j \text{ and } \|(x_j)\|_p = \left(\sum_{j=1}^{\infty} \|x_j\|^p\right)^{\frac{1}{p}} < \infty\right\}.$$

We also define

$$\left(\sum_{j=1}^{\infty} X_j\right)_{\infty} = \{(x_j)_{j=1}^{\infty}\colon x_j \in X_j \text{ and } \|(x_j)\|_{\infty} = \sup_j \|x_j\| < \infty\}$$

and

$$\left(\sum_{j=1}^{\infty} X_j\right)_0 = \left\{(x_j)_{j=1}^{\infty} \in \left(\sum_{j=1}^{\infty} X_j\right)_{\infty}\colon \|x_j\| \to 0\right\}.$$

In all the above sums the algebraic operations are performed coordinatewise. It is routine to check that all the above sums are Banach spaces.

Very similarly to the case of l_p spaces one checks that

$$\left(\sum_{j=1}^{\infty} X_j\right)_0^* = \left(\sum_{j=1}^{\infty} X_j^*\right)_1,$$

$$\left(\sum_{j=1}^{\infty} X_j\right)_p^* = \left(\sum_{j=1}^{\infty} X_j^*\right)_q \quad \text{if} \quad 1 \le p < \infty \quad \text{and} \quad \frac{1}{p} + \frac{1}{q} = 1.$$

In particular if the X_j's are all reflexive and $1 < p < \infty$ then $\left(\sum_{j=1}^{\infty} X_j\right)_p$ is reflexive.

22 Examples. (a) $L_p[0,1]$ is isometric to $\left(\sum_{n=1}^{\infty} L_p[0,1]\right)_p$ for $1 \le p \le \infty$

(b) l_p is isometric to $\left(\sum_{n=1}^{\infty} l_p\right)_p$ for $1 \le p \le \infty$ and c_0 is isometric to $\left(\sum_{n=1}^{\infty} c_0\right)_0$.

(c) $C(\Delta)$ is isomorphic to $\left(\sum_{n=1}^{\infty} C(\Delta)\right)_0$ where Δ is the Cantor set.

To see (a) is relatively easy. We put $A_n = [(n+1)^{-1}, n^{-1}]$ and identifying $f \in L_p[0,1]$ with the sequence $(f|A_n)_{n=1}^{\infty}$ we see $L_p[0,1] = \left(\sum_{n=1}^{\infty} L_p(A_n)\right)_p$. Since $L_p(A_n) \cong L_p[0,1]$ we have the claim. (b) is analogous. In order to prove (c) we start with the same idea. Let us fix an increasing sequence $t_n \nearrow 1$ such that $t_1 < 0, t_2 > 0$ and $\Delta \cap (t_n, t_{n+1}) = \Delta_n$ is homeomorphic to Δ. Identifying the n-th summand of $\left(\sum_{n=1}^{\infty} C(\Delta)\right)_0$ with $C(\Delta_n)$ we can identify $\left(\sum_{n=1}^{\infty} C(\Delta)\right)_0$ with the subspace $C_0(\Delta) = \{f \in C(\Delta) : f(1) = 0\}$. So we have to prove that $C_0(\Delta) \sim C(\Delta)$. This follows from the following

23 Lemma. *The space $C(\Delta)$ contains a complemented subspace isomorphic to c_0.*

Proof: With Δ_n as above we put $\xi_n = \sup \Delta_n$. The map $Pf = \sum_{n=1}^{\infty} [f(\xi_n) - f(1)]\chi_{\Delta_n}$ is a projection from $C(\Delta)$ onto a subspace of $C_0(\Delta)$ of functions constant on each Δ_n. One easily checks that this subspace is isometric to c_0. ∎

Now we can write $C(\Delta) \sim Z \oplus c_0$. Since $C_0(\Delta)$ has codimension 1 in $C(\Delta)$ Proposition 3 gives that $C_0(\Delta)$ is isomorphic to $Z \oplus V$ where

$V \subset c_0, V = \{(\alpha_j)_{j=1}^{\infty} \in c_0 \colon \alpha_1 = 0\}$. Since V is clearly isometric to c_0 we have

$$C_0(\Delta) \sim Z \oplus V \sim Z \oplus c_0 \sim C(\Delta). \qquad \blacksquare$$

24. Arguments like that used in **22.**(c) are used quite often in proving that two Banach spaces are isomorphic. In the above case it is possible to write an explicit formula for the isomorphism, but in more complicated cases it is practically impossible. The following theorem is the most common variant of the 'decomposition method' which is the elaboration of such arguments.

Theorem. *Let X and Y be two Banach spaces such that X is isomorphic to a complemented subspace of Y and Y is isomorphic to a complemented subspace of X. Assume moreover that X is isomorphic to $\left(\sum_{n=1}^{\infty} X\right)_p$ for some p with $1 \le p \le \infty$ or 0. Then $X \sim Y$.*

Proof: Let us write $X \sim Z \oplus Y$ and $Y \sim V \oplus X$. Then

$$Y \sim V \oplus X \sim V \oplus \left(\sum_{n=1}^{\infty} X\right)_p \sim V \oplus X \oplus \left(\sum_{n=1}^{\infty} X\right)_p \sim Y \oplus X,$$

and also

$$X \sim \left(\sum_{n=1}^{\infty} X\right)_p \sim \left(\sum_{n=1}^{\infty} Z \oplus Y\right)_p \sim \left(\sum_{n=1}^{\infty} Z\right)_p \oplus \left(\sum_{n=1}^{\infty} Y\right)_p$$

$$\sim \left(\sum_{n=1}^{\infty} Z\right)_p \oplus \left(\sum_{n=1}^{\infty} Y\right)_p \oplus Y \sim X \oplus Y.$$

So $X \sim Y$. One easily checks that each step is justisfied. $\blacksquare$

Notes and Remarks.

Much of the material in this chapter is folklore of Banach space theory. The notion of isomorphic embedding appeared in the early 30's, and various cases of embedding and impossibility of embedding were studied by Banach and his collaborators. Those results are summarized in Banach [1932] where in particular *Theorem 4* is proved and in Banach-Mazur [1933]. The comparison between Banach [1931] and Banach [1932] shows the crystallization of the notion.

The general notion of a Schauder basis was introduced in Schauder [1927]. By now the subject has grown enormously. An encyclopaedic treatment is in Singer [1970] and [1981] and a very good, concise presentation in Lindenstrauss-Tzafriri [1977]. The Haar system, introduced in Haar [1910], is one of the most important orthonormal systems (see Kashin-Saakian [1984]). It is also a prototype of the general notion of martingale difference sequence. Haar studied his system in order to show that the expansion with respect to this system of every $f \in C[0,1]$ is uniformly convergent (Exercise 7(a)). *Proposition 10.(b)* was proved by Schauder [1928]. The classical Faber-Schauder system was introduced and shown to be a basis in $C[0,1]$ in Faber [1910]. The general case was independently treated in Schauder [1927]. The notion of a basic sequence and a block basis emerged in the late 50's in the work of Pełczyński and Bessaga. The main ideas of *Proposition 15* are in Krein-Milman-Rutman [1940]. They were rediscovered and applied in Bessaga-Pełczyński [1958]. This paper contains also *Proposition 17* and *Corollary 18*. Direct sums appear already in Banach [1932]. *Theorem 24* is taken from Pełczyński [1960] but germs of such ideas are due to Borsuk (see Borsuk [1933] and Banach [1932] chap XI. §7).

Exercises

1. Show that, if X_1 and X_2 are closed subspaces of a Banach space X, both of codimension n, then $X_1 \sim X_2$.

2. A norm $||| \cdot |||$ on a Banach space $(X, \| \cdot \|)$ is called an equivalent norm if for some $C > c > 0$ we have $c\|x\| \leq |||x||| \leq C\|x\|$ for all $x \in X$.

    (a) Suppose $Y \subset X$ and $d(Y, Z) < c$ (here $d(Y, Z)$ is the Banach-Mazur distance defined in II.E.6) for some Banach space Z. Construct an equivalent norm $||| \cdot |||$ on X such that $c^{-1}\|x\| \leq |||x||| \leq \|x\|$ for all $x \in X$ and $(Y, ||| \cdot |||)$ is isometric to Z.

    (b) For $f \in C[0,1]$ put $|||f||| = \|f\|_\infty + \|f\|_2$. Show that $||| \cdot |||$ is an equivalent norm on $C[0,1]$ and ℓ_1 is not isometric to any subspace of $(C[0,1], ||| \cdot |||)$.

    (c) A norm is strictly convex if for all x, y, with $x \neq y$ and $\|x\| = \|y\| = 1$ we have $\|x + y\| < 2$. Show that every separable Banach space has an equivalent strictly convex norm.

    (d) If Γ is a set of continuum cardinality then $\ell_\infty(\Gamma)$ does not have any equivalent strictly convex norm.

3. Suppose that $(x_n)_{n=1}^\infty$ is a basis in X. Show that the biorthogonal functionals $(x_n^*)_{n=1}^\infty$ form a basic sequence in X^*. This need not be a basis (even if X^* is separable).

4. Let $(x_n)_{n=1}^{\infty} \subset X$ be a bounded sequence which is not norm relatively compact. Show that there are sequences n_k and m_k such that $(x_{n_k} - x_{m_k})_{k=1}^{\infty}$ is a basic sequence.

5. Suppose X and Y are subspaces of a Banach space Z, such that $X \cap Y = \{0\}$. Show that, if the algebraic sum $X + Y$ is not closed in Z, then there exist a basic sequence $(x_n)_{n=1}^{\infty} \subset X$ and a basic sequence $(y_n)_{n=1}^{\infty} \subset Y$ such that $\|x_n\| = \|y_n\| = 1$ for $n = 1, 2, \ldots$ and $\|x_n - y_n\| < 4^{-n}$ for $n = 1, 2, \ldots$. In particular X and Y have isomorphic subspaces.

6. Prove the following.

    (a) The space $C[0, 1]$ has a basis $(f_n)_{n=1}^{\infty}$ consisting of polynomials.

    (b) There exists a sequence of polynomials which is a basis in all $L_p[0, 1]$, $1 \le p < \infty$.

7. (a) Show that for $f \in C[0, 1]$ the Haar-Fourier expansion converges uniformly.

    (b) Construct a basis in $C(\Delta)$.

8. Let $(\varphi_n(t))_{n=1}^{\infty}$ be the Faber-Schauder system. Show that $f = \sum_{n=1}^{\infty} a_n \varphi_n \in Lip_\alpha[0, 1], 0 < \alpha < 1$ if and only if $|a_n| = 0(n^{-\alpha})$, so $Lip_\alpha[0, 1]$ is isomorphic to ℓ_∞.

9. Show the following isomorphisms:

    (a) $H_p(\mathbb{T}) \sim L_p(\mathbb{T})$ for $1 < p < \infty$;

    (b) $Lip_1[0, 1] \sim L_\infty[0, 1]$;

    (c) $C^k[0, 1] \sim C[0, 1]$ for $k = 1, 2, 3, \ldots$;

    (d) $W_p^k[0, 1] \sim L_p[0, 1]$ for $k = 1, 2, 3, \ldots$ and $1 \le p < \infty$.

10. Suppose $S \subset \Delta$ is a closed subset. Prove the following.

    (a) If S has continuum cardinality then $C(S) \sim C(\Delta)$.

    (b) If S is countable then $C(S)$ is not isomorphic to $C(\Delta)$.

11. Show the following isomorphisms.

    (a) $C[0, 1] \sim (\Sigma C[0, 1])_0$;

    (b) $C[0, 1] \oplus c_0 \sim C[0, 1]$;

    (c) $L_p[0, 1] \oplus \ell_p \sim L_p[0, 1]$.

12. Prove the following.

    (a) Every isometric embedding of ℓ_p into $L_p(\mu)$ (μ-arbitrary measure), $1 \leq p < \infty, p \neq 2$, maps unit vectors onto disjointly supported, norm-one functions.

    (b) Every subspace of $L_p(\mu)$ isometric to ℓ_p, $1 \leq p < \infty$ is complemented.

13. Suppose $x_n \xrightarrow{\omega} 0$ in X. Show that for every $\varepsilon > 0$ and every $N \in \mathbb{N}$ there exists a finite sequence $(\lambda_j)_{j=N}^{N+M}$ such that $\sum_{j=N}^{N+M} \lambda_j = \sum_{j=N}^{N+M} |\lambda_j| = 1$ and $\left\| \sum_{j=N}^{N+M} \varepsilon_j \lambda_j x_j \right\| < \varepsilon$ for every choice of ε_j with $|\varepsilon_j| = 1$.

14. Let Σ be a σ-algebra of subsets of Ω and let μ be a probability measure on Σ. Let Σ_1 be a sub-σ-algebra of Σ. Prove the following.

    (a) For every $f \in L_1(\Sigma, \mu)$ there exists a unique function Pf which is Σ_1-measurable, such that for every $B \in \Sigma_1$ we have $\int_B f d\mu = \int_B Pf d\mu$.

    (b) P is a linear map.

    (c) If $f \geq 0$ then $Pf > 0$.

    (d) P is a norm-one projection in $L_p(\Sigma, \mu)$ for $1 \leq p \leq \infty$.

    (e) If $g \in L_\infty(\Sigma_1, \mu)$ and $f \in L_1(\Sigma, \mu)$ then $P(gf) = gPf$.

15. Prove the following.

    (a) Every separable Banach space X is a quotient of ℓ_1.

    (b) If Γ is a set of continuum cardinality then $\ell_1(\Gamma)$ is isometric to a subspace of ℓ_∞.

    (c) If Γ is a set of continuum cardinality then $\ell_p(\Gamma)$, $1 < p < \infty$ and $c_0(\Gamma)$ are not isomorphic to any subspace of ℓ_∞.

16. (The weak basis theorem). Suppose that $(x_n)_{n=1}^\infty \subset X$ is such that for every $x \in X$ there exists a unique sequence of scalars $(t_n)_{n=1}^\infty$ such that the series $\sum_{n=1}^\infty t_n x_n$ converges weakly to x. Then $(x_n)_{n=1}^\infty$ is a Schauder basis in X.

II.C. Weak Compactness

The sets compact in the $\sigma(X, X^*)$-topology are important in many applications. We study such sets in this section. The main result is the Eberlein-Šmulian theorem which says that weak compactness of a set is determined by properties of sequences, even when the $\sigma(X, X^*)$-topology on this set is not metrizable. We apply this to study weakly compact operators, i.e. operators such that the image of any ball is contained in a weakly compact set. We show that each weakly compact operator factorizes through a reflexive space, and use this to investigate properties of such operators.

1. This section is devoted to the study of weakly compact sets in Banach spaces, i.e. subsets $A \subset X$ which are compact in the $\sigma(X, X^*)$-topology. We say that the set $A \subset X$ is relatively weakly compact if its $\sigma(X, X^*)$-closure in X is weakly compact. From Theorem II.A.14 we infer that every bounded subset of a reflexive space is relatively weakly compact. Also by Theorem II.A.4 and II.A.14 we get that every convex, bounded, norm-closed subset of a reflexive space is weakly compact. Also if X is a reflexive space and if $T: X \rightarrow Y$ is a continuous linear operator, then $T(B_X)$ is a weakly compact set.

2. We have

Lemma. *A subset $A \subset X$ is relatively weakly compact if and only if it is bounded and the $\sigma(X^{**}, X^*)$-closure of $i(A)$ in X^{**} is contained in $i(X)$.*

Proof: Immediate from Alaoglu's theorem II.A.9 and the remarks after Proposition II.A.10.

3. We have seen (Exercises II.A.1 and 2) that the weak topology (even on a weakly compact set) need not be metrizable. Thus it seems that we are not permitted to use sequential arguments. Actually it is not so. The following important and quite surprising theorem says essentially that sequential arguments about weakly compact sets are permissible.

Theorem. (Eberlein-Šmulian). *A set $A \subset X$ is relatively weakly compact if and only if every sequence $(a_n) \subset A$ has a weakly convergent subsequence.*

Proof: Assume A is relatively weakly compact in X and fix $(a_n)_{n=1}^\infty \subset A$. Denote $V = \text{span}\{a_n\}_{n=1}^\infty$. Clearly V is a separable subspace of X. Let us fix a sequence $(x_n^*)_{n=1}^\infty \subset X^*$ such that if $x \in V$ and $x_n^*(x) = 0$ for $n = 1, 2, \ldots$ then $x = 0$. (There are many ways to construct such a sequence, e.g. take $(v_n)_{n=1}^\infty$ dense in V and use the Hahn-Banach theorem to get $x_n^* \in X^*$ such that $\|x_n^*\| = 1$ and $x_n^*(v_n) = \|v_n\|$ for $n = 1, 2, 3, \ldots$.) Using a standard diagonal argument we find a subsequence $(a_{n_k})_{k=1}^\infty$ such that $\lim_{k \to \infty} x_n^*(a_{n_k})$ exists for every $n = 1, 2, \ldots$. Let $\bar{y} \in X$ be any weak cluster point of the set $\{a_{n_k}\}_{k=1}^\infty \subset A$; thus $x_n^*(\bar{y}) = \lim_{k \to \infty} x_n^*(a_{n_k})$ for $n = 1, 2, \ldots$. Since V is a closed subspace, it is weakly closed, so $\bar{y} \in V$. Thus $\bar{y}$ is the unique cluster point of the sequence $(a_{n_k})_{k=1}^\infty$. We have to show that $a_{n_k} \xrightarrow{\omega} \bar{y}$ as $k \to \infty$, i.e. that for every $x^* \in X^*$ we have $x^*(a_{n_k}) \to x^*(\bar{y})$ as $k \to \infty$. But if for some $x_0^* \in X^*$ we have $x^*(a_{n_{k_s}}) \longrightarrow \alpha \neq x_0^*(\bar{y})$, then there exists a cluster point of $\{a_{n_{k_s}}\}_{s=1}^\infty$, thus a cluster point of $\{a_{n_k}\}_{k=1}^\infty$, different from $\bar{y}$. This contradicts the fact that $\bar{y}$ was the unique cluster point of $\{a_{n_k}\}_{k=1}^\infty$. So in fact $a_{n_k} \xrightarrow{\omega} \bar{y}$.

To prove the converse let us assume that $A \subset X$ is not relatively weakly compact. We have to produce a sequence $(a_n)_{n=1}^\infty \subset A$ without a weakly convergent subsequence. Using Lemma 2 we find $F \in X^{**} \backslash i(X)$ such that F is in the $\sigma(X^{**}, X^*)$ closure of A. Let $\theta = \text{dist}(F, i(X)) > 0$.

We will inductively construct $(a_n, g_n)_{n=1}^\infty \subset A \times X^*$ such that

(1) $g_n \in X^*$ and $\|g_n\| = 1$,

(2) $a_n \in A$,

(3) $\text{Re}\, F(g_n) > \frac{3}{4}\theta$ for $n = 1, 2, 3 \ldots$,

(4) $|g_n(a_j)| < \frac{1}{4}\theta$ for $j < n$,

(5) $\text{Re}\, g_n(a_j) > \frac{3}{4}\theta$ for $j \geq n$.

Such $(a_n)_{n=1}^\infty$ has no weakly convergent subsequence, because if $(a_{n_k}) \xrightarrow{\omega} a$ then by Corollary II.A.5 there is a convex combination

$$\sum_{k=N}^M \alpha_k a_{n_k} \qquad \text{such that} \qquad \left\| \sum_{k=N}^M \alpha_k a_{n_k} - a \right\| < \frac{1}{4}\theta.$$

From (4) we see that for $n > n_M$ we have $|g_n(\sum_{k=N}^M \alpha_k a_{n_k})| < \frac{1}{4}\theta$. Thus we infer that for $n > n_M$ we have $|g_n(a)| < \frac{1}{2}\theta$. On the other hand (5) implies $\text{Re}\, g_n(a) \geq \frac{3}{4}\theta$ for $n = 1, 2, \ldots$, a contradiction.

Digression. Let us consider the following simple example: $A = B_{c_0} \subset c_0$. This is clearly not a weakly compact set. Put $a_n = \sum_{j=1}^n e_j$. Then $F = (1, 1, \ldots) \in \ell_\infty = c_0^{**}, \theta = 1$ and $g_n = e_n^*$. Conditions (1)-(5) are clearly satisfied. Our construction is an attempt to imitate this example in every set not weakly compact. End of digression.

Inductive construction. Since $\|F\| \geq \theta$ there exists g_1 satisfying (1) and (3). Since $F \in \overline{i(A)}^{\sigma(X^{**}, X^*)}$ there exists $a_1 \in A$ such that $|F(g_1) - g_1(a_1)|$ is small enough so that (5) holds.

Suppose we have $(a_j, g_j)_{j=1}^n$ satisfying (1)-(5). Use the Hahn-Banach theorem to obtain $\varphi \in X^{***}$ such that $\varphi(a_j) = 0$ for $j = 1, 2, \ldots, n$ and $\varphi(F) > \frac{3}{4}\theta$ and $\|\varphi\| < 1$. The Goldstine theorem II.A.13 gives $g_{n+1} \in X^*$ such that (1), (3) and (4) hold. Next we choose $a_{n+1} \in A$ which approximates F on $g_1, \ldots, g_{n+1}$ so well that (3) gives (5). ∎

4. Now we would like to introduce and study the concept of a weakly compact operator. An operator $T \colon X \to Y$ is said to be weakly compact if the set $T(B_X)$ is relatively weakly compact. We infer from Theorem 3 that this is equivalent to the condition that for every bounded sequence $(x_n)_{n=1}^\infty \subset X$ the sequence $(T(x_n))_{n=1}^\infty$ has a weakly convergent subsequence. From Theorem II.A.14 we get easily that $T \colon X \to Y$ is weakly compact if X or Y is reflexive. On the other hand the operator $T \colon L_1[0, 1] \to C[0, 1]$ defined as $Tf(x) = \int_0^x f(t)dt$ is not weakly compact since $T(2n(\chi_{[\frac{1}{2} - \frac{1}{n}, \frac{1}{2}]} - \chi_{[\frac{1}{2}, \frac{1}{2} + \frac{1}{n}]}))$ has no weakly convergent subsequence. We should note that the requirement that $T(B_X)$ be weakly compact is far too strong. For example the identity $id \colon C[0, 1] \to L_2[0, 1]$ is weakly compact but $id(B_{C[0,1]})$ is not norm-closed in $L_2[0, 1]$. We can also find a functional on $C[0, 1]$, e.g. $\varphi(f) = \int_0^{\frac{1}{2}} f(t)dt - \int_{\frac{1}{2}}^1 f(t)dt$, such that $\varphi(B_{C[0,1]})$ is not closed in $\mathbb{R}$.

5. Our investigation of weakly compact operators is based on the following:

Theorem. *An operator $T \colon X \to Y$ is weakly compact if and only if T factors through a reflexive space, i.e. there exist a reflexive space R and operators $\alpha \colon X \to R$ and $\beta \colon R \to Y$ such that $T = \beta \cdot \alpha$. Moreover such a factorization can be chosen such that $\|\alpha\| \cdot \|\beta\| \leq 4 \cdot \|T\|$.*

The 'if' part is obvious from **4**, so we will discuss the 'only if' part.

Proof: We can assume that T is 1-1. (If not consider $\widetilde{T} \colon X/\ker T \longrightarrow Y$ defined by $\widetilde{T}([x]) = Tx$.) Let us put $T(B_X) = W$. The set W is convex and weakly compact. We define

$$\mathcal{U}_n = 2^n W + 2^{-n} B_Y$$
$$= \{y_0 \in Y \colon y_0 = 2^n w + 2^{-n} y \text{ with } w \in W \text{ and } \|y\| \leq 1\}.$$

Each $\mathcal{U}_n$ is convex and $2^{-n}B_Y \subset \mathcal{U}_n \subset (2^n\|T\| + 2^{-n})B_Y$. Thus $\mathcal{U}_n$ induces a norm $\|\cdot\|_n$ on Y such that

$$(2^n\|T\| + 2^{-n})^{-1}\|y\| \le \|y\|_n \le 2^n\|y\| \qquad \text{for all} \qquad y \in Y.$$

We define $R = \{y \in Y: \|\|y\|\| = (\sum_{n=1}^{\infty}\|y\|_n^2)^{\frac{1}{2}} < \infty\}$. Obviously $(R, \|\|\cdot\|\|)$ is a closed subspace of $(\sum_{n=1}^{\infty}(Y, \|\cdot\|_n))_2$ so it is a Banach space. One checks (see II.B.21) that R^{**} can be canonically identified with $\{y^{**} \in Y^{**}: (\sum_{n=1}^{\infty}\|y^{**}\|_n^2)^{\frac{1}{2}} < \infty\}$. Since W is weakly compact the unit ball in $(Y^{**}, \|\cdot\|_n)$ equals $2^nW + 2^{-n}B_{Y^{**}}$. Thus for $y^{**} \in Y^{**}\backslash i(Y)$ we have $\|y^{**}\|_n \ge 2^n \inf\{\|y^{**} - y\|: y \in Y\}$. This shows that for $y^{**} \in Y^{**}\backslash i(Y)$ the series $\sum_{n=1}^{\infty}\|y^{**}\|_n^2$ diverges, in other words $R^{**} \subset (\sum_{n=1}^{\infty}(Y, \|\cdot\|_n))_2$ so R is reflexive. Now we define $\alpha: X \to R$ as $\alpha(x) = T(x)$. Since $Tx \in \|x\| \cdot W$ we have $\|Tx\|_n \le 2^{-n}$ so $Tx \in R$ and $\|\alpha\| \le 2$. We define $\beta(y) = y$ for $y \in R$. Clearly $\|\beta\| \le 2\|T\|$.   ∎

We will now use Theorem 5 to deduce the fundamental properties of weakly compact operators from corresponding properties of reflexive spaces.

6 Theorem. *(a) Suppose $T: X \to Y$ is weakly compact and $U: X_1 \to X$ and $V: Y \to Y_1$ are continuous. Then VTU is weakly compact.*
(b) An operator $T: X \to Y$ is weakly compact if and only if $T^*: Y^* \to X^*$ is weakly compact.
(c) An operator $T: X \to Y$ is weakly compact if and only if $T^{**}(X^{**}) \subset i(Y)$.
(d) The set of all weakly compact operators from X into Y is a norm-closed linear subspace of $L(X, Y)$.

Proof: (a) follows immediately from the fact that norm-continuous linear operators are also weakly continuous and the well known fact that a continuous image of a compact set is compact.
(b) Follows from Theorem 5 and Theorem II.A.14.
(c)   If T is weakly compact then from Theorem 5 we get that $T^{**}(X^{**}) \subset \beta^{**}(R^{**}) = \beta(R) \subset Y$. Conversely, suppose $T^{**}(B_{X^{**}}) \subset i(Y)$. Since $T^{**}(B_{X^{**}})$ is $\sigma(Y^{**}, Y^*)$-compact (see Theorem II.A.9) we infer (cf. remarks after Proposition II.A.10) that $i^{-1}(T^{**}(B_{X^{**}})$ is $\sigma(Y, Y^*)$-compact. Obviously $T(B_X) \subset i^{-1}(T^{**}(B_{X^{**}}))$ so $T(B_X)$ is relatively weakly compact, so T is weakly compact.
(d) The fact that the set of weakly compact operators is linear can be seen easily from (c). To show that it is complete it is enough to show

that if $T = \sum_{n=1}^{\infty} T_n$ with T_n weakly compact and $\|T_n\| \leq 4^{-n}$ then T is weakly compact. For each T_n let us take a factorization as in Theorem 5 with R_n reflexive and $\|\alpha_n\| \leq 2 \cdot 2^{-n}$ and $\|\beta_n\| \leq 2 \cdot 2^{-n}$. We define $R = (\sum_{n=1}^{\infty} R_n)_2$ and $\alpha \colon X \to R$ by $\alpha(x) = (\alpha_n(x))_{n=1}^{\infty}$ and $\beta(R) \to Y$ by $\beta((r_n)_{n=1}^{\infty}) = \sum_{n=1}^{\infty} \beta_n(r_n)$. The space R is reflexive (II.B.21) and $\beta\alpha(x) = T(x)$ so T is weakly compact. ∎

7. A more elementary proof of the fact that the norm limit of weakly compact operators is weakly compact follows from the following

Lemma. *A subset $A \subset X$ is weakly compact provided for every $\varepsilon > 0$ there exists a weakly compact set $A_\varepsilon \subset X$ such that $A \subset A_\varepsilon + \varepsilon B_X$.*

Proof: Since the $\sigma(X^{**}, X^*)$-closure of $A_\varepsilon + \varepsilon B_X$ in X^{**} equals $A_\varepsilon + \varepsilon B_{X^{**}}$ (use Lemma 2) we infer that the $\sigma(X^{**}, X^*)$-closure of A is contained in $\bigcap_{\varepsilon > 0} A_\varepsilon + \varepsilon B_{X^{**}} \subset X$. Lemma 2 gives that A is weakly compact. ∎

The reader undoubtedly observed that the above arguments have already been used in the proof of Theorem 5.

8 Corollary. *If $A \subset X$ is a relatively weakly compact set then $\overline{\mathrm{conv}}\, A$ is weakly compact.*

Proof: Assume first that X is separable. Replacing A by its weak closure we can assume that A is actually weakly compact. Let $\ell_1(A)$ denote the space of functions $x(a)$ defined on A such that $\sum_{a \in A} |x(a)| < \infty$. This is clearly a Banach space. Define an operator $T \colon \ell_1(A) \to X$ by $T((x(a))_{a \in A}) = \sum_{a \in A} x(a) \cdot a$. Since A is bounded (Lemma 2) the operator T is bounded. The adjoint operator $T^* \colon X^* \to \ell_\infty(A)$ acts as $T^*(x^*)(a) = x^*(a)$; thus actually $T^*(X^*) \subset C(A, \sigma(X, X^*))$. Let S denote the same map as T^* but acting from X^* into $C(A)$. Let $(x_n^*) \subset B_{X^*}$. Since X is separable, passing to a subsequence we can assume that $x_n^* \to x_0^*$ in the $\sigma(X^*, X)$-topology, so $S(x_n^*) \longrightarrow S(x_0^*)$ pointwise in $C(A)$. Thus for every $\mu \in C(A)^*$, the Riesz representation Theorem I.B.11 and the Lebesgue theorem give $\int_A S(x_n^*)d\mu \longrightarrow \int_A S(x_0^*)d\mu$, so $S(x_n^*)$ converges weakly to $S(x_0^*)$. Theorem 3 gives that $S(B_{X^*})$ is relatively weakly compact, so T^* is weakly compact so by Theorem 6 the operator T is weakly compact. Since $\overline{\mathrm{conv}}\, A \subset \overline{T(B_{\ell_1(A)})}$ we see that $\overline{\mathrm{conv}}\, A$ is weakly compact.

For non-separable X we argue as follows. If $\overline{\mathrm{conv}}\ A$ is not weakly compact then by Theorem 3 there is a sequence in $\overline{\mathrm{conv}}\ A$ that is not relatively weakly compact. Each element of such a sequence is the norm limit of some convex combinations of a countable subset of A. Thus we get a separable subspace $Y \subset X$ such that $\overline{\mathrm{conv}}(Y \cap A)$ is not weakly compact. This is impossible in view of the first part of our argument. ∎

Notes and remarks.

The fundamental *Theorem 3* is, as usual, the result of much work by many mathematicians. The standard references to papers where various pieces appeared in full generality are Eberlein [1947] and Šmulian [1940]. Our proof of sufficiency is taken from James [1981]. Other proofs can be found in Dunford-Schwartz [1958], Pełczyński [1964], Whitley [1967]. Basic facts about weakly compact operators (*Theorem 6 (a,b,c)*) have been proved in Gantmacher [1940]. *Theorem 5* which is our basis for discussing weakly compact operators was proved by Davis-Figiel-Johnson-Pełczyński [1974]. This paper contains numerous applications of the technique of the proof to problems of the geometry of Banach spaces. *Theorem 5* is the first instance of an application in this book of a factorization idea. The reader will encounter many more examples later on. It should also be noted that the proof is a very elementary application of an interpolation argument. *Corollary 7* was proved by M. Krein-V. Šmulian [1940].

Exercises

1. Let X be a Banach space. Show that $i(X^*)$ is complemented in X^{***}.

2. Let $f \in L_1(\mu)$ and $f \geq 0$. Show that $\{g \in L_1(\mu) : |g| \leq f\}$ is weakly compact, and if $\mu|\mathrm{supp}f$ is not purely atomic then it is not norm-compact.

3. Let X and Y be Banach spaces and let $T: X \to Y$ be an operator that is not weakly compact. Show that there exist $S: \ell_1 \to X$ and $U: Y \to \ell_\infty$ such that $UTS = \sigma$ where $\sigma((\xi_n)_{n=1}^\infty) = \left(\sum_{n=1}^k \xi_n\right)_{k=1}^\infty$.

4. Let μ be a measure on $\mathbb{T}$ and let $T_\mu: L_1(\mathbb{T}) \to L_1(\mathbb{T})$ be given by $T_\mu(f) = f * \mu$. Show that the following conditions are equivalent:

   (a) T_μ is weakly compact;

   (b) T_μ is compact;

   (c) $\mu \in L_1(\mathbb{T})$.

5. Suppose that $T_K f(x) = \int_0^1 K(x,y)f(y)dy$ where $K(x,y)$ is a measurable function on $[0,1]\times[0,1]$. Show that if $\text{supess}_x \int_0^1 |K(x,y)|^2 dy < \infty$, then T_K maps $L_\infty[0,1]$ into $L_\infty[0,1]$ and is weakly compact. Show that for $K_0(x,y) = x^{-1}\chi_{[0,x]}(y)$ the operator T_{K_0} maps $C[0,1]$ into itself but is not weakly compact.

6. Let $S_N(\mu) = \sum_{-N}^N \hat\mu(n)e^{in\theta}$. Show that if μ is a measure on $\mathbb{T}$ such that $\sup_N \|S_N(\mu)\|_1 < \infty$ then $\hat\mu(n) \to 0$ as $|n| \to \infty$. Let X be the space of all measures μ such that $|||\mu||| = \sup_N \|S_N(\mu)\|_1 < \infty$. Show that the Fourier transform $\hat{} : X \to c_0$ is a weakly compact operator.

7. Let μ be a measure on $\mathbb{T}$ such that $\hat\mu(n) = 0$ or 1 for $n = 0,\pm1,\pm2,\ldots$. Show that $\{n \in \mathbb{N}: \hat\mu(n) = 1\} = (F_1 \cup V)\backslash F_2$ with F_1, F_2 finite sets and V a periodic set, i.e. a finite sum of arithmetic progressions.

8. A basis $(x_n)_{n=1}^\infty$ in a Banach space X is called boundedly complete if for every sequence of scalars $(a_n)_{n=1}^\infty$ such that $\sup_N \|\sum_{n=1}^N a_n x_n\| < \infty$ the series $\sum_{n=0}^\infty a_n x_n$ converges. A basis $(x_n)_{n=1}^\infty$ is called shrinking if for every $x^* \in X^*, \lim_{N\to\infty} \|x^* \mid \text{span}(x_n)_{n\geq N}\| = 0$.

    (a) Show that if $(x_n)_{n=1}^\infty$ is a shrinking basis in X then $(x_n^*)_{n=1}^\infty$ is a boundedly complete basis in X^*.

    (b) Let $(x_n)_{n=1}^\infty$ be a boundedly complete basis in X and let $Y = \text{span}(x_n^*)_{n=1}^\infty \subset X^*$. Show that X is naturally isometric to Y^*.

    (c) Show that every basis in a reflexive space is both shrinking and boundedly complete.

    (d) Show that if $(x_n)_{n=1}^\infty$ is a shrinking basis in X then X^{**} can be naturally identified with the set $\{(a_n)_{n=1}^\infty : \sup_N \|\sum_{n=1}^N a_n x_n\| < \infty\}$.

9. (James space). Let us define the following norm on sequences $(\xi_j)_{j=1}^\infty$:

$$\|(\xi_j)\|_J = \sup\left\{\left(\sum_{s=1}^{n-1}|\xi_{j_{s+1}} - \xi_{j_s}|^2\right)^{\frac{1}{2}} : 1 \leq j_1 < j_2 < \cdots < j_n\right\}.$$

The James space $J = \{(\xi_j)_{j=1}^\infty \in c_0 : \|(\xi_j)\|_J < \infty\}$.

    (a) Show that J is a Banach space.

(b) Show directly that the unit ball in J is not relatively weakly compact.

(c) Show that the unit vectors form a basis in J and that this basis is shrinking.

(d) Show that J^{**} can be naturally identified with $\{(\xi_j)_{j=1}^\infty \in c\colon \|(\xi_j)\|_J < \infty\}$. Thus $\dim J^{**}/i(J) = 1$.

(e) Show that J is isomorphic to J^{**}.

(f) Show that J is not isomorphic to $J \oplus J$.

10. Apply the construction of Theorem 5 to the operator $\Sigma\colon \ell_1 \to c$ defined by $\Sigma\left((\xi)_{j=1}^\infty\right) = \left(\sum_{j=1}^n \xi_j\right)_{n=1}^\infty$. Show that the resulting space X have the property that $\dim X^{**}/i(X) = 1$.

II.D. Convergence Of Series

We start this section with various characterizations of unconditionally and weakly unconditionally convergent series in Banach spaces. We prove the classical Orlicz theorem that every unconditionally convergent series in L_p, $1 \leq p \leq 2$, has norms square summable. We also introduce the notion of an unconditional Schauder basis and show that $C[0,1]$ and $L_1[0,1]$ are not subspaces of any space with an unconditional basis. We conclude with the proof that the Haar system is an unconditional basis in $L_p[0,1]$ for $1 < p < \infty$.

1. This section deals with various types of convergence of series of elements of a Banach space X. The series $\sum_{n=1}^{\infty} x_n$ is said to converge absolutely if $\sum_{n=1}^{\infty} \|x_n\| < \infty$. It is an obvious consequence of the triangle inequality that absolutely convergent series converge. The series $\sum_{n=1}^{\infty} x_n$ is said to converge unconditionally if the series $\sum_{n=1}^{\infty} \varepsilon_n x_n$ converges for all ε_n with $\varepsilon_n = \pm 1$ for $n = 1, 2, 3, \ldots$.

When X is finite dimensional then the classical Riemann theorem asserts that absolute and unconditional convergence coincide. This assertion actually characterizes finite dimensional Banach spaces (Exercise III.F.8). For the time being let us note that $\sum_{n=1}^{\infty} \xi_n e_n$ converges unconditionally in ℓ_p, $1 < p < \infty$, whenever $(\xi_n)_{n=1}^{\infty} \in \ell_p$ but converges absolutely only when $(\xi_n)_{n=1}^{\infty} \in \ell_1$.

The following characterize unconditional convergence.

2 Theorem. *For a series $\sum_{n=1}^{\infty} x_n$ in a Banach space X the following conditions are equivalent:*

(a) *the series $\displaystyle\sum_{n=1}^{\infty} x_n$ is unconditionally convergent;*

(b) *the series $\displaystyle\sum_{n=1}^{\infty} a_n x_n$ is convergent for every $(a_n)_{n=1}^{\infty} \in \ell_\infty$;*

(c) *there exists a compact operator $T\colon c_0 \to X$ such that $T(e_n) = x_n$ for $n = 1, 2, 3, \ldots$;*

(d) *for every permutation σ of the integers the series $\sum_{n=1}^{\infty} x_{\sigma(n)}$ converges;*

(e) *for every increasing sequence of integers $(n_k)_{k=1}^{\infty}$ the series $\sum_{n=1}^{\infty} x_{n_k}$ converges.*

Proof: We will prove the following implications:

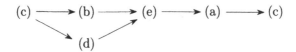

(c)$\Rightarrow$(b). For every N we put $v_N = \sum_{n=1}^{N} a_n e_n \in c_0$. Since v_N is weakly Cauchy $T(v_N)$ is norm-Cauchy, so $\sum_{n=1}^{\infty} a_n x_n$ converges.

(b)$\Rightarrow$(e). Take appropriate sequence of zeros and ones.

(e)$\Rightarrow$(a). Given a sequence $\varepsilon_n = \pm 1$ we define n_k in such a way that $\varepsilon_{n_k} = 1$ for $k = 1, 2, 3, \ldots$ and $\varepsilon_n = -1$ if $n \neq n_k$ for $k = 1, 2, \ldots$. Then $\sum_{n=1}^{\infty} \varepsilon_n x_n = 2 \sum_{k=1}^{\infty} x_{n_k} - \sum_{n=1}^{\infty} x_n$. Since both series on the right hand side converge the left hand side series also converges.

(a)$\Rightarrow$(c). We will give the proof for real spaces. The changes for complex spaces are straightforward. Every vector in c_0 with a finite number of non-zero coefficients is a convex combination of vectors taking values $1, -1, 0$. Thus for every N we have $\| T \mid \operatorname{span}(e_n)_{n \geq N} \| \leq \sup_{\varepsilon_n = \pm 1} \| \sum_{n=N}^{\infty} \varepsilon_n x_n \|$. Note that $\lim_{N \to \infty} \sup_{\varepsilon_n = \pm 1} \| \sum_{n=N}^{\infty} \varepsilon_n x_n \| = 0$ since otherwise we can inductively produce a sequence of signs $(\varepsilon_n)_{n=1}^{\infty}$ such that $\sum_{n=1}^{\infty} \varepsilon_n x_n$ is not Cauchy. This shows that T is compact.

(c)$\Rightarrow$(d). The permutation σ induces an isometry of c_0 defined as $I_\sigma(\xi_n) = (\xi_{\sigma(n)})$. The operator $T \circ I_\sigma \colon c_0 \to X$ is compact and satisfies $(T \circ I_\sigma)(e_n) = x_{\sigma(n)}$ so the already proven implication (c)$\Rightarrow$(a) gives that $\sum_{n=1}^{\infty} x_{\sigma(n)}$ converges.

(d)$\Rightarrow$(e). If (e) does not hold then we can find $\varepsilon > 0$ and increasing sequences $(n_k)_{k=1}^{\infty}$ and $(N_s)_{s=0}^{\infty}$ with $N_0 = 0$ such that $\| \sum_{k=N_s+1}^{N_{s+1}} x_{n_k} \| \geq \varepsilon$ for $k = 1, 2, 3, \ldots$. Let m_s be an increasing enumeration of $\mathbb{N} \setminus \{n_k\}_{k=1}^{\infty}$. One checks that for the permutation σ defined by

$$(1, 2, 3, \ldots) \leftrightarrow (n_1, n_2, \ldots, n_{k_{N_1}}, m_1, n_{k_{N_1}+1}, \ldots, n_{k_{N_2}}, m_2, n_{k_{N_2}+1}, \ldots)$$

the series $\sum_{n=1}^{\infty} x_{\sigma(n)}$ diverges.   ■

3.   There is also a related notion of weakly unconditionally convergent series. The series $\sum_{n=1}^{\infty} x_n$ is said to be weakly unconditionally convergent if for every functional $x^* \in X^*$ the scalar series $\sum_{k=1}^{\infty} x^*(x_n)$ is unconditionally convergent. Actually the name 'weakly unconditionally convergent' is a bit misleading, because such series need not converge (even weakly). As an example take $\sum_{n=1}^{\infty} e_n$ in c_0.

More generally, the series $\sum_{n=0}^{\infty} f_n$ in the space $C(K)$, K compact, is weakly unconditionally convergent if and only if there is a c such that

$\sum_{n=0}^{\infty} |f_n(k)| \leq c < \infty$ for every $k \in K$. This can be easily checked using the Riesz representation theorem (I.B.11).

4 Proposition. *For a series $\sum_{n=1}^{\infty} x_n$ in a Banach space X the following conditions are equivalent:*

(a) *the series $\sum_{n=1}^{\infty} x_n$ is weakly unconditionally convergent;*

(b) *there is a constant C such that*

$$\sum_{n=1}^{\infty} |x^*(x_n)| \leq C\|x^*\| \text{ for every } x^* \in X^*;$$

(c) *there exists a constant C such that for every $(t_n)_{n=1}^{\infty} \in \ell_\infty$*

$$\sup_N \left\| \sum_{n=1}^{N} t_n x_n \right\| \leq C\|(t_n)\|_\infty;$$

(d) *there exists an operator $T\colon c_0 \to X$ such that $T(e_n) = x_n$.*

Proof: (a)$\Rightarrow$(b). We define $S\colon X^* \to \ell_1$ by $S(x^*) = (x^*(x_n))$. The closed graph theorem implies that S is continuous, thus (b) holds.
(b)$\Rightarrow$(c). For every N and every $(t_n)_{n=1}^{\infty} \in \ell_\infty$ we have

$$\left\| \sum_{n=1}^{N} t_n x_n \right\| = \sup_{\|x^*\| \leq 1} \left| x^* \left(\sum_{n=1}^{N} t_n x_n \right) \right|$$

$$= \sup_{\|x^*\| \leq 1} \left| \sum_{n=1}^{N} t_n x^*(x_n) \right| \leq \|(t_n)\|_\infty \sup_{\|x^*\| \leq 1} \sum_{n=1}^{\infty} |x^*(x_n)|$$

$$\leq C\|(t_n)\|_\infty.$$

(c)$\Rightarrow$(d). Obvious.
(d)$\Rightarrow$(a). For $x^* \in X^*$ we have

$$\sum_{n=1}^{\infty} |x^*(x_n)| = \sum_{n=1}^{\infty} |x^*(Te_n)| = \sum_{n=1}^{\infty} |T^*(x^*)(e_n)|.$$

Since $T^*(x^*) \in \ell_1 = c_0^*$ this is finite.   ∎

5.   Now we will investigate weakly unconditionally convergent series which are not convergent. It turns out that the example we gave in **3** is a canonical one. More precisely we have

Proposition.   *Suppose that $\sum_{n=1}^{\infty} x_n$ is a non-convergent weakly unconditionally convergent series. Then there exist increasing sequences of integers p_k, r_k with $p_1 < r_1 < p_2 < r_2 < \cdots$ such that the vectors*

$$u_k = \sum_{n=p_k}^{r_k} x_n \tag{1}$$

form a basic sequence equivalent to the unit vector basis in c_0.

Proof: Since $\sum_{n=1}^{\infty} x_n$ does not converge we can find sequences $(p_k)_{k=1}^{\infty}$ and $(r_k)_{k=1}^{\infty}$ as above such that the vectors $(u_k)_{k=1}^{\infty}$ given by (1) satisfy $\|u_k\| \geq \varepsilon > 0$ for some ε. From Proposition 4(d) we infer that $u_k \xrightarrow{\omega} 0$. Corollary II.B.18 yields a subsequence (which we will still denote u_k) which is basic. Since $(u_k)_{k=1}^{\infty}$ is basic, there exists a constant C such that for all finite sequences (t_k) we have

$$\|(t_k)\|_{\infty} \leq C \left\| \sum_k t_k u_k \right\|.$$

On the other hand Proposition 4(d) yields a constant C_1 such that

$$\left\| \sum_k t_k u_k \right\| \leq C \|(t_k)\|_{\infty}$$

for all $(t_k) \in c_0$. This completes the proof.   ■

6.   We would like now to investigate the unconditionally convergent series in $L_p, 1 \leq p \leq 2$. We have the following

Theorem. (Orlicz).   *If the series $\sum_{n=1}^{\infty} x_n$ converges unconditionally in $L_p(\Omega, \mu)$ with $1 \leq p \leq 2$ and μ is a probability measure then $\sum_{n=1}^{\infty} \|x_n\|^2 < \infty$.*

Proof: Let $(r_n(t))_{n=1}^{\infty}$ denote the Rademacher functions. It follows from Proposition 4(e) that there exists a constant such that

$\sup_N \sup_{t\in[0,1]} \|\sum_{n=1}^{N} r_n(t)x_n\|_p \leq C$. Thus we have

$$C^p \geq \int_0^1 \left\|\sum_{n=1}^{N} r_n(t)x_n\right\|_p^p dt = \int_0^1 \int_\Omega \left|\sum_{n=1}^{N} r_n(t)x_n(\omega)\right|^p d\mu(\omega)dt \quad (2)$$

$$= \int_\Omega \int_0^1 \left|\sum_{n=1}^{N} r_n(t)x_n(\omega)\right|^p dt d\mu(\omega).$$

The Khintchine inequality I.B.8 and the fact that $p \leq 2$ yield

$$C^p \geq K_p^p \int_\Omega \left(\sum_{n=1}^{N} |x_n(\omega)|^2\right)^{p/2} d\mu(\omega)$$

$$= K_p^p \int_\Omega \left(\sum_{n=1}^{N} (|x_n(\omega)|^p)^{\frac{2}{p}}\right)^{\frac{p}{2}} d\mu(\omega) \quad (3)$$

$$\geq K_p^p \left(\sum_{n=1}^{N} \left(\int_\Omega |x_n(\omega)|^p d\mu(\omega)\right)^{\frac{2}{p}}\right)^{\frac{p}{2}}$$

$$= K_p^p \left(\sum_{n=1}^{N} \|x_n\|_p^2\right)^{\frac{p}{2}}.$$

Since N was arbitrary the theorem follows. ∎

The important technical feature of the above proof is the use of the Rademacher functions to represent all possible choices of signs, each occurring with the same probability. This will be used extensively in the sequel.

7 Remark. We will discuss this type of question in more detail in III.A. Let us note that what we actually proved is the following inequality: There exists a constant C such that

$$\left(\sum_{n=1}^{N} \|x_n\|_p^2\right)^{\frac{1}{2}} \leq C \sup_{\varepsilon_n = \pm 1} \left\|\sum_{n=1}^{N} \varepsilon_n x_n\right\|_p \quad (4)$$

for all finite $(x_n)_{n=1}^{N} \subset L_p(\Omega, \mu)$, $1 \leq p \leq 2$, and μ a probability measure. A general Banach space X in which (4) holds is said to have the Orlicz property.

8.   A Schauder basis $(x_n)_{n=1}^\infty$ with biorthogonal functionals $(x_n^*)_{n=1}^\infty$ in a Banach space X (see II.B.5) is called an unconditional basis if for every $x \in X$ the series $\sum_{n=1}^\infty x_n^*(x)x_n$ converges unconditionally. A basis which is not unconditional is called conditional. Clearly the unit vectors in $\ell_p, 1 \leq p < \infty$, or in c_0 form an unconditional basis. One easily checks that if $(x_n)_{n=1}^\infty$ is an unconditional basis in X then $(x_n^*)_{n=1}^\infty$ is an unconditional basic sequence in X^*.

9 Proposition.   *The trigonometric system $(e^{in\theta})_{n=-\infty}^{+\infty}$ is an unconditional basis in $L_p(\mathbb{T})$ only when $p = 2$.*

Proof: The trigonometric system, being complete and orthonormal is clearly an unconditional basis in $L_2(\mathbb{T})$. By duality it is enough to consider only the case $1 \leq p < 2$. Suppose that the series $\sum_{-\infty}^{+\infty} a_n e^{in\theta}$ represents a function $f \in L_p(\mathbb{T})$ and converges unconditionally. Then Theorem 6 gives $\sum_{-\infty}^{+\infty} |a_n|^2 < \infty$, thus $f \in L_2(\mathbb{T})$.   ∎

In view of Proposition 9 the question arises whether L_p, $p \neq 2$, has an unconditional basis at all. The answer is positive for $1 < p < \infty$ and negative for $p = 1$. We have the following.

10 Theorem.   *The space $L_1[0,1]$ does not embed into any space with an unconditional basis.*

The proof of this theorem relies on the following general

11 Proposition.   *A block basic sequence of an unconditional basis is an unconditional basic sequence.*   ∎

Proof of Theorem 10.   Let $(r_n(t))_{n=1}^\infty$ be the sequence of Rademacher functions. Note that for every $x \in L_1[0,1]$

$$x \cdot r_n \xrightarrow{\omega} 0 \quad \text{as} \quad n \to \infty \tag{5}$$

and

$$\|x + xr_n\|_1 \longrightarrow \|x\|_1 \text{ as } n \to \infty. \tag{6}$$

Assume that $L_1[0,1]$ embeds into a Banach space Y with an unconditional basis $(y_n)_{n=1}^\infty$. Starting with $x_1 = 1$ we define

$$x_n = (x_1 + \cdots + x_{n-1})r_{k_n}$$

where k_n increases so fast that

$$\frac{1}{2} \leq \|x_n\|_1 = \|x_1 + \cdots + x_{n-1}\|_1 \leq 2 \qquad (7)$$

the sequence $(x_n)_{n=1}^{\infty}$ considered in Y is equivalent $\qquad$ (8)
to a block-basic sequence of the basis $(y_n)_{n=1}^{\infty}$.

Condition (8) can be ensured using (5) and arguments like those in II.B.17. Condition (7) follows from (6) because $x_1 + \cdots + x_{n-1} = (x_1 + \cdots + x_{n-2}) + (x_1 + \cdots + x_{n-2})r_{k_{n-1}}$. Conditions (7), (8) and Proposition 11 show that for some constant C (depending on the norm of the embedding of $L_1[0,1]$ into Y) and for every choice of $\varepsilon_n = \pm 1$ and for every N

$$\left\| \sum_{n=1}^{N} \varepsilon_n x_n \right\|_1 \leq C.$$

But this and (7) clearly contradict the inequality (4). $\qquad\blacksquare$

12 Corollary. *The space $C[0,1]$ does not embed into any space with an unconditional basis.*

Proof: It is a special case of II.B.4 that $L_1[0,1]$ embeds into $C[0,1]$, so the proof follows from Theorem 10. $\qquad\blacksquare$

The existence of an unconditional basis in the spaces $L_p[0,1], 1 < p < \infty$, is contained in the following theorem, which actually gives more precise information.

13 Theorem. *Let $1 < p < \infty$. If $(a_k)_{k=0}^{\infty}$ and $(b_k)_{k=0}^{\infty}$ are complex numbers with $|b_k| \leq |a_k|$ for $k = 0, 1, 2$, then for all $n \geq 0$*

$$\left\| \sum_{k=0}^{n} b_k h_k \right\|_p \leq (p^* - 1) \left\| \sum_{k=0}^{n} a_k h_k \right\|_p \qquad (9)$$

where $(h_k)_{k=0}^n$ is the Haar system and $p^* = \max(p, p/(p-1))$.

Let us recall that the Haar system was defined in II.B.9.

Proof: One easily checks using duality that it is enough to consider only the case $2 < p < \infty$ (when $p^* = p$). Let $v : \mathbb{C} \times \mathbb{C} \to \mathbb{R}$ be defined as

$$v(x,y) = |y|^p - (p-1)^p |x|^p. \qquad (10)$$

The crucial part of the proof is the following

14 Lemma.   *There exists a function* $u: \mathbb{C} \times \mathbb{C} \rightarrow \mathbb{R}$ *such that if* $a, b, x, y \in \mathbb{C}$ *and* $|b| \leq |a|$ *then*

$$v(x, y) \leq u(x, y), \tag{11}$$

$$u(x, y) = u(-x, -y), \tag{12}$$

$$u(0, 0) = 0, \tag{13}$$

$$u(x + a, y + b) + u(x - a, y - b) \leq 2u(x, y). \tag{14}$$

Assuming this lemma for a moment we complete the proof as follows. Put $f_n = \sum_{k=0}^{n} a_k h_k$ and $g_n = \sum_{k=0}^{n} b_k h_k$. Assume also that $(h_k)_{k=0}^{\infty}$ is normalized so that $\|h_k\|_{\infty} = 1$ for k=0,1,... . This clearly does not influence the theorem but some formulas are a bit shorter. Then by (10) and (11)

$$\|g_n\|_p^p - (p-1)^p \|f_n\|_p^p = \int v(f_n(t), g_n(t))dt \leq \int u(f_n(t), g_n(t))dt. \tag{15}$$

Since both f_{n-1} and g_{n-1} are constant on the $I_n = \text{supp } h_n$ we get from (14)·

$$
\begin{aligned}
\int_0^1 u(f_n(t), g_n(t))dt = &\int_{[0,1]\backslash I_n} u(f_{n-1}(t), g_{n-1}(t))dt \\
&+ \int_{\{h_n>0\}} u(f_{n-1} + a_n, g_{n-1} + b_n)dt \\
&+ \int_{\{h_n<0\}} u(f_{n-1} - a_n, g_{n-1} - b_n)dt \qquad (16) \\
\leq &\int_{[0,1]\backslash I_n} u(f_{n-1}, g_{n-1})dt \\
&+ \frac{1}{2}\int_{I_n} [u(f_{n-1} + a_n, g_{n-1} + b_n) \\
&+ u(f_{n-1} - a_n, g_{n-1} - b_n)]dt \\
\leq &\int_0^1 u(f_{n-1}, g_{n-1})dt
\end{aligned}
$$

and thus inductively

$$\|g_n\|_p^p - (p-1)^p \|f_n\|_p^p \leq \int u(f_0, g_0)dt = u(a_0, b_0)$$

$$= \frac{1}{2}(u(a_0, b_0) + u(-a_0, -b_0)) \leq u(0,0) = 0. \blacksquare$$

Proof of Lemma 14. The desired function $u(x,y)$ is given by

$$u(x,y) = \alpha_p(|x|+|y|)^{p-1}(|y|-(p-1)|x|) \text{ where } \alpha_p = p\left(1-\frac{1}{p}\right)^{p-1}. \quad (17)$$

Conditions (12) and (13) are clearly satisfied. To prove (11), by homogeneity it is enough to assume $|x|+|y| = 1$. Letting $|x| = s$, (11) reduces then to the inequality

$$F(s) = \alpha_p(1-ps) - (1-s)^p + (p-1)^p s^p \geq 0 \quad \text{for} \quad 0 \leq s \leq 1 \text{ and } p \geq 2. \quad (18)$$

One checks by a direct computation that

$$F(0) > 0, \ F(1) > 0, \ F\left(\frac{1}{p}\right) = F'\left(\frac{1}{p}\right) = 0, \ F''\left(\frac{1}{p}\right) > 0$$

and F'' has only one zero in $[0,1]$. This is enough to see that (18) holds. To prove (14) it is enough to assume that x and a are linearly independent over $\mathbb{R}$ and the same is true for y and b. Under this assumption the function $G(t) = u(x+ta, y+tb)$ is infinitely differentiable. A routine but rather tedious computation gives

$$G''(0) = \alpha_p\{$$
$$- p(p-1)(|a|^2 - |b|^2)(|x|+|y|)^{p-2}$$
$$- p(p-2)[|b|^2 - Re < \frac{y}{|y|}, b >^2]|y|^{-1}(|x|+|y|)^{p-1} \quad (19)$$
$$- p(p-1)(p-2)[Re < \frac{x}{|x|}, a > +Re < \frac{y}{|y|}, b >]^2|x|(|x|+|y|)^{p-3}\}.$$

Since $|b| \leq |a|$, the Cauchy-Schwarz inequality gives $G''(0) \leq 0$. Since this holds for all x, y, a and b with $|b| \leq |a|$ we see that in general $G''(t) \leq 0$. This clearly implies (14). $\blacksquare$

Notes and remarks.
The interplay between conditional and unconditional, i.e. absolute convergence for scalar series, was already the subject of research in the

nineteenth century, and probably even earlier. Thus, with the emergence of a theory of Banach spaces, investigation of series in Banach spaces became an important topic. In this section we only scratch the surface of this vast area. We will present further development of these ideas in some of the subsequent chapters, most notably in III.A and III.F.

The fundamental early paper on the subject is Orlicz [1933]. It contains our *Theorem 2* (actually it contains the proof of $(d) \Leftrightarrow (e)$ and condition (a) is used without comment in the proofs) and *Theorem 6* with basically the same proofs. The notion of weakly unconditionally convergent series also appeared quite early. Already in Orlicz [1929] the following theorem is proved.

Theorem. *Every weakly unconditionally convergent series in a weakly sequentially complete space is unconditionally convergent.*

Let us point out that a space X is weakly sequentially complete if every weakly Cauchy sequence is weakly convergent. This theorem is a forerunner of our *Proposition 5* which is due to Bessaga-Pełczyński [1958]. Unconditional bases (under the name of absolute bases) were already investigated in Karlin [1948] where among other results *Proposition 9* and a weak version of *Corollary 12* were proved. *Theorem 10* was shown in Pełczyński [1961] and the proof we present was given in Milman [1971]. *Theorem 13* except for the constant p^{-1} is a classical result of Marcinkiewicz [1937] but the main work was done by Paley [1932]. This theorem was the starting point of much of the later work on orthogonal series and martingale inequalities. The proof we present is due to Burkholder [1988]. It has the advantage of being rather elementary and it gives the best constant in the inequality (9); see Exercise 9.

Exercises

1.   (a) Show that if $f \in C[0,1]$ then the series $\sum_{n=0}^{\infty} \langle f, h_n \rangle h_n$, where $(h_n)_{n=0}^{\infty}$ is the Haar system, converges uniformly to f.

     (b) Show that if $\omega(\delta)$ is the modulus of continuity of $f \in C[0,1]$ then $\left\| \sum_{n=1}^{N} \langle f, h_n \rangle h_n - f \right\|_\infty \le C\omega(\frac{2}{N})$.

2.   Show that $\sum_{n=1}^{\infty} a_n \varphi_n(t)$, where $(\varphi_n)_{n=1}^{\infty}$ is the Faber-Schauder system, converges unconditionally in $C[0,1]$ if $\sum_k \sup_{2^k < n \le 2^{k+1}} |a_n| < \infty$. Show that this is not a necessary condition.

3. (Orlicz-Pettis theorem). Suppose that for every subsequence (n_s) of the integers the series $\sum_{s=1}^{\infty} x_{n_s}$ converges weakly to an element in X. Show that the series $\sum_{n=1}^{\infty} x_n$ converges unconditionally.

4. For $f \in L_p[0,1]$, $1 < p \leq 2$ with $f = \sum_{n=0}^{\infty} \langle f, h_n \rangle h_n$, where $(h_n)_{n=0}^{\infty}$ is the Haar system, define $M(f) = \sum_{n=1}^{\infty} \langle f, h_n \rangle n^{\frac{1}{2} - \frac{1}{p}} h_n$. Show that $M: L_p[0,1] \to L_2[0,1]$.

5. Show that every linear operator $T: C(K) \to \ell_1$ is compact.

6. Show that $\ell_p, p > 2$, does not embed isomorphically into any space $L_q(\mu)$, $1 \leq q \leq 2$.

7. Let X be a Banach space and let $\sum_{n=1}^{\infty} x_n$ be a conditionally convergent series in X with $\sum_{n=1}^{\infty} x_n = 0$. Let

    $$\mathcal{U}(x_n) = \left\{ x \in X: \text{ there exists a permutation } \sigma \text{ of } \mathbb{N} \text{ such that} \right.$$

    $$\left. x = \sum_{n=1}^{\infty} x_{\sigma(n)} \right\}$$

    (a) (Steinitz). Let X be an n-dimensional Banach space and let $(x_j)_{j=1}^{m} \subset X$ be such that $\sum_{j=1}^{m} x_j = 0$ and $\|x_j\| \leq 1$ for $j = 1, 2, \ldots, m$. Show that there exists a permutation σ of numbers $\{1, 2, \ldots, m\}$ such that

    $$\sup_k \left\| \sum_{j=1}^{k} x_{\sigma(j)} \right\| \leq C(n)$$

    and the constant $C(n)$ depends only on the dimension n.

    (b) (Steinitz). Show that if X is finite dimensional then $\mathcal{U}(x_n)$ is a linear subspace.

    (c) Show that in $L_2[0,1]$ there exists a conditionally convergent series $\sum_{n=1}^{\infty} x_n$ such that $\mathcal{U}(x_n)$ is not a linear subspace.

    (d) Show that in $L_2[0,1]$ there exists a conditionally convergent series $\sum_{n=1}^{\infty} x_n$ such that $\mathcal{U}(x_n)$ consists of exactly two points.

8. Suppose that X is a Banach space with an unconditional basis $(x_n)_{n=1}^{N}$ (where N is finite or infinite). We define the unconditional

basis constant of this basis

$$ubc(x_n) = \sup \left\{ \left\| \sum_{n=1}^{N} \varepsilon_n a_n x_n \right\| : |\varepsilon_n| = 1 \text{ for } n = 1, 2, \ldots, N \text{ and} \right.$$

$$\left. \left\| \sum_{n=1}^{N} a_n x_n \right\| = 1 \right\}.$$

We define the unconditional basis constant of the space X as

$$ubc(X) = \inf\{ubc(x_n) : (x_n)_{n=1}^{N} \text{ is an unconditional basis in } X\}.$$

(a) Suppose that (y_n) is a block basis of the basis $(x_n)_{n=1}^{N}$. Show that $ubc(y_n) \leq ubc(x_n)$.

(b) Suppose that $g_{n,k}$, $n = 0, 1, 2, \ldots$, $k = 1, 2, \ldots, 2^n$, is a system of functions on $[0,1]$ such that

(1) supp $g_{n,k} = A_{n,k}$ and for every n, the sets $A_{n,k}, k = 1, 2, \ldots, 2^n$, are disjoint and each has measure 2^{-n},

(2) $A_{n+1,2k-1} = \{t : g_{n,k} = 2^{\frac{n}{2}}\}$ and $A_{n+1,2k} = \{t : g_{n,k} = -2^{\frac{n}{2}}\}$.

Show that $\{g_{n,k}\}$ is an orthonormal system. Show that in $L_p[0,1]$, $1 \leq p \leq \infty$ it is a basic sequence equivalent to the Haar system in $L_p[0,1]$.

(c) Show that every basis in $L_p[0,1]$, $1 \leq p < \infty$, has a block-basic sequence equivalent to the Haar system.

(d) Show that for every complete orthonormal system $(\varphi_n)_{n=1}^{\infty}$ on $[0,1]$ consisting of bounded functions, there exists an $f \in L_1[0,1]$ such that $\max_N \left| \sum_{n=1}^{N} \langle f, \varphi_n \rangle \varphi_n \right| \notin L_1[0,1]$.

9. (a) Let $1 < p \leq 2$ and $x > 0$. Let $\omega > p$ satisfy $x^p + p\omega^{p-1} - \omega^p = 0$. Let us set $\theta = 1 - \frac{1}{\omega} = \frac{1}{\omega'}$ and $\beta_k = 1 - \frac{\omega\delta}{(x+k\delta)}$ for $k = 1, 2, \ldots$, where $0 < \delta < \frac{x}{\omega}$. Let us define

$$d_1 = x \cdot \chi_{[0,1]},$$
$$d_2 = \delta\chi_{[0,\beta_1]} + [\theta(x+\delta) - x]\chi_{[\beta_1,1]},$$
$$d_3 = \delta\chi_{[0,\beta_1\cdot\beta_2]} + [\theta(x+2\delta) - (x+\delta)]\chi_{[\beta_1\beta_2,\beta_1]},$$
$$d_4 = \delta\chi_{[0,\beta_1\beta_2\beta_3]} + [\theta(x+3\delta) - (x+2\delta)]\chi_{[\beta_1\beta_2\beta_3,\beta_1\beta_2]}, \text{etc.}$$

(a) Show that $\lim_{x\to 0} \lim_{\delta\to 0} \lim_{n\to\infty} \left\| \sum_{k=1}^{m} (-1)^k d_k \right\|_p = 1$ and $\lim_{x\to 0} \lim_{\delta\to 0} \lim_{n\to\infty} \left\| \sum_{k=1}^{n} d_k \right\|_p = p - 1$.

(b) Show that the constant $(p^* - 1)$ in (9) is the smallest possible.

(c) Show that $ubc(L_p[0,1]) = (p^* - 1)$.

II.E. Local Properties

In this chapter we present some results and notions concerning finite dimensional Banach spaces and the relation between an infinite dimensional Banach space and its finite dimensional subspaces. We start with a discussion of the bounded approximation property and the π_λ-spaces. We also prove the local reflexivity principle which connects the local properties of X and X^{**}. We prove the Auerbach lemma which allows a good identification of an n-dimensional Banach space with $\mathbb{R}^n$ or $\mathbb{C}^n$. We also study the concept of Banach-Mazur distance.

1. By local properties of a Banach space we mean the properties which depend on the structure of finite dimensional subspaces of the space. Some examples of such properties will be pointed out in this chapter and many more will be encountered in the sequel.

  The basic aim of this chapter is to provide an elementary understanding of local phenomena. Even at this early stage it is apparent that one needs a clarification of two points:

(a) how the general Banach space is built up from finite dimensional subspaces;

(b) what are the relevant properties of finite dimensional spaces.

  Let us start with some definitions and examples which explain point (a) a little . What we are really thinking about in (a) is the approximation problem: how well can we approximate the identity operator on the space X by finite dimensional operators?

2. A Banach space X is said to have the bounded approximation property (b.a.p.) if there exists a constant C such that for every finite subset $\{x_1, \ldots, x_n\} \subset X$ and for every $\varepsilon > 0$ there exists a finite dimensional operator $T \colon X \to X$ such that $\|Tx_j - x_j\| \leq \varepsilon$, for $j = 1, \ldots, n$ and $\|T\| \leq C$.

3. A Banach space X is said to be a π_λ-space if there exists a family $(X_\gamma)_{\gamma \in \Gamma}$ of finite dimensional subspaces of X such that

for every $\gamma_1, \gamma_2 \in \Gamma$ there exists $\gamma_3 \in \Gamma$ such that $X_{\gamma_1} \subset X_{\gamma_3}$ and $X_{\gamma_2} \subset X_{\gamma_3}$, \hfill (1)

$\bigcup_{\gamma \in \Gamma} X_\gamma$ is dense in X,   (2)

for every $\gamma \in \Gamma$ there is a projection $P_\gamma \colon X \overset{\text{onto}}{\longrightarrow} X_\gamma$ with $\|P_\gamma\| \leq \lambda$.   (3)

Many similar properties can be thought of and have been investigated in the literature but we will discuss only these and the existence of a basis. We have

4 Proposition. *A Banach space with a Schauder basis is a π_λ-space for some λ. A π_λ-space has the bounded approximation property.*

Proof: Let $(x_n)_{n=1}^\infty$ be a Schauder basis for X. We let $X_n = \text{span}(x_k)_{k=1}^n$ and P_n be the partial sum projection. One obviously has (1)-(3). Now let X be a π_λ space and $x_1, \ldots, x_n$ be a fixed finite subset of X. Given $\delta > 0$ we use (1) and (2) to find $\gamma \in \Gamma$ such that $\text{dist}(x_j, X_\gamma) < \delta$ for $j = 1, \ldots, n$. Fix $\bar{x}_j \in X_\gamma$ such that $\|\bar{x}_j - x_j\| < \delta$ for $j = 1, \ldots, n$. Then we have

$$\|x_j - P_\gamma x_j\| \leq \|x_j - \bar{x}_j\| + \|\bar{x}_j - P_\gamma x_j\|$$
$$\leq \delta + \|P_\gamma(\bar{x}_j - x_j)\| \leq (\lambda + 1)\delta.$$

This shows that X has the b.a.p.   ∎

5 Examples. (a) Let (Ω, μ) be a probability measure space. Then $L_p(\Omega, \mu), 1 \leq p \leq \infty$ is a π_1-space. We define Γ as the set of all finite partitions of Ω into sets of positive measure. For a partition $\gamma = (\Omega_1, \ldots, \Omega_n)$ we put $X_\gamma = \text{span}\{\chi_{\Omega_j}\}_{j=1}^n$ and $P_\gamma(f) = \sum_{j=1}^n \mu(\Omega_j)^{-1} \int f \chi_{\Omega_j} d\mu \cdot \chi_{\Omega_j}$. The properties (1)-(3) are clearly satisfied (see the proof of II.B.10(b) or Exercise II.B.14). Observe that $L_p(\Omega, \mu), 1 \leq p \leq \infty$, may not be separable and then it does not have a Schauder basis. Note also that for each γ we have $\text{Im} P_\gamma \cong \ell_p^n$.

(b) The disc algebra $A(\mathbb{D})$ has the bounded approximation property. To see this, it is enough to consider operators $\mathcal{F}_n(f) = f * \mathcal{F}_n$ where $\mathcal{F}_n$ is the n-th Fejér kernel (I.B.16). It also has a Schauder basis (see III.E.17) but is not a π_1-space (Exercise III.E.9).

(c) Every $C(K)$, K compact, is a $\pi_{1+\varepsilon}$-space for every $\varepsilon > 0$. (Actually it is a π_1-space, but this requires more care.) This can be proved directly. The alternative argument is to note that $C(K)^{**} = L_\infty(\mu)$ for some (usually non-σ-finite) measure μ. From this and the principle of local reflexivity (Theorem 15) we infer that there exists an increasing net (X_γ) of finite dimensional subspaces of $C(K)$ such that $\bigcup X_\gamma$ is dense

in $C(K)$ and $d(X_\gamma, \ell_\infty^{\dim X_\gamma}) \leq 1 + \varepsilon$. (The notion $d(X, Y)$ is defined in
6.) It follows from the Hahn-Banach theorem that there is a projection
$P_\gamma \colon C(K) \xrightarrow{\text{onto}} X_\gamma$ with $\|P_\gamma\| \leq 1 + \varepsilon$.

6. Since all n-dimensional spaces are isomorphic, in order to investigate
finite dimensional spaces more precisely we will use a more quantitative
notion.

Let X and Y be two isomorphic Banach spaces. The Banach-Mazur
distance between X and Y denoted as $d(X, Y)$ is defined as

$$d(X, Y) = \inf \{ \|T\| \cdot \|T^{-1}\| \colon \quad T \colon X \xrightarrow{\text{onto}} Y \text{ is an isomorphism} \}. \qquad (4)$$

If the spaces X and Y are not isomorphic we set $d(X, Y) = \infty$.

The Banach-Mazur distance has the following, almost obvious,
properties:
$$d(X, Y) \leq d(X, Z) \cdot d(Y, Z); \qquad (5)$$

$$\text{if } X \text{ and } Y \text{ are isometric then } d(X, Y) = 1. \qquad (6)$$

Thus the Banach-Mazur distance is not a metric in the geometrical sense,
but its logarithm is. We follow, however, the long established custom,
and use the definition (4) and the name 'Banach-Mazur distance'.

We have the following converse to (6).

7 Proposition. *If X is finite dimensional and $d(X, Y) = 1$ then X is
isometric to Y.*

Proof: Let us take $T_n \colon X \to Y$ such that $\|T_n\| \cdot \|T_n^{-1}\| \to 1$. Multiplying
T_n by an appropriate scalar we can assume $\|T_n\| = 1$ for $n = 1, 2, \ldots$.
Since $L(X, Y)$ and $L(Y, X)$ are finite dimensional Banach spaces, passing
to a subsequence we can assume $\|T_n - T\| \longrightarrow 0$ as $n \to \infty$ and $\|T_n^{-1} - S\| \longrightarrow 0$ as $n \to \infty$ for some $T \in L(X, Y)$ and $S \in L(Y, X)$. Obviously
T is invertible and $T^{-1} = S$. Moreover $\|T\| = \|S\| = 1$ so T is the
desired isometry.   ∎

8. As an easy example of a computation of the Banach-Mazur distance
we offer the following

Proposition. *For every $n = 1, 2, 3, \ldots$ and p such that $1 \leq p \leq \infty$ we
have $d(\ell_p^n, \ell_2^n) = n^{|\frac{1}{p} - \frac{1}{2}|}$.*

Proof. Note that for reflexive spaces X and Y, $d(X, Y) = d(X^*, Y^*)$;
thus it is enough to consider $2 \leq p \leq \infty$. The upper estimate is obvious

if we take $id: \ell_p^n \to \ell_2^n$. In order to check the lower estimate take any operator $T: \ell_p^n \to \ell_2^n$ with $\|T\| \le 1$. Analogously as in the proof of II.D.(4) we get

$$n^{\frac{2}{p}} = \sup_{\pm} \left\| \sum_{j=1}^n \pm e_j \right\|^2 \ge \sup_{\pm} \left\| \sum_{j=1}^n \pm T e_j \right\|^2 \ge \int \left\| \sum_{j=1}^n r_j(t) T e_j \right\|^2 dt$$

$$= \int \sum_{k=1}^n \left| \sum_{j=1}^n r_j(t) T e_j(k) \right|^2 dt$$

$$= \sum_{k=1}^n \sum_{j=1}^n \left| T e_j(k) \right|^2$$

$$= \sum_{j=1}^n \left\| T e_j \right\|^2 . \qquad \blacksquare$$

Thus $\inf_j \|T e_j\|_2 \le n^{\frac{1}{p}-\frac{1}{2}}$ and $\|T^{-1}\| \ge n^{\frac{1}{2}-\frac{1}{p}}$.

9.   Let us consider the space T_n^p of trigonometric polynomials of order n with L_p-norm. More precisely the elements of $T_n^p, 1 \le p \le \infty$, are trigonometric polynomials of the form

$$h(\theta) = \sum_{k=-n}^n a_k e^{ik\theta} \quad \text{with the norm} \quad \|h\|_p = \left(\frac{1}{2\pi} \int_0^{2\pi} |h(\theta)|^p d\theta \right)^{\frac{1}{p}} .$$

We have

Theorem.   *There is a constant C such that*

$$d(T_n^p, \ell_p^{2n+1}) \le C \frac{p^2}{p-1} .$$

Proof: Let $\theta_k = e^{i \cdot \frac{2\pi k}{(2n+1)}}$ for $k = 0, 1, \ldots, 2n$. The theorem follows from the following two estimates valid for every $h \in T_n^p$.

$$\left(\frac{1}{2n+1} \sum_{k=0}^{2n} |h(\theta_k)|^p \right)^{\frac{1}{p}} \le 3\|h\|_p, \quad 1 \le p \le \infty, \tag{7}$$

$$\|h\|_p \le C \frac{p^2}{p-1} \left(\frac{1}{2n+1} \sum_{k=0}^{2n} |h(\theta_k)|^p \right)^{\frac{1}{p}}, \quad 1 < p < \infty. \tag{8}$$

Proof of (7):   Let $\mathcal{V}_n$ be the de la Valle-Poussin kernel (cf. I.B.17.). Clearly $h * \mathcal{V}_n = h$ for $h \in T_n^p$. Thus

$$\frac{1}{3}|h(\theta)| \leq \frac{2}{3}|h * \mathcal{F}_{2n-1}(\theta)| + \frac{1}{3}|h * \mathcal{F}_{n-1}(\theta)|$$

so from the properties of the Fejér kernel I.B.16 we get

$$\left(\frac{1}{3}|h(\theta)|\right)^p \leq \frac{2}{3}|h * \mathcal{F}_{2n-1}(\theta)|^p + \frac{1}{3}|h * \mathcal{F}_{n-1}(\theta)|^p$$

$$\leq \frac{2}{3}|h|^p * \mathcal{F}_{2n-1}(\theta) + \frac{1}{3}|h|^p * \mathcal{F}_{n-1}(\theta).$$

Summing these inequalities we get

$$\frac{1}{3^p(2n+1)}\sum_{k=0}^{2n}|h(\theta_k)|^p$$

$$\leq \frac{2}{3(2n+1)}\sum_{k=0}^{2n}\frac{1}{2\pi}\int_0^{2\pi}|h(\theta)|^p\mathcal{F}_{2n-1}(\theta - \theta_k)d\theta$$

$$+ \frac{1}{3(2n+1)}\sum_{k=0}^{2n}\frac{1}{2\pi}\int_0^{2\pi}|h(\theta)|^p\mathcal{F}_{n-1}(\theta - \theta_k)d\theta. \qquad (9)$$

Let us observe that for every trigonometric polynomial $\varphi(\theta) = \sum_{-m}^{m}\alpha_k e^{ik\theta}$ with $m \leq 2n$ we have

$$\frac{1}{2n+1}\sum_{k=0}^{2n}\varphi(\theta_k) = \alpha_0 = \frac{1}{2\pi}\int_0^{2\pi}\varphi(\theta)d\theta. \qquad (10)$$

Interchanging the order of summation and integration in (9) and applying (10) for $\mathcal{F}_{2n-1}$ and $\mathcal{F}_{n-1}$ we get (7).

Proof of (8).   The condition (8) follows by duality from (7). Let P_n be the natural projection from $L_p(\mathbb{T})$ onto T_n^p. We know (see II.B.11) that $\|P_n\| \leq \frac{cp^2}{(p-1)}$. Take $g \in L_q(\mathbb{T}), \frac{1}{p} + \frac{1}{q} = 1, \|g\|_q = 1$ such that

$(2\pi)^{-1}\int_0^{2\pi} g(\theta)h(\theta)d\theta = \|h\|_p$. Then, using (10) we get

$$\|h\|_p = \frac{1}{2\pi}\int_0^{2\pi} g(\theta)h(\theta)d\theta$$

$$= \frac{1}{2\pi}\int_0^{2\pi} P_n(g)(\theta)h(\theta)d\theta$$

$$= \frac{1}{2n+1}\sum_{k=0}^{2n} P_n(g)(\theta_k)h(\theta_k).$$

Applying Hölder's inequality and (7) we get (8).   ∎

10.   One of the most powerful tools available for investigating finite dimensional spaces which does not exist in the infinite dimensional situation is volume. Since we can treat an n-dimensional space as $\mathbb{R}^n$ (or $\mathbb{C}^n$) with some norm it is clear what the volume is. The normalization is arbitrary but in most cases it does not matter. As an illustration of it let us prove

Proposition.   Let X be an n-dimensional Banach space. If $\{x_j\}_{j=1}^N$ is an ε-net in B_X, then $N \geq \varepsilon^{-n}$ for real X and $N \geq \varepsilon^{-2n}$ for complex X. There exists an ε-net in B_X, $\{x_j\}_{j=1}^N$ with $N \leq \left(\frac{1+\varepsilon}{\varepsilon}\right)^n$ for real X and $N \leq \left(\frac{1+\varepsilon}{\varepsilon}\right)^{2n}$ for complex X.

Proof: Let us consider real spaces first. If $\{x_j\}_{j=1}^N$ is an ε-net in B_X then $\bigcup_{j=1}^N B(x_j,\varepsilon) \supset B_X$ so

$$\text{vol } B_X \leq \text{vol } \bigcup_{j=1}^N B(x_j,\varepsilon)$$

$$\leq \sum_{j=1}^N \text{vol } B(x_j,\varepsilon) = N \cdot \text{vol } \varepsilon B_X = N \cdot \varepsilon^n \cdot \text{vol } B_X.$$

Thus $N \geq \varepsilon^{-n}$. In order to find an ε-net of a small cardinality let us fix a maximal 2ε-separated set, i.e. a maximal set $\{x_j\}_{j=1}^N \subset B_X$ such that $\|x_j - x_i\| > 2\varepsilon$ for $i \neq j$. Such a set is obviously an ε-net in B_X. Since $B(x_j,\varepsilon) \cap B(x_i,\varepsilon) = \emptyset$ for $i \neq j$ and $B(x_j,\varepsilon) \subset (1+\varepsilon)B_X$ for $j = 1,\ldots,N$ we have

$$N\cdot\varepsilon^n\cdot\text{vol } B_X = \text{vol }\left(\bigcup_{j=1}^N B(x_j,\varepsilon)\right) \leq \text{vol}((1+\varepsilon)B_X) = (1+\varepsilon)^n \text{ vol } B_X$$

so $N \leq \left(\frac{1+\varepsilon}{\varepsilon}\right)^n$. Since an n-dimensional complex Banach space is a $2n$-dimensional real space, the claim for the complex spaces also follows. ∎

Another application of volume considerations is the following.

11 Lemma. (Auerbach). *Let X be an n-dimensional Banach space. There exists a biorthogonal system $(x_j, x_j^*)_{j=1}^n$ in $X \times X^*$ with $\|x_j\| = 1$ and $\|x_j^*\| = 1$ for $j = 1, \ldots, n$.*

Proof: Let $(y_j, y_j^*)_{j=1}^n$ be any biorthogonal system in X. The function $V(z_1, \ldots, z_n) = \det(y_j^*(z_i))_{i,j=1}^n$ is a continuous function on the n-fold product of B_X. Let $V(x_1, \ldots, x_n) = \max\{V(z_1, \ldots, z_n) : \|z_j\| \leq 1$ for $j = 1, \ldots, n\}$. Clearly $\|x_j\| = 1$ for $j = 1, \ldots, n$. We define

$$x_j^*(x) = V(x_1, \ldots, x_{j-1}, x, x_{j+1}, \ldots, x_n) \cdot V(x_1, \ldots, x_n)^{-1}.$$

A moment's reflection gives that $(x_j, x_j^*)_{j=1}^n$ is the desired biorthogonal system. ∎

As an application of Auerbach's lemma we show certain general perturbation results.

12 Proposition. (a) *Let X be a Banach space and let $(X_\gamma)_{\gamma \in \Gamma}$ be a family of subspaces satisfying (1) and (2). Let X_0 be a finite dimensional subspace of X. For every $\varepsilon > 0$ there exists a $\gamma \in \Gamma$ and $X_0' \subset X_\gamma$ such that $d(X_0, X_0') < 1 + \varepsilon$.*

(b) *Let X have the b.a.p. Then there exists a constant C such that for every finite dimensional subspace $X_0 \subset X$ there exists an operator $T: X \to X$ with $\|T\| \leq C$ and $T \mid X_0 = id_{X_0}$.*

Proof: Let $(x_j, x_j^*)_{j=1}^n \subset X_0 \times X_0^*$ be given by Auerbach's lemma. In order to show (a) fix $\gamma \in \Gamma$ such that there are $\bar{x}_j \in X_\gamma$ with $\|x_j - \bar{x}_j\| \leq \frac{\delta}{n}$. Define $T: X_0 \to X_\gamma$ by $T(x_j) = \bar{x}_j$ and put $X_0' = T(X_0)$. For $x \in X_0$ we have

$$\|x\| = \left\| \sum_{j=1}^n x_j^*(x) x_j \right\| = \left\| \sum_{j=1}^n x_j^*(x) \bar{x}_j + \sum_{j=1}^n x_j^*(x)(x_j - \bar{x}_j) \right\|.$$

Since

$$\left\| \sum_{j=1}^n x_j^*(x)(x_j - \bar{x}_j) \right\| \leq \sum_{j=1}^n |x_j^*(x)| \, \|x_j - \bar{x}_j\| \leq \sum_{j=1}^n \|x_j^*\| \cdot \|x\| \cdot \|x_j - \bar{x}_j\|$$
$$\leq \delta \|x\|$$

we get

$$\|Tx\| - \delta\|x\| \le \|x\| \le \|Tx\| + \delta\|x\|.$$

If δ is suitably chosen this gives (a). To show (b) we use the definition of b.a.p. and we fix an operator $T_1: X \to X$ such that $\|T_1(x_j) - x_j\| \le \frac{\delta}{n}$ and $\|T\| \le C_1$. We define $T(x) = T_1(x) + \sum_{j=1}^{n} \tilde{x}_j^*(x)(x_j - Tx_j)$ where $\tilde{x}_j^*$ is a Hahn-Banach extension of x_j^* to a functional on X, $j = 1, \ldots, n$. Clearly $T(x_j) = x_j$ for $j = 1, \ldots, n$ so $T \mid X_0 = id_{X_0}$. Also

$$\|Tx\| \le \|T_1 x\| + \sum_{j=1}^{n} |\tilde{x}_j^*(x)| \, \|x_j - Tx_j\|$$

$$\le (C_1 + \delta)\|x\|.$$

Thus actually (b) holds with any constant C bigger than the constant given by (2). ∎

13 Proposition. *For every n-dimensional Banach space X and for every $\varepsilon > 0$ there exists an embedding $i: X \to \ell_\infty^N$ with $(1 - \varepsilon)\|x\| \le \|i(x)\| \le \|x\|$ where $N \le \left(\frac{1+\varepsilon}{\varepsilon}\right)^n$ if X is a real space and $N \le \left(\frac{1+\varepsilon}{\varepsilon}\right)^{2n}$ if X is a complex space.*

Proof: Let $(x_j^*)_{j=1}^N$ be an ε-net in B_{X^*} given by Proposition 10. We define $i: X \to \ell_\infty^N$ by $i(x) = (x_j^*(x))_{j=1}^N$. Given $x \in X$ let $x^* \in B_{X^*}$ be such that $\|x\| = x^*(x)$, and let j be such that $\|x_j^* - x^*\| < \varepsilon$. Then $|x_j^*(x)| = |x^*(x) + x_j^*(x) - x^*(x)| \ge |x^*(x)| - |(x_j^* - x^*)(x)| \ge (1 - \varepsilon)\|x\|$. Thus $\|i(x)\| \ge (1 - \varepsilon)\|x\|$. Obviously $\|i\| \le 1$. ∎

Remark: This result is a finite dimensional or 'local' version of Theorem II.B.4. Actually without an estimate for N this result easily follows from Theorem II.B.4.

14. Now we will return to the interplay between local and global properties of Banach spaces. We start with the important

Theorem. (Principle of local reflexivity.) *Let X be a Banach space and let $E \subset X^{**}$ and $F \subset X^*$ be finite dimensional subspaces. Given $\varepsilon > 0$ there exists an operator $T: E \to X$ such that*

$$\|T\| \cdot \|T^{-1} \mid T(E)\| \le 1 + \varepsilon, \tag{11}$$

$$T \mid E \cap X = id, \tag{12}$$

$$f(Te) = e(f) \text{ for all } f \in F \text{ and } e \in E. \tag{13}$$

This theorem asserts that finite dimensional subspaces of X^{**} are basically the same as finite dimensional subspaces of X. The proof relies on the following

15 Lemma. Let $(A_j)_{j=1}^N$ be bounded, norm-open convex subsets of X and let $\tilde{A}_j$ denote the norm interior of the $\sigma(X^{**}, X^*)$-closure of A_j in X^{**}.

(a) If $\bigcap_{j=1}^N \tilde{A}_j \neq \emptyset$ then $\bigcap_{j=1}^N A_j \neq \emptyset$.

(b) If we have a map $T: X \to Y$ with Y a finite dimensional Banach space then $T^{**}(\bigcap_{j=1}^N \tilde{A}_j) = T(\bigcap_{j=1}^N A_j)$.

Proof: (a) Let X_N be the direct sum of N copies of X. The set

$$A = \{(x_j)_{j=1}^N \in X_N : x_j \in A_j, j = 1, \ldots, N\}$$

is a bounded, norm-open, convex subset of X_N. If $\bigcap_{j=1}^N A_j = \emptyset$ then $A \cap V = \emptyset$ where $V = \{(x_j)_{j=1}^N \in X_N : x_j = x_1 \text{ for } j = 1, 2, \ldots, N\}$. Let

$$\tilde{A} = \{(x_j^{**})_{j=1}^N \in X_N^{**} : x_j^{**} \in \tilde{A}_j, \ j = 1, \ldots, N\}$$

and

$$V^{**} = \{(x_j^{**})_{j=1}^N \in X_N^{**} : x_j^{**} = x_1^{**} \text{ for } j = 1, \ldots, N\}.$$

If $A \cap V = \emptyset$ then there exists $\phi = (\phi_1, \ldots, \phi_N) \in X_N^*$ such that $\phi|V = 0$ and $\phi(a) > 0$ for all $a \in A$. Since A is $\sigma(X^{**}, X^*)$-dense in $\tilde{A}$ (use Goldstine's theorem II.A.13) we get $\phi(a^{**}) \geq 0$ for all $a^{**} \in \tilde{A}$. But $\tilde{A}$ is open so $\phi(\tilde{A})$ is open. This yields $\phi(a^{**}) > 0$ for all $a^{**} \in \tilde{A}$, so $\tilde{A} \cap V^{**} = \emptyset$, i.e. $\bigcap_{j=1}^N \tilde{A}_j = \emptyset$.

(b) Clearly $T^{**}(\bigcap_{j=1}^N \tilde{A}_j)$ and $T(\bigcap_{j=1}^N A_j)$ are open, convex sets and $T(\bigcap_{j=1}^N A_j) \subset T^{**}(\bigcap_{j=1}^N \tilde{A}_j)$. If they are not equal, then there exists a point $p \in T^{**}(\bigcap_{j=1}^N \tilde{A}_j)$ and a functional ϕ on Y and numbers $\alpha > \beta$ such that

$$\phi(p) > \alpha > \beta > \phi\left(T(\bigcap_{j=1}^N A_j)\right).$$

Let $x^* \in X^*$ equal $T^*(\phi)$. Let us consider sets $A_j^+ = A_j \cap \{x \in X : x^*(x) > \alpha\}$. Since $\tilde{A}_j^+ = \tilde{A}_j \cap \{x^{**} \in X^{**} : x^{**}(X^*) > \alpha\}$ we see that $\bigcap_{j=1}^N \tilde{A}_j^+ \neq \emptyset$ but $\bigcap_{j=1}^N A_j^+$ is empty. This contradicts (a) and so proves (b). ∎

Proof of Theorem 14. Let $\dim E = n$ and $\dim E \cap X = n - k$. Fix a biorthogonal system $(x_j^{**}, e_j^*)_{j=1}^n$ in $E \times E^*$ such that $\text{span}(x_j^{**})_{j=k+1}^n = E \cap X$, and $\|x_j^{**}\| = 1$. The identity $id \colon E \to X^{**}$ can be written as $id(e) = \sum_{j=1}^n e_j^*(e) x_j^{**}$. We want to find $x_1, \ldots, x_k$ in X such that the map $T \colon E \to X$ defined as $T(e) = \sum_{j=1}^k e_j^*(x) x_j + \sum_{k+1}^n e_j^*(e) x_j^{**}$ will have the desired properties. Property (12) is satisfied with this definition. Let Z be the direct sum of k copies of X and let $\delta > 0$ be a small number. Fix the following finite sets:

$\{f_j\}_{j=1}^M$ is a basis in F,

$\{x_j^*\}_{j=1}^R \subset B_{X^*}$ is such that for every $e \in E$

$$\|e\| \le (1 + \delta) \sup\{|x_j^*(e)| \colon j = 1, \ldots, R\},$$

$\{e_j\}_{j=1}^N$ is a δ-net in B_E.

We have

$$e_j = \sum_{r=1}^n \lambda_r^j x_r^{**}.$$

Now we form the following subsets of Z:

$$C_j = \{(x_s)_{s=1}^k \colon \left\| \sum_{s=1}^k \lambda_s^j x_s + \sum_{s=k+1}^n \lambda_s^j x_s^{**} \right\| < \|e_j\|$$

$$\text{and } \|x_s\| < 1 + \delta, \; s = 1, \ldots k\}, \; j = 1, \ldots, N.$$

Those subsets of Z are norm-open, bounded and convex. Since $(x_s^{**})_{s=1}^k \in \bigcap_{j=1}^N \tilde{C}_j \subset Z^{**}$ (where $\tilde{}$ has the same meaning as in Lemma 15) Lemma 15 gives $(x_s)_{s=1}^k \in \bigcap_{j=1}^N C_j \ne \emptyset$.

Let us consider an operator $S \colon Z \to \mathbb{R}^{M \cdot k} \oplus \mathbb{R}^{R \cdot k}$ (or into $\mathbb{C}^{M \cdot k} \oplus \mathbb{C}^{R \cdot k}$), defined as

$$S\big((x_s)_{s=1}^k\big) = \big(f_j(x_s), x_k^*(x_s)\big)_{j=1,\ldots,M, \; k=1 \ldots, R, \; s=1,\ldots,k}.$$

From Lemma 15 we infer that there exists $(x_s)_{s=1}^k$ such that

$$(x_s)_{s=1}^k \in \bigcap_{j=1}^N C_j$$

and

$$S\big((x_s)_{s=1}^k\big) = S^{**}\big((x_s^{**})_{s=1}^k\big).$$

With this choice of $(x_s)_{s=1}^k$ we clearly get (13). Also we get

$$\|Te\| \ge \sup_{j=1,\ldots,R} |x_j^*(Te)| = \sup_{j=1,\ldots,R} |x_j^*(e)| \ge (1 + \delta)^{-1} \|e\|. \qquad (14)$$

Given $e \in B_E$ let us fix e_j with $\|e - e_j\| \leq \delta$. Since $(x_s)_{s=1}^k \in C_j$ we have $\|Te_j\| \leq \|e_j\|$. So

$$\|Te\| \leq \|Te_j\| + \|T(e - e_j)\| \leq \|e_j\| + \|T\| \cdot \delta \leq \|e\| + \delta + \delta\|T\|. \quad (15)$$

A very crude estimate for $\|T\|$ is $(1 + \delta) \sum_{j=1}^n \|e_j^*\| \leq 2 \sum_{j=1}^n \|e_j^*\|$, so if δ is small enough (11) follows from (14) and (15). ∎

15. We say that a Banach space X is finitely representable in a Banach space Y if there exists a constant C such that for every finite dimensional subspace $X_1 \subset X$ there exists $Y_1 \subset Y$ with $d(X_1, Y_1) \leq C$. In other words X is finitely representable in Y if finite dimensional subspaces of X are subspaces of Y. Note that Proposition 12 yields that every Banach space X is finitely representable in c_0. From Proposition 12(a) and Example 5(a) we get that $L_p(\Omega, \mu)$ is finitely representable in ℓ_p for $1 \leq p \leq \infty$. On the other hand from II.D.(4) we easily get that ℓ_p for $p > 2$ is not finitely representable in any $L_q(\Omega, \mu)$ for $1 \leq q \leq 2$. Theorem 14 shows that X^{**} is finitely representable in X.

Notes and remarks.
The general idea of approximation of a separable Banach space by finite dimensional ones is quite old and was around in Lwów in the thirties. Banach [1932] asked the question if every separable Banach space has a basis. A notion of approximation property (a.p.), the concept even weaker than b.a.p., was also invented then. We say that a Banach space X has the approximation property if for every norm-compact set $K \subset X$ and for every $\varepsilon > 0$ there exists a finite dimensional operator $T: X \to X$ such that $\sup\{\|Tk - k\|: k \in K\} \leq \varepsilon$. The first deep study of a.p. and b.a.p. is contained in Grothendieck [1955]. The notion of π_λ-space emerged in the sixties (see Lindenstrauss [1964] and Michael-Pełczyński [1967]). The real breakthrough in the study of approximation properties come with Enflo's [1973] example of a Banach space without a.p. Many examples differentiating various approximation properties have been produced later. We refer the interested reader to Lindenstrauss-Tzafriri [1977] and [1979], Pisier [1986] and Szarek [1987]. All this is a very fascinating subject but beyond the scope of our book.

As a first result of the local theory of Banach spaces one can consider the following well known fact proved in Jordan-von Neumann [1935]. Actually the analogous three dimensional characterisation was given earlier by Frechet [1935].

A Banach space X is isometric to a Hilbert space if and only if

$$\|x + y\|^2 + \|x - y\|^2 = 2(\|x\|^2 + \|y\|^2) \qquad \text{for all } x, y \in X.$$

Note that this result implies that X is isometric to a Hilbert space if and only if every two-dimensional subspace of X is isometric to a Hilbert space. For an isomorphic version of this see Exercise 9(b) . The local theory of Banach spaces gathered momentum in the sixties with the study of L_1-preduals (see Lacey [1974]) and p-absolutely summing operators (III. F). Today it is a vast subject having connections with operator theory, harmonic analysis, geometry of convex bodies etc. Some of it will be presented later. For a more detailed presentation of different aspects of the theory, the reader should consult Milman-Schechtman [1986] or Tomczak-Jaegermann [1989] or Beauzamy [1985].

The notion of Banach-Mazur distance is in Banach [1932]. *Theorem 9* is a classical result of Marcinkiewicz [1937a] (see also Zygmund [1968] chapter X §7). *Proposition 8* is an easy special case of a result of Gurarii-Kadec-Macaev [1965]. The Auerbach lemma is mentioned without proof in Banach [1932]. The principle of local reflexivity (*Theorem 14*) was proved in Lindenstrauss-Rosenthal [1969].

Exercises

1. A real Banach space X is called uniformly convex if there exists a function $\varphi(\varepsilon) > 0$ for $\varepsilon > 0$ (called the modulus of convexity) such that if $x, y \in X, \|x\| = \|y\| = 1$ and $\|x - y\| > 2\varepsilon$ then $\|\frac{(x+y)}{2}\| \le 1 - \varphi(\varepsilon)$ (draw the picture).

    (a) Let X be uniformly convex and let $x^* \in X^*, \|x^*\| = 1$. Show that

    $$\text{diam}\{x \in X \colon \|x\| \le 1 \text{ and } x^*(x) > 1 - \varepsilon\} \to 0 \text{ as } \varepsilon \to 0.$$

    (b) Show that uniformly convex spaces are reflexive.

    (c) Show that $L_p[0, 1]$, $1 < p < \infty$, is uniformly convex.

    (d) Show that if $\sum_{n=1}^{\infty} x_n$ converges unconditionally in X then

    $$\sum_{n=1}^{\infty} \varphi(\|x_n\|) < \infty.$$

    (e) Suppose that X is a uniformly convex space and that $Y \subset X$ is a closed subspace. Show that for every $x \in X$, there exists a unique $y \in Y$ such that $\|x - y\| = \text{dist}(x, Y)$.

2. Let X be a complex Banach space. We say that X is complexly uniformly convex if there exists a function $\varphi(\varepsilon) > 0$ for $\varepsilon > 0$ (called the complex modulus of convexity) such that if $x, y \in X$ with $\|y\| \geq \varepsilon$ and $\|x + e^{i\theta}y\| \leq 1$ for all θ then $\|x\| \leq 1 - \varphi(\varepsilon)$.

   (a) Show that $L_1[0,1]$ is complexly uniformly convex.

   (b) Show that if φ is a complex modulus of convexity of X and the $\sum_{n=1}^{\infty} x_n$ converges unconditionally in X then $\sum_{n=1}^{\infty} \varphi(\|x_n\|) < \infty$.

3. Find two Banach spaces X and Y such that $d(X, Y) = 1$ but X and Y are not isometric.

4. Let T_n^∞ be the space of trigonometric polynomials of the form $\sum_{-n}^{n} d_k e^{ik\theta}$ with the sup-norm.

   (a) Show that $d(T_n^\infty, \ell_\infty^{2n+1}) \leq C \log(n+2)$ for $n = 1, 2, 3, \ldots$.

   (b) Show that T_n^∞ contains a subspace isometric to ℓ_∞^{n+1}

5. Let $T_{N,2}^p$ be the space of trigonometric polynomials of degree at most N in two variables (i.e. $f(\theta, t) \in T_{N,2}^p$ if and only if $f(\theta, t) = \sum_{n,m=-N}^{N} a_{n,m} e^{in\theta} e^{imt}$), equipped with the norm $\|f\|_p = \left((4\pi^2)^{-1} \int_0^{2\pi} \int_0^{2\pi} |f(\theta, t)|^p d\theta dt\right)^{\frac{1}{p}}$. Show that $d(T_{N,2}^p, \ell_p^{(2N+1)^2}) < \frac{cp^2}{(p-1)^2}$ for $1 < p < \infty$.

6. Let X be an infinite dimensional Banach space. Show that there is no translation invariant Borel measure μ on X such that $\mu(U) > 0$ for every open set U and such that $\mu(U_1) < \infty$ for some open set U_1. Translation invariant means that $\mu(A + x) = \mu(A)$ for every $x \in X$.

7. Suppose that $T: X \xrightarrow{\text{onto}} Y$. Let $(Y_\alpha)_{\alpha \in \Gamma}$ be a net of finite dimensional subspaces of Y, ordered by inclusion and such that $\overline{\bigcup_{\alpha \in \Gamma} Y_\alpha} = Y$. Assume that there is a C such that for each $\alpha \in \Gamma$ there is an operator $S_\alpha: Y_\alpha \to X$ such that $\|S_\alpha\| \leq C$ and $TS_\alpha = id_{Y_\alpha}$. Show that $T^*(Y^*)$ is complemented in X^*.

8. (a) Show that ℓ_2 is isomorphic to a complemented subspace of $\left(\sum_{n=1}^{\infty} \ell_2^n\right)_\infty$.

   (b) Let M_n denote the set of all n-dimensional Banach spaces (up to isometry, i.e. we identify isometric spaces). Show that for every $n \in \mathbb{N}$ the set M_n with the Banach-Mazur distance (or rather, its logarithm) is a compact space.

(c) Let $(B_k)_{k=1}^{\infty}$ be a sequence of finite dimensional spaces such that $\{B_k\}_{k=1}^{\infty} \cap M_n$ is dense in M_n for all $n \in \mathbb{N}$. Show that for every separable Banach space X the space $\left(\sum_{k=1}^{\infty} B_k\right)_{\infty}$ contains a complemented copy of X^*.

9. (a) Show that if X is a Banach space finitely representable in a uniformly convex Banach space Y (Exercise 1), then X has an equivalent uniformly convex norm.

   (b) Show that if a Banach space X is finitely representable in ℓ_2 then X is isomorphic to a Hilbert space.

10. A Banach space X has the uniform approximation property $(u.a.p.)$ if there exist a constant C and a function $\varphi(\varepsilon, n), n \in \mathbb{N}, \varepsilon > 0$, such that for all $x_1, x_2, \ldots, x_n \in X$ there exists an operator $T: X \to X$ such that $\|Tx_j - x_j\| \leq \varepsilon\|x_j\|$ for $j = 1, 2, \ldots, n$ and $\|T\| \leq C$ and $\dim T(X) \leq \varphi(\varepsilon, n)$. Show that $L_p[0,1]$, $1 \leq p \leq \infty$, have the u.a.p.

11. Show that ℓ_2^n is isometric to a subspace of ℓ_∞ but is not isometric to any subspace of c_0.

Part III

Selected Topics

III.A L_p-Spaces; Type And Cotype.

In this chapter we investigate the $L_p(\mu)$-spaces, $1 < p < \infty$. We start by proving the isomorphisms of some natural spaces to spaces $L_p(\mu)$. We show that the Sobolev space $W_p^1(\mathbb{T}^2)$ is isomorphic to $L_p(\mathbb{T}^2)$ for $1 < p < \infty$ and that the Bergman space $B_p(\mathbb{D})$ is isomorphic to ℓ_p for $1 \leq p < \infty$. Along the way we prove a useful criterion for the boundedness of integral operators on $L_p(\Omega, \mu)$ (Proposition 9). Later we borrow from probability theory and show the existence and basic properties of stable laws. These provide isometric embeddings of ℓ_p into L_q, $q \leq p \leq 2$. We continue the line of thought started in II.D.6 and introduce the general notion of type and cotype of a Banach space. In order to study these notions efficiently we prove the vector valued generalization of Khintchine inequality (Kahane's inequality). A generalization of a classical result of Carleman from the theory of orthonormal series is also presented. We conclude this chapter with the Banach-Saks theorem and its generalizations to almost everywhere convergence.

1. We start this chapter with some general observations.

Proposition. *A separable space* $L_p(\Omega, \mu), 1 \leq p < \infty$, *is isometric to one of the following spaces:* $\ell_p^n, n = 1, 2, \ldots,$ $\ell_p, L_p[0, 1], (L_p[0, 1] \oplus \ell_p^n)_p, n = 1, 2, \ldots, (L_p[0, 1] \oplus \ell_p)_p$.

The proof is an immediate consequence of the characterization of non-atomic, separable measure spaces given in I.B.1. Let us also note that the above list contains at most two non-isomorphic infinite dimensional spaces, namely ℓ_p and $L_p[0, 1]$ (see II.B Exercise. 11). That for $p \neq 2, 1 \leq p < \infty$ these spaces are really non-isomorphic follows from Propositions 5 and 7. Thus there are rather few separable L_p-spaces. Some questions about non-separable L_p-spaces can be reduced to the separable case using

2 Proposition. *Every separable subspace $X \subset L_p(\Omega, \mu), 1 \le p < \infty$, is contained in a separable $Y \subset L_p(\Omega, \mu)$ isometric to some $L_p(\Omega, \mu_1)$.*

Proof: Let us fix a countable dense subset $(x_j)_{j=1}^{\infty}$ in X, and consider sets $A_{j,a,b} = \{\omega \in \Omega\colon a < x_j(\omega) < b\}$ where a, b are rational numbers. Let Σ be the σ-algebra generated by all sets $A_{j,a,b}$. The space $L_p(\Omega, \Sigma, \mu)$ of all Σ-measurable, p-integrable functions is the desired Y. ∎

3. One of the reasons why L_p-spaces are important is that many other spaces common in analysis are isomorphic to L_p-spaces (see II.B.Exercise 9). We want to present one more example of this type.

Proposition. *The space $W_p^1(\mathbb{T}^2)$ is isomorphic to $L_p(\mathbb{T}^2), 1 < p < \infty$.*

For a doubly indexed sequence of numbers $(a_{n,m})_{n,m=-\infty}^{+\infty}$ we say that $(a_{n,m})$ is a multiplier on $L_p(\mathbb{T}^2)$ if the map $\sum \hat{f}(n, m)e^{in\theta_1}e^{im\theta_2} \mapsto \sum a_{n,m}\hat{f}(n, m)e^{in\theta_1}e^{im\theta_2}$ extends to a continuous operator from $L_p(\mathbb{T}^2)$ into $L_p(\mathbb{T}^2)$.

4 Lemma. *The doubly indexed sequences $\left(\frac{1}{(1+|n|+|m|)}\right)_{n,m=-\infty}^{\infty}$, $\left(\frac{n}{(1+|n|+|m|)}\right)_{n,m=-\infty}^{\infty}$ and $\left(\frac{m}{(1+|n|+|m|)}\right)_{n,m=-\infty}^{\infty}$ are multipliers on $L_p(\mathbb{T}^2)$ for $1 < p < \infty$.*

This Lemma is a special case of the multidimensional multiplier Theorem I.B.32.

Proof of Proposition 3. Let $T\colon L_p(\mathbb{T}^2) \to W_p^1(\mathbb{T}^2)$ be defined by the multiplier $\left(\frac{1}{(1+|n|+|m|)}\right)_{n,m=-\infty}^{\infty}$. It follows easily from Lemma 4 that T is continuous. We define $\Sigma\colon W_p^1(\mathbb{T}^2) \to L_p(\mathbb{T}^2)$ by $\Sigma(f) = f + \tilde{\partial}_1 f + \tilde{\partial}_2 f$ where

$$\tilde{\partial}_1\left(\sum_{n,m} \hat{f}(n, m)e^{in\theta_1}e^{im\theta_2}\right) = \sum_{n,m} |n|\hat{f}(n, m)e^{in\theta_1}e^{im\theta_2}$$

and $\tilde{\partial}_2$ is defined analogously. Since $f \in W_p^1(\mathbb{T}^2)$ the function $\sum_{n,m} n\hat{f}(n, m)e^{in\theta_1}e^{im\theta_2} \in L_p(\mathbb{T}^2)$ and using the Riesz projection (see Theorem I.B.20) in the variable θ_1 we get that $\tilde{\partial}_1\colon W_p^1(\mathbb{T}^2) \to L_p(\mathbb{T}^2)$. Analogously for $\tilde{\partial}_2$, so Σ is continuous. A routine calculation shows that $\Sigma T f = f$ for $f \in L_p(\mathbb{T}^2)$. Since Σ is clearly 1-1 we get the desired isomorphism. ∎

5. Now we will investigate the spaces ℓ_p.

Proposition. *Let X be an infinite dimensional subspace of ℓ_p, $1 \leq p < \infty$, or of c_0. Then X contains a subspace Y such that $Y \sim \ell_p$ (or c_0) and is complemented in ℓ_p (or c_0).*

Proof: Let $(z_n)_{n=1}^{\infty}$ be a block-basic sequence of the unit vector basis in ℓ_p (or c_0), so $z_n = \sum_{k_n+1}^{k_{n+1}} \alpha_j e_j$ with k_n increasing to ∞ and $\|z_n\| = 1$. One checks that $\text{span}\{z_n\}_{n=1}^{\infty}$ is isometric to ℓ_p. Let z_n^* be a functional on ℓ_p (or c_0) such that $z_n^*(z_n) = 1 = \|z_n^*\|$ and $z_n^* = \sum_{k_n+1}^{k_{n+1}} \beta_j e_j^*$, $n = 1, 2, 3, \dots$. We define $P \colon \ell_p \to \ell_p$ by $P(x) = \sum_{n=1}^{\infty} z_n^*(x) z_n$. P is algebraically a projection onto $\text{span}\{z_n\}_{n=1}^{\infty}$ and

$$\|P(x)\|_p = \left(\sum_{n=1}^{\infty} |z_n^*(x)|^p \right)^{\frac{1}{p}} \leq \left(\sum_{n=1}^{\infty} \left| z_n^* \left(\sum_{k_n+1}^{k_{n+1}} x(j) e_j \right) \right|^p \right)^{\frac{1}{p}}$$

$$\leq \left(\sum_{n=1}^{\infty} \left\| \sum_{k_n+1}^{k_{n+1}} x(j) e_j \right\|^p \right)^{\frac{1}{p}}$$

$$\leq \left(\sum_{j=1}^{\infty} |x(j)|^p \right)^{\frac{1}{p}}$$

$$= \|x\|_p$$

so $\|P\| = 1$. From II.B.17 we infer that X contains a sequence $(x_n)_{n=1}^{\infty}$ very close to such a $(z_n)_{n=1}^{\infty}$ and II.B.15 gives that $\text{span}(x_n)_{n=1}^{\infty}$ is complemented in ℓ_p (or c_0) and isomorphic to $\text{span}(z_n)_{n=1}^{\infty}$ so to ℓ_p (or c_0). ∎

Using the above Proposition 5 and Theorem II.B.24 we get

6 Theorem. *Every infinite dimensional complemented subspace of ℓ_p, $1 \leq p < \infty$, or of c_0 is isomorphic to the whole space.* ∎

7. This simple structure of complemented subspaces of ℓ_p is rather exceptional. For example $L_p[0,1]$, $1 < p < \infty$, contains complemented subspaces isomorphic to Hilbert space as well as those isomorphic to ℓ_p or $L_p[0,1]$ itself (see II.B.2(b)).

Proposition. *Let $(r_n)_{n=1}^{\infty}$ be Rademacher functions. Then the space $\text{span}(r_n)_{n=1}^{\infty} \subset L_p[0,1]$ is isomorphic to ℓ_2 for $1 \leq p < \infty$ and is complemented for $1 < p < \infty$.*

Proof: The first claim is just the Khintchine inequality I.B.8. For the second let $P: L_2[0,1] \to L_2[0,1]$ be an orthonormal projection onto span$(r_n)_{n=1}^\infty$. For $\infty > p \geq 2$ the Khintchine inequality gives

$$\|Pf\|_p \leq C_p\|Pf\|_2 \leq C_p\|f\|_2 \leq C_p\|f\|_p \qquad \text{for} \quad f \in L_p[0,1].$$

By duality we get $P: L_p[0,1] \to L_p[0,1]$ for $1 < p < \infty$.   ∎

Remark: The same property also holds for span$(e^{in_k\theta})_{k=1}^\infty$, for any lacunary sequence $(n_k)_{k=1}^\infty$, i.e. any n_k such that $\inf_k(n_{k+1}/n_k) > 1$. The proof is exactly the same, only uses the analogue of the Khintchine inequality for lacunary sequences of characters (see I.B.8).

8.   Let $\mathbb{D} = \{z \in \mathbb{C}: |z| < 1\}$ and ν be Lebesgue measure on $\mathbb{D}$. $B_p \subset L_p(\mathbb{D}, d\nu)$ denotes the Bergman space $B_p(\mathbb{D})$ (see I.B.28 for definitions).

Theorem.   *For every $s > 0$ the operator*

$$P_s f(z) = \frac{s+1}{\pi} \int\limits_{\mathbb{D}} \frac{(1-|w|^2)^s f(w)}{(1-z\bar{w})^{2+s}} d\nu(w) \tag{1}$$

is a continuous projection from $L_p(\mathbb{D}, d\nu)$ onto $B_p, 1 \leq p < \infty$.

Note that this theorem is false for $p = \infty$. Obviously $B_\infty = H_\infty$ and there is no continuous linear projection from L_∞ onto H_∞; see remarks after Proposition III.E.15. In the proof we will need the following criterion for the boundedness of integral operators on $L_p(\Omega, \mu)$.

9 Proposition.   *Let (Ω, μ) be a measure space and let $K(\omega_1, \omega_2)$ be a measurable function on $\Omega \times \Omega$. Let us define*

$$Tf(\omega_2) = \int_\Omega K(\omega_1, \omega_2) f(\omega_1) \, d\mu(\omega_1).$$

Then

(a) *if* supess$_{\omega_1 \in \Omega} \int\limits_\Omega |K(\omega_1, \omega_2)| \, d\mu(\omega_2) < \infty$ *then*

$$T: L_1(\Omega, \mu) \to L_1(\Omega, \mu),$$

(b) *if* supess$_{\omega_2 \in \Omega} \int\limits_\Omega |K(\omega_1, \omega_2)| d\mu(\omega_1) < \infty$ *then*

$$T: L_\infty(\Omega, \mu) \to L_\infty(\Omega, \mu),$$

(c) if $1 < p < \infty$ and there exists a measurable positive function g on Ω and constants a, b such that for $\frac{1}{p} + \frac{1}{p'} = 1$ we have

$$\int_\Omega |K(\omega_1, \omega_2)| g(\omega_1)^{p'} d\mu(\omega_1) \le [ag(\omega_2)]^{p'} \quad \mu - a.e.$$

and

$$\int_\Omega |K(\omega_1, \omega_2)| g(\omega_2)^p d\mu(\omega_2) \le [bg(\omega_1)]^p \quad \mu - a.e.$$

then $T: L_p(\Omega, \mu) \to L_p(\Omega, \mu)$.

Proof: The argument for (a) and (b) is obvious, so we will prove (c). We have

$$|Tf(\omega_2)|$$
$$\le \int_\Omega |K(\omega_1, \omega_2)| \, |f(\omega_1)| d\mu(\omega_1)$$
$$= \int_\Omega [|K(\omega_1, \omega_2)|^{\frac{1}{p'}} g(\omega_1)] \cdot [|K(\omega_1, \omega_2)|^{\frac{1}{p}} |f(\omega_1)| g(\omega_1)^{-1}] d\mu(\omega_1)$$
$$\le ag(\omega_2) \cdot \left\{ \int_\Omega |K(\omega_1, \omega_2)| (|f|/g)^p(\omega_1) d\mu(\omega_1) \right\}^{\frac{1}{p}}.$$

Hence

$$\|Tf\|_p \le a \left[\int_\Omega \left(\frac{|f|}{g} \right)^p (\omega_1) \int_\Omega g^p(\omega_2) |K(\omega_1, \omega_2)| d\mu(\omega_2) d\mu(\omega_1) \right]^{\frac{1}{p}}$$
$$\le ab \left[\int_\Omega (|f|/g)^p(\omega_1) g^p(\omega_1) d\mu(\omega_1) \right]^{\frac{1}{p}}$$
$$= ab \left[\int_\Omega |f(\omega_1)|^p d\mu(\omega_1) \right]^{\frac{1}{p}}. \qquad \blacksquare$$

10. The other fact we will use is the following

Lemma. For every α and s such that $-1 < \alpha < 1$ and $s > 0$, there exists a constant $C_{\alpha,s}$ such that for $|w| < 1$ we have

$$\int_D \frac{(1 - |z|^2)^\alpha}{|1 - z\bar{w}|^{2+s}} d\nu(z) \le C_{\alpha,s} (1 - |w|^2)^{\alpha - s}. \tag{2}$$

Proof: There exists a constant $\gamma > 0$ such that for all ρ with $0 \le \rho \le 1$ and all θ with $|\theta| < \pi$ we have $|1 - \rho e^{i\theta}| \ge \gamma(1 - \rho + |\theta|)$. This gives

$$\int_0^{2\pi} \frac{d\theta}{|1 - \rho e^{i\theta}|^{2+s}} \le \gamma^{-2-s} \int_{-\pi}^{\pi} (1 - \rho + |\theta|)^{-2-s} d\theta$$

$$\le C_s \int_{1-\rho}^{5} \theta^{-2-s} d\theta \tag{3}$$

$$\le C_s (1 - \rho)^{-1-s}.$$

Since the integral in (2) depends only on $\rho = |w|$, passing to polar coordinates and using (3) we get

$$\int_{\mathbb{D}} \frac{(1 - |z|^2)^\alpha}{|1 - z\bar{w}|^{2+s}} d\nu(z) = \frac{1}{2\pi} \int_0^{2\pi} \int_0^1 r \frac{(1 - r^2)^\alpha}{|1 - \rho r e^{i\theta}|^{2+s}} dr d\theta \tag{4}$$

$$\le \frac{1}{2\pi} \int_0^1 (1 - r^2)^\alpha \left(\int_0^{2\pi} \frac{d\theta}{|1 - \rho r e^{i\theta}|^{2+s}} \right) dr$$

$$\le C \int_0^1 (1 - r)^\alpha (1 - \rho r)^{-1-s} dr.$$

Integrating the last integral by parts we see that it equals

$$C_1 + C_2 \int_0^1 (1 - r)^{1+\alpha} (1 - \rho r)^{-2-s} dr$$

$$\le C_1 + C_2 \int_0^1 (1 - \rho r)^{1+\alpha} (1 - \rho r)^{-2-s} dr$$

$$\le C_{\alpha,s} (1 - \rho)^{\alpha-s},$$

so we get (2). ∎

Proof of Theorem 8. First we show that (1) defines a bounded operator on $L_1(\mathbb{D}, \nu)$. The adjoint operator is given by the formula

$$P_s^*(f)(z) = \frac{s+1}{\pi} (1 - |z|^2)^s \int_{\mathbb{D}} \frac{g(z) d\nu(z)}{(1 - z\bar{w})^{2+s}} \tag{5}$$

so we infer from Proposition 9(b) and Lemma 10 that P_s^* is bounded on $L_\infty(\mathbb{D}, \nu)$ so P_s is bounded on $L_1(\mathbb{D}, \nu)$.

For $1 < p < \infty$ we apply Proposition 9(c) for $g(z) = (1 - |z|^2)^{-\frac{1}{pq}}$ where $\frac{1}{p} + \frac{1}{q} = 1$. Lemma 10 yields

$$\int_{\mathbb{D}} \left| \frac{(1 - |z|^2)^s}{(1 - z\bar{w})^{2+s}} \right| (1 - |w|^2)^{-\frac{1}{p}} d\nu(w) \leq C (1 - |z|^2)^s (1 - |z|^2)^{-\frac{1}{p} - s}$$

$$\leq [Cg(z)]^q$$

and analogously

$$\int_{\mathbb{D}} \left| \frac{(1 - |z|^2)^s}{(1 - z\bar{w})^{2+s}} \right| (1 - |z|^2)^{-\frac{1}{q}} d\nu(z) \leq [Cg(w)]^p$$

so P_s is bounded. Clearly for $f \in L_1(\mathbb{D}, \nu)$, $P_s(f)$ is analytic in $\mathbb{D}$. Moreover for $n = 0, 1, 2, \ldots$ we have

$$P_s(z^n)(z) = \frac{s+1}{\pi} \int_{\mathbb{D}} \frac{(1 - |w|^2)^s w^n}{(1 - z\bar{w})^{2+s}} d\nu(w)$$

$$= \frac{s+1}{\pi} \int_{\mathbb{D}} (1 - |w|^2)^s w^n \sum_{k=0}^{\infty} \frac{\Gamma(k + 2 + s)}{k! \Gamma(2 + s)} (z\bar{w})^k d\nu(w)$$

$$= \frac{s+1}{\pi} \frac{\Gamma(n + 2 + s)}{n! \Gamma(2 + s)} z^n \int_{\mathbb{D}} (1 - |w|^2)^s |w|^{2n} d\nu(w).$$

Evaluating the last integral in polar coordinates and using the well known properties of Euler's beta function we get that $P_s(z^n) = z^n$ for $n = 0, 1, 2, \ldots$. This shows that P_s is a projection onto $B_p(\mathbb{D})$.   ∎

As an application of Theorem 8 we show

11 Theorem.   *The space B_p is isomorphic to $\ell_p, 1 \leq p < \infty$.*

The proof of this theorem will be based upon the following

12 Lemma.   *Every compact operator $T: X \to L_p$ admits a factorization*

with $\|\alpha\| \cdot \|\beta\| \leq 8\|T\|$.

Proof: Since T is compact and L_p is a π_1-space (II.E.5(a)) there exists a sequence of norm-one projections $P_n\colon L_p \to L_p$ such that $\|T - P_nT\| \leq 4^{-n}$ and $d(ImP_n, \ell_p^{\dim(ImP_n)}) = 1$, for $n = 1, 2, \ldots$. We identify ℓ_p with $\left(\sum_{n=1}^{\infty} ImP_n\right)_p$ and we define

$$\alpha(x) = (P_1Tx, 2(P_2Tx - P_1Tx), \ldots, 2^{n-1}(P_nTx - P_{n-1}Tx), \ldots)$$

and $\beta(f_n) = \sum_{n=1}^{\infty} 2^{-n+1} f_n$. One checks the desired properties. ∎

Proof of Theorem 11.   Let us fix an increasing sequence of numbers $(r_n)_{n=1}^{\infty}$ tending to 1 with $r_1 > 0$. Let us put

$$\mathbb{D}_0 = \{z \in \mathbb{D}\colon |z| \leq r_1\} \quad \text{and} \quad \mathbb{D}_n = \{z \in \mathbb{D}\colon r_n < |z| \leq r_{n+1}\}.$$

Let $I_n\colon B_p \to L_p(\mathbb{D}_n, d\nu)$ be the natural restriction operator. From I.B.28 and a standard normal family argument we infer that each I_n is compact. Let (α_n, β_n) be a factorization of I_n given by Lemma 12, with $\|\alpha_n\| = 1, \|\beta_n\| \leq 8$.

Thus we have the commutative diagram

$$
\begin{array}{ccccc}
B_p & \xrightarrow{\ \ I\ \ } & L_p(\mathbb{D}, d\nu) & \xrightarrow{\ \ P\ \ } & B_p \\[2mm]
{\scriptstyle \alpha}\big\downarrow & & \big\uparrow{\scriptstyle \Sigma} & & \\[2mm]
(\Sigma\ell_p)_p & \xrightarrow{\ \ \beta\ \ } & (\Sigma L_p(D_n, d\nu))_p & &
\end{array}
$$

where $\alpha(f) = (\alpha_n(f|\mathbb{D}_n))_{n=1}^{\infty}, I$ denotes the identity embedding, $\beta((x_n)_{n=1}^{\infty}) = (\beta_n(x_n))_{n=1}^{\infty}$ and $\Sigma((f_n)_{n=1}^{\infty}) = \sum_{n=1}^{\infty} f_n$ and P is any projection onto B_p (see Theorem 8). Since $PI = id_{B_p}$ we get that α is an isomorphic embedding of B_p into $(\Sigma\ell_p)_p \cong \ell_p$ and $P\Sigma\beta$ is a projection onto $\alpha(B_p)$. Theorem 6 gives the claim. ∎

Remark: It is easy to see from (5) that $Im(P_s^*) \subset L_\infty(\mathbb{D}, \nu)$ is exactly

$$\{f(z)\colon f(z) = (1 - |z^2|)^s \cdot g(z) \text{ with } g(z) \text{ analytic}\}$$

so we infer from Theorem 11 that the space X_s of all analytic functions $f(z)$ such that $\sup_{|z|<1}(1 - |z|)^s|f(z)| < \infty$ is isomorphic to ℓ_∞, for $s > 0$.

13.   Our aim now is to introduce the so-called stable laws. These are well known probability distributions, but because of their importance in Banach space theory we will discuss them here in some detail. Let us recall some general notions. To each real valued random variable f on a probability space (Ω, P) there corresponds a probability measure μ_f on $\mathbb{R}$, called the distribution of f, determined by the relations $P\{\omega: f(\omega) < \lambda\} = \mu_f((-\infty, \lambda)), \lambda \in \mathbb{R}$. Conversely for each probability measure μ on $\mathbb{R}$ there exist random variables f such that $\mu_f = \mu$. It is clear and well known that the integrability properties of f are reflected in properties of μ_f. More precisely we have the following formula:

$$\int\limits_{\Omega} F(f(\omega)) dP(\omega) = \int\limits_{-\infty}^{+\infty} F(x) d\mu_f(x), \tag{6}$$

valid for every bounded continuous or positive continuous function $F: \mathbb{R} \to \mathbb{R}$.

14 Theorem.   *For every $0 < p \leq 2$ there exists a distribution μ_p such that*

$$\int\limits_{-\infty}^{\infty} e^{i\alpha x} d\mu_p(\alpha) = e^{-|x|^p}. \tag{7}$$

Every function (random variable) whose distribution equals μ_p is called p-stable. Those variables do not exhaust the class of all p-stable variables considered in probability theory. This is a simple special case, but sufficient for our purposes.

Proof: Note that (7) defines what in probability theory is called the characteristic function and in harmonic analysis the Fourier transform of the measure μ. Its basic properties are well known and can be found in many books, e.g. Katznelson [1968] Chapter VI. Let $\mathcal{B}$ denote the class of functions on $\mathbb{R}$ which are Fourier transforms of positive measures on $\mathbb{R}$. This class satisfies the following properties:

if $f_1, f_2 \in \mathcal{B}, a_1, a_2 \geq 0$ then $a_1 f_1 + a_2 f_2 \in \mathcal{B}$; \hfill (8)

if $f_1, f_2 \in \mathcal{B}$ then $f_1 \cdot f_2 \in \mathcal{B}$; \hfill (9)

if $(f_n)_{n=1}^{\infty} \subset \mathcal{B}$ and f_n converges almost uniformly on $\mathbb{R}$ to f

then $f \in \mathcal{B}$. \hfill (10)

Our aim is to show that $e^{-|x|^p} \in \mathcal{B}$, $0 < p \leq 2$, because then we get from (7) that $\mu_p(\mathbb{R}) = 1$, so μ_p is a probability measure on $\mathbb{R}$. The case $p = 2$ is the classical Gaussian (normal) distribution so $d\mu_2(x) = (2\pi)^{\frac{1}{2}} e^{-\frac{x^2}{2}} dx$. From now on we assume $0 < p < 2$. From the formula

$$|x|^p = C_p \int_0^\infty \frac{\alpha^{p-1}}{1 + (\frac{\alpha}{x})^2} d\alpha \qquad (11)$$

which is easy to check using the substitution $\frac{\alpha}{x} = u$ we get

$$e^{-|x|^p} = \exp -C_p \int_0^\infty \frac{\alpha^{p-1}}{1 + (\frac{\alpha}{x})^2} d\alpha.$$

Approximating this integral we see that $e^{-|x|^p}$ is an almost uniform limit of the functions

$$\exp -\sum_{v=1}^N \frac{a_v}{1 + (\frac{b_v}{x})^2} = \prod_{v=1}^N \exp -\frac{a_v}{1 + (\frac{b_v}{x})^2} \quad \text{with} \quad a_v \geq 0.$$

So it is enough to check (see (9) and (10)) that $\exp\left(-\frac{a^2}{1+(\frac{b}{x})^2}\right) \in \mathcal{B}$. But

$$\exp -\frac{a^2}{1 + (\frac{b}{x})^2} = \exp \frac{-a^2 x^2}{x^2 + b^2}$$

$$= \exp\left(\frac{a^2 b^2}{x^2 + b^2} - a^2\right)$$

$$= C \cdot \exp \frac{a^2 b^2}{x^2 + b^2}$$

$$= C \sum_{n=0}^\infty \frac{(ab)^{2n}}{n!} (x^2 + b^2)^{-n}.$$

Since the convergence is almost uniform, from (8), (9) and (10) we see that it is enough to check that $(x^2 + b^2)^{-1} \in \mathcal{B}$. This follows from the formula

$$(x^2 + b^2)^{-1} = \frac{1}{2b} \int_{-\infty}^{+\infty} e^{i\alpha x} e^{-b|\alpha|} d\alpha$$

$$= \frac{1}{b} \int_0^\infty \cos \alpha x \, e^{-b\alpha} d\alpha \qquad (12)$$

which is easily verified using integration by parts twice.   ∎

15 Proposition. *Let f be a p-stable function on a probability measure space (Ω, μ). Then*

(a) *if $p = 2$ then $f \in L_q(\Omega, \mu)$ for $0 < q < \infty$,*

(b) *if $p < 2$ then $f \in L_q(\Omega, \mu)$ for $0 < q < p$.*

Proof: The case (a) easily follows from (6), since we know that $d\mu_2(x) = (2\pi)^{\frac{1}{2}} e^{-\frac{x^2}{2}} dx$. Let $p < 2$ and f be a p-stable variable with the distribution μ_p satisfying (7). We have to estimate $\int_{-\infty}^{+\infty} |x|^q d\mu_p(x)$ (see (6)). Since $|x|^q = C_q \int_0^\infty (1 - \cos xt) t^{-1-q} dt$ (substitute $xt = u$ here the condition $q < 2$ is important) we get using (7)

$$\int_{-\infty}^{+\infty} |x|^q d\mu_p(x) = C_q \int_{-\infty}^{+\infty} \int_0^\infty \frac{1 - \cos xt}{t^{1+q}} dt d\mu_p(x)$$

$$= C_q \int_0^\infty \frac{1}{t^{1+q}} \int_{-\infty}^{+\infty} (1 - \cos xt) d\mu_p(x) dt$$

$$= C_q \int_0^\infty \frac{1}{t^{1+q}} \int_{-\infty}^{+\infty} (1 - Re\ e^{ixt}) d\mu_p(x) dt$$

$$= C_q \int_0^\infty \frac{1}{t^{1+q}} [1 - e^{-t^p}] dt.$$

Substituting in the last integral $t^p = u$ we get

$$\int_0^\infty \frac{1 - e^{-t^p}}{t^{1+q}} dt = \frac{1}{p} \int_0^\infty \frac{1 - e^{-u}}{u^{\frac{1+q}{p}}} du,$$

which is finite for $0 < q < p$.   ∎

16.   Suppose now that $(f_n)_{n=1}^\infty$ is a sequence of independent p-stable functions on a probability measure space (Ω, μ). Let $(a_n)_{n=1}^\infty$ be a finite sequence of real numbers with $\sum_{n=1}^\infty |a_n|^p = 1$ and put $f = \sum_{n=1}^\infty a_n f_n$.

If μ_f is the distribution of f then we have

$$\int\limits_{-\infty}^{+\infty} e^{i\alpha x} d\mu_f(\alpha) = \int\limits_{\Omega} e^{ixf(\omega)} d\mu(\omega)$$

$$= \int\limits_{\Omega} \exp\Big(ix \sum_{n=1}^{\infty} a_n f_n(\omega)\Big) d\mu(\omega)$$

$$= \int\limits_{\Omega} \prod_{n=1}^{\infty} e^{ixa_n f_n(\omega)} d\mu(\omega)$$

$$= \prod_{n=1}^{\infty} \int\limits_{\Omega} e^{ixa_n f_n(w)} d\mu(w)$$

$$= \prod_{n=1}^{\infty} e^{-|a_n x|^p}$$

$$= \exp -\Big(\sum_{n=1}^{\infty} |a_n|^p\Big)|x|^p = e^{-|x|^p}.$$

Thus f is also p-stable. In particular (see Proposition 15) we have

Corollary. Let $(f_n)_{n=1}^{\infty}$ be a sequence of independent p-stable functions, $0 < p \leq 2$. The $\mathrm{span}(f_n)_{n=1}^{\infty}$ in real $L_q(\Omega, \mu)$ is isometric to ℓ_p if $q < p$ and $p < 2$. For $p = 2$ it is isometric to ℓ_2 for $0 < q < \infty$. ∎

17.   We now wish to return to the circle of ideas connected with unconditional convergence of series in Banach spaces which were discussed in II.D.7. Motivated by Orlicz's theorem and in particular by II.D.(4) we introduce the following definitions.

Definition. *A Banach space X is said to have cotype $p, 2 \leq p \leq \infty$, if there exists a constant C such that for all finite sets $(x_j)_{j=1}^{n} \subset X$*

$$C \int\limits_0^1 \Big\| \sum_{j=1}^n r_j(t)x_j \Big\| dt \geq \Big(\sum_{j=1}^n \|x_j\|^p \Big)^{\frac{1}{p}} \tag{13}$$

A Banach space is said to have type $p, 1 \leq p \leq 2$, if there exists a constant C such that for all finite sets $(x_j)_{j=1}^{n} \subset X$

$$\int\limits_0^1 \Big\| \sum_{j=1}^n r_j(t)x_j \Big\| dt \leq C \Big(\sum_{j=1}^n \|x_j\|^p \Big)^{\frac{1}{p}}. \tag{14}$$

Recall that $(r_j)_{j=1}^\infty$ are the Rademacher functions.

A few comments about these definitions are in order.

(a) Since scalars satisfy neither (13) for $p < 2$ nor (14) for $p > 2$ (see Khintchine's inequality I.B.8) we see that the above restrictions for p are essential if we hope to get non-trivial concepts.

(b) If a Banach space X has type p and cotype q and Y is finitely representable in X then Y also has type p and cotype q. In particular X and X^{**} have the same type and cotype; cf. principle of local reflexivity II.E.14.

(c) Every Banach space X has type 1 and cotype ∞. Also if X has cotype p it has also cotype q for $q > p$ and if X has type p it has also type q for $q < p$.

(d) The smallest constant for which (13) holds for a given space X is called the cotype p constant of X and is denoted $C_p(X)$. Similarly we define the type p constant of X, denoted by $T_p(X)$.

The following vector valued generalization of the Khintchine inequality is a fundamental tool for investigating types and cotypes.

18 Theorem. (Kahane's inequality) *There exist constants $C_p, 1 \leq p < \infty$ such that for every Banach space X the inequality*

$$\int_0^1 \left\| \sum_{j=0}^n r_j(t)x_j \right\| dt \leq \left(\int_0^1 \left\| \sum_{j=0}^n r_j(t)x_j \right\|^p dt \right)^{\frac{1}{p}} \leq C_p \int_0^1 \left\| \sum_{j=1}^n r_j(t)x_j \right\| dt \tag{15}$$

holds for every finite sequence $(x_j)_{j=1}^n \subset X$.

The proof follows from the following distributional inequality.

19 Proposition. *Let $V(t) = \left\| \sum_{j=0}^n r_j(t)x_j \right\|$. Then for every $\alpha > 0$ we have $|\{t\colon V(t) > 2\alpha\}| \leq 4|\{t\colon V(t) > \alpha\}|^2$.*

Proof: For $k \leq n$ let us put $V_k(t) = \left\| \sum_{j=0}^k r_j(t)x_j \right\|$ and let us define the following sets:

$$A_m = \{t\colon V_k(t) \leq \alpha, \ k = 0, \ldots, m-1, \ V_m(t) > \alpha\},$$

$$A = \bigcup_{m=1}^n A_m = \{t\colon \sup_k V_k(t) > \alpha\},$$

$$B = \{t\colon V(t) > \alpha\},$$

$$C = \{t\colon V(t) > 2\alpha\},$$

$$C_m = \{t\colon \left\| \sum_{j=m}^n r_j(t)x_j \right\| > \alpha\}.$$

Let us write $\sum_{j=0}^{n} r_j(t)x_j = \sum_{j=0}^{m} r_j(t)x_j + \sum_{j=m+1}^{n} r_j(t)x_j = a(t) + b(t)$. It follows from properties of the Rademacher functions that $b(t)$ is symmetric on every set where $a(t)$ is constant. Since $\|x\| \leq \max(\|x+y\|, \|x-y\|)$ for every $x, y \in X$, we see that at least on half of the set A_m we have $\|a\| \leq \|a+b\|$. Thus

$$|A_m| \leq 2|B \cap A_m| \quad \text{so} \quad |A| \leq 2|B|. \tag{16}$$

Analogously

$$|C_m| \leq 2|B|. \tag{17}$$

We put $A_m = (A_m \cap \{t: r_m(t) = 1\}) \cup (A_m \cap \{t: r_m(t) = -1\}) = A_m^+ \cup A_m^-$ and $C_m = (C_m \cap \{t: r_m(t) = 1\}) \cup (C_m \cap \{t: r_m(t) = -1\}) = C_m^+ \cup C_m^-$. The independence of the Rademacher functions gives

$$|A_m^+ \cap C_m^+| = 2|A_m^+| \cdot |C_m^+|$$

and

$$|A_m^- \cap C_m^-| = 2|A_m^-| \cdot |C_m^-|.$$

Since $|A_m^+| = |A_m^-| = \frac{1}{2}|A_m|$ and $|C_m^+| = |C_m^-| = \frac{1}{2}|C_m|$ we have

$$|A_m \cap C_m| = 4|A_m^+| \, |C_m^+| = |A_m| \cdot |C_m|. \tag{18}$$

Since obviously $C \subset B \subset A$, from (18), (17) and (16) we get

$$|C| \leq \sum |A_m \cap C| \leq \sum |A_m \cap C_m| = \sum |A_m| \, |C_m|$$

$$\leq \sup_m |C_m| \cdot \sum |A_m| \leq 2|B| \cdot |A| \leq 4|B|^2. \qquad \blacksquare$$

Proof of Theorem 18.   The left hand side inequality is obvious, while the right hand side inequality is a standard passage from a distributional inequality to an integral one. We can assume $\int_0^1 V(t)dt = 1$, so $|\{t: V(t) > 8\}| \leq \frac{1}{8}$. Applying Proposition 19 inductively for $k = 1, 2, 3, \ldots$ we obtain

$$|\{t: V(t) > 2^k \cdot 8\}| \leq 4^{2^k} 8^{-2^k} = 2^{-2^k}.$$

This gives

$$\int_0^1 V(t)^p dt \leq 8 + \sum_{k=1}^{\infty} (2^k \cdot 8)^p \cdot |\{t: \, 2^{k-1} \cdot 8 < V(t) \leq 2^k \cdot 8\}|$$

$$\leq 8 + \sum_{k=1}^{\infty} (2^k \cdot 8)^p \cdot 2^{-2^k} = C_p^p. \qquad \blacksquare$$

20 Remarks. (a) An obvious and immediate consequence of Theorem 18 is that in (13) and (14) instead of $\int \|\Sigma r_j x_j\|$ we can use $(\int \|\Sigma r_j x_j\|^q)^{\frac{1}{q}}$ for every $q, 1 \leq q < \infty$.

(b) For some applications the magnitude of the constant C_p is important. Our proof, as can be easily verified, gives $C_p \leq C \cdot p$ for $p \geq 2$. The correct order of magnitude is $C_p \leq C\sqrt{p}$ (see e.g. Milman-Schechtman [1986]).

21. The following repeats arguments from II.D.6.

Proposition. *If X has cotype p and $\sum_{n=1}^{\infty} x_n$ is an unconditionally convergent series in X, then $\sum_{n=1}^{\infty} \|x_n\|^p < \infty$.*

Proof. We have for every N

$$\left(\sum_{n=1}^{N} \|x_n\|^p \right)^{\frac{1}{p}}$$

$$\leq C_p(X) \int_0^1 \left\| \sum_{n=1}^{N} r_n(t) x_n \right\| dt \leq C_p(X) \sup_t \left\| \sum_{n=1}^{N} r_n(t) x_n \right\| \leq C. \;\blacksquare$$

22. A part of the relation between type and cotype is explained by the following.

Proposition. *If X has type p then X^* has cotype q, where $\frac{1}{p} + \frac{1}{q} = 1$.*

Proof: For arbitrary $x_1, \ldots, x_n \in X$ and $x_1^*, \ldots, x_n^* \in X^*$ we have

$$\sum_{i=1}^{n} x_i^*(x_i) = \int_0^1 \left(\sum_{i=1}^{n} r_i(t) x_i^* \right) \left(\sum_{i=1}^{n} r_i(t) x_i \right) dt$$

$$\leq \int_0^1 \left\| \sum_{i=1}^{n} r_i(t) x_i^* \right\| \cdot \left\| \sum_{i=1}^{n} r_i(t) x_i \right\| dt \qquad (19)$$

$$\leq \left(\int_0^1 \left\| \sum_{i=1}^{n} r_i(t) x_i^* \right\|^2 dt \right)^{\frac{1}{2}} \left(\int_0^1 \left\| \sum_{i=1}^{n} r_i(t) x_i \right\|^2 \right)^{\frac{1}{2}}.$$

Since

$$\left(\sum_{i=1}^{n} \|x_i^*\|^q \right)^{\frac{1}{q}} = \sup \left\{ \sum_{i=1}^{n} x_i^*(x_i) : x_i \in X, \sum_{i=1}^{n} \|x_i\|^p \leq 1 \right\}$$

from (19) we get

$$\left(\sum_{i=1}^{n}\|x_i^*\|^q\right)^{\frac{1}{q}} \leq \left(\int_0^1 \|\sum_{i=1}^{n} r_i(t)x_i^*\|^2\right)^{\frac{1}{2}}$$

$$\cdot \sup\left\{\left(\int_0^1 \|\sum_{i=1}^{n} r_i(t)x_i\|^2\right)^{\frac{1}{2}} : \sum_{i=1}^{n}\|x_i\|^p \leq 1\right\}.$$

Using Theorem 18 we get the claim.   ∎

23.   The types and cotypes of L_p-spaces are as follows.

Theorem.   *The space $L_p(\Omega, \mu), 1 \leq p < \infty$, is of type $\min(2, p)$ and of cotype $\max(2, p)$.*

Proof: Clearly $L_\infty(\Omega, \mu)$ has type 1 and cotype ∞. For $1 \leq p < \infty$ exactly like in II.D.6 we obtain for every $(x_j) \subset L_p(\Omega, \mu)$

$$\left(\int_0^1 \left\|\sum_j r_j(t)x_j\right\|_p^p dt\right)^{\frac{1}{p}}$$

$$= \left(\int_\Omega \int_0^1 \left|\sum_j r_j(t)x_j(\omega)\right|^p dt d\mu(\omega)\right)^{\frac{1}{p}} \qquad (20)$$

$$\sim \left(\int_\Omega \left(\sum_j |x_j(\omega)|^2\right)^{\frac{p}{2}} d\mu(\omega)\right)^{\frac{1}{p}}$$

$$= \left(\int_\Omega \left[\sum_j (|x_j(\omega)|^p)^{\frac{2}{p}}\right]^{\frac{p}{2}} d\mu(\omega)\right)^{\frac{1}{p}}$$

where '$\sim$' indicate, that there are inequalities in both directions with constants independent of the set (x_j). If $1 \leq p \leq 2$ then Theorem 18 and (20) give

$$\int_0^1 \left\|\sum_j r_j(t)x_j\right\|_p dt \leq C\left(\int_\Omega \sum_j |x_j(\omega)|^p d\mu(\omega)\right)^{\frac{1}{p}} \leq C\left(\sum_j \|x_j\|_p^p\right)^{\frac{1}{p}}$$

and

$$\int_0^1 \Big\| \sum_j r_j(t)x_j \Big\|_p dt \geq C\Big(\sum_j \Big(\int_\Omega |x_j(\omega)|^p d\mu(\omega) \Big)^{\frac{2}{p}} \Big)^{\frac{1}{2}}$$

$$= C\Big(\sum_j \|x_j\|_p^2 \Big)^{\frac{1}{2}}.$$

For $2 \leq p < \infty$ Theorem 18 and (20) give

$$\int_0^1 \Big\| \sum_j r_j(t)x_j \Big\|_p dt \leq C\Big(\int_\Omega \Big(\sum_j |x_j(\omega)|^2 \Big)^{\frac{p}{2}} d\mu(\omega) \Big)^{\frac{1}{p}}$$

$$\leq C\Big(\sum_j \Big(\int_\Omega |x_j(\omega)|^p d\mu(\omega) \Big)^{\frac{2}{p}} \Big)^{\frac{1}{2}}$$

$$= C\Big(\sum_j \|x_j\|_p^2 \Big)^{\frac{1}{2}}$$

and

$$\int_0^1 \Big\| \sum_j r_j(t)x_j \Big\|_p dt \geq C\Big(\int_\Omega \Big(\sum_j |x_j(\omega)|^2 \Big)^{\frac{p}{2}} d\mu(\omega) \Big)^{\frac{1}{p}} \geq$$

$$\geq C\Big(\int_\Omega \sum_j |x_j(\omega)|^p d\mu(\omega) \Big)^{\frac{1}{p}} = C\Big(\sum_j \|x_j\|_p^p \Big)^{\frac{1}{p}} \blacksquare$$

24. As we said earlier (**17**) our Definition 17 was motivated by the Orlicz property (II.D.(4)). There is however a more formal connection between these two notions.

Proposition. *Suppose the Banach space X has the Orlicz property and suppose that $X \sim \big(\sum_{n=1}^\infty X \big)_p$ for some $p, 1 \leq p \leq 2$. Then X has cotype 2.*

Proof: Let us take a finite set $(x_j)_{j=1}^m \subset X$. Let $\tilde{r}_j(k)$, $k = 1, 2, \ldots, 2^m$, $j = 1, \ldots, m$ be the 'Rademacher type' functions on the set $\{1, 2, \ldots, 2^m\}$. Let us define

$$\tilde{x}_j = 2^{-\frac{m}{p}} \sum_{k=1}^{2^m} \tilde{r}_j(k)x_j^k$$

where x_j^k denotes the vector x_j considered in the k-th summand of $\left(\sum_{k=1}^{\infty} X\right)_p$. Using the isomorphism between X and $\left(\sum_{k=1}^{\infty} X\right)_p$ and the Orlicz property we get

$$\left(\sum_{j=1}^{m} \|x_j\|^2\right)^{\frac{1}{2}} = \left(\sum_{j=1}^{m} \|\tilde{x}_j\|^2\right)^{\frac{1}{2}} \le C \sup_{t\in[0,1]} \left\|\sum_{j=1}^{m} r_j(t)\tilde{x}_j\right\|$$

$$= C \sup_{t\in[0,1]} \left\|2^{-\frac{m}{p}}\sum_{j=1}^{m} r_j(t)\sum_{k=1}^{2^m} \tilde{r}_j(k)x_j^k\right\| =$$

$$= C \sup_{t\in[0,1]} 2^{-\frac{m}{p}}\left(\sum_{k=1}^{2^m} \left\|\sum_{j=1}^{m} r_j(t)\tilde{r}_j(k)x_j\right\|^p\right)^{\frac{1}{p}}$$

$$= C \sup_{t\in[0,1]} \left(\int_0^1 \left\|\sum_{j=1}^{m} r_j(t)r_j(u)x_j\right\|^p du\right)^{\frac{1}{p}} =$$

$$= C\left(\int \left\|\sum_{j=1}^{m} r_j(u)x_j\right\|^p du\right)^{\frac{1}{p}}.$$

We see from Remark 20 that this completes the proof.   ■

25.   As a simple application of these ideas we will present the generalization of a classical result of Carleman.

Theorem.   *For every complete orthonormal system $(\varphi_n)_{n=1}^{\infty} \subset L_2[0,1]$ there exists an $f \in C[0,1]$ such that $\sum_{n=1}^{\infty} |\langle f, \varphi_n\rangle|^p = \infty$ for every $p < 2$.*

Proof: Suppose it is not so. A standard category argument or the closed graph theorem applied to the space $\bigcup_{p<2} \ell_p$ yields $p < 2$ such that $\sum_{n=1}^{\infty} |\langle f, \varphi_n\rangle|^p < \infty$ for all $f \in C[0,1]$. Thus we have a commutative diagram

$$C[0,1] \xrightarrow{\;id\;} L_2[0,1]$$

with φ on the lower-left and Σ on the lower-right, meeting at ℓ_p.

where operators φ and Σ are defined by $\varphi(f) = (\langle f, \varphi_n\rangle)_{n=1}^{\infty}$ and $\Sigma(\xi_n) = \sum_{n=1}^{\infty} \xi_n\varphi_n$. Since $id\colon C[0,1] \to L_2[0,1]$ is clearly a non-compact operator the following lemma gives the contradiction.

26 Lemma.   *Every operator from $C(K)$ into $\ell_p, 1 < p < 2$, is compact.*

Proof: Let $T: C(K) \to \ell_p$. Take $T^*: \ell_q \to M(K), \frac{1}{p}+\frac{1}{q} = 1$. If T is not compact, nor is T^*, thus there exists a sequence $(x_n)_{n=1}^{\infty} \subset \ell_q, \|x_n\| \leq 1$ such that $\|T^*x_n - T^*x_m\| \geq \delta$ for $n \neq m$. Since ℓ_q is reflexive we can pass to a subsequence such that $x_{n_k} \xrightarrow{\omega} x_\infty$ and to another subsequence such that $y_k = x_{n_k} - x_\infty$ is equivalent to the block basis of the unit vector basis in ℓ_q (apply II.B.17) and thus to the unit vector basis in ℓ_q. Since $\|T^*y_k\| \geq \delta$ and $M(K)$ has cotype 2, for $N = 1, 2, \ldots$ we get

$$\sqrt{N} \cdot \delta \leq \left(\sum_{k=1}^{N} \|T^*y_k\|^2 \right)^{\frac{1}{2}} \leq C \int_0^1 \left\| \sum_{k=1}^{N} r_n(t)T^*y_k \right\| dt$$

$$\leq C\|T\| \int_0^1 \left\| \sum_{k=1}^{N} r_n(t)y_k \right\| dt \leq C\|T\| \left(\sum_{k=1}^{N} \|y_k\|^q \right)^{\frac{1}{q}} \leq CN^{\frac{1}{q}}.$$

Since $q > 2$ this is a contradiction. ∎

This lemma is also true for ℓ_1 (Exercise II.D.5).

27. We know from II.A.5 that for every weakly convergent sequence there exists a sequence of convex combinations convergent in norm. For L_p spaces, $1 < p < \infty$, this can be improved.

Theorem. (Banach-Saks). *Every bounded sequence of functions $(x_n)_{n=1}^{\infty} \subset L_p(\Omega, \mu), 1 < p < \infty$, contains a subsequence $(x_{n_k})_{k=1}^{\infty}$ such that $N^{-1} \sum_{k=1}^{N} x_{n_k}$ converges in norm.*

This theorem is an obvious consequence of the following.

28 Proposition. *Every bounded sequence $(x_n)_{n=1}^{\infty} \subset L_p(\Omega, \mu), 1 < p < \infty$, contains a subsequence $(x_{n_k})_{k=1}^{\infty}$ such that for some $x \in L_p(\Omega, \mu)$*

$$\left\| \sum_{k \in A} (x_{n_k} - x) \right\| \leq C \left(\sum_{k \in A} \|x_{n_k} - x\|^s \right)^{\frac{1}{s}} \tag{21}$$

for every finite subset of integers A and for $s = \min(2, p)$.

Proof: Propositions 1 and 2 show that it suffices to consider $L_p[0, 1]$. First we use reflexivity to choose the subsequence (x_{n_k}) such that $x_{n_k} \xrightarrow{\omega} x$. If for some further subsequence (still call it x_{n_k}) we have $\|x_{n_k} - x\| \to 0$ we take once more a subsequence such that $\|x_{n_k} - x\| \leq 2^{-k}$ so (21) holds. Otherwise $\|x_{n_k} - x\| \geq \delta$ for $k = 1, 2, \ldots$ so $(x_{n_k} - x)$

has a further subsequence equivalent to the block basis of the Haar system (II.B.17), thus unconditional (see II.D.11 and II.D.13). Theorem 23 easily gives (21) in this case. ∎

29.   We also wish to present a similar result for almost everywhere convergence. Because of the application in III.C.8. we formulate it for countable family of sequences.

Theorem *Suppose that for every $m \in \mathbf{Z}(x_n^m)_{n=1}^\infty$ is a bounded sequence in $L_p(\Omega, \mu), 1 < p < \infty$. Then there exists an increasing sequence of integers $(n_k)_{k=1}^\infty$ such that for every $m \in \mathbf{Z}$ there exists $x^m \in L_p(\Omega, \mu)$ such that*

$$\frac{1}{N} \sum_{k=1}^N x_{n_{\sigma(k)}}^m \to x^m \qquad \mu\text{-a.e.} \tag{22}$$

for every $m \in \mathbf{Z}$ and every permutation σ of natural numbers.

Proof: A standard diagonal procedure and Proposition 28 shows that there exists an increasing sequence of integers $(n_k)_{k=1}^\infty$ such that for each $m \in \mathbf{Z}$ the sequence $(x_{n_k}^m)_{k=1}^\infty$ satisfies (21) with some constant depending on m. We will show that any sequence satisfying (21) satisfies (22). Clearly it suffices to consider each sequence $(x_{n_k}^m)_{k=1}^\infty$ separately, so in the rest of the proof we will omit the superscript m.

Denote $x_{n_{\sigma(k)}} - x = h_k$ and $N^{-1} \sum_{k=1}^N h_k = H_N$. We have to show that $\sum_{N=1}^\infty (H_{N+1} - H_N)$ converges μ-a.e.

Since for $(x_n) \subset L_p(\Omega, \mu)$ we have $\|\Sigma|x_n|\|_p \leq \Sigma\|x_n\|$, we see that absolutely convergent series converge μ-a.e. We write $H_{N+1} - H_N = \left(\frac{-1}{N(N+1)}\right) H_N + \frac{h_{(N+1)}}{N+1}$. From (21) we get $\sum_N \left(\frac{1}{N(N+1)}\right)\|H_N\| < \infty$ so $\sum_N \left(\frac{-1}{N(N+1)}\right) H_N$ converges μ-a.e. It remains to show that

$$\sum_{n=2}^\infty \frac{h_n}{n} \qquad \text{converges} \quad \mu\text{-a.e.} \tag{23}$$

Fix numbers α and β such that

$$0 < \alpha < 1, \ \beta > 1, \ \alpha\beta < 1, \ \alpha s > 1 \text{ and } s\beta + 1 > s + \beta.$$

For each integer K we have from (21)

$$\left\| \sum_{n=[K^\beta]}^{[(K+1)^\beta]} \frac{h_n}{n} \right\| \leq C \left(\sum_{n=[K^\beta]}^{[(K+1)^\beta]} (K^{-\beta})^s \right)^{\frac{1}{s}} \leq C \left(\frac{K^{\beta-1}}{K^{\beta s}} \right)^{\frac{1}{s}}.$$

Since $\beta > 1$ we infer that

$$\lim_{K \to \infty} \sum_{n=1}^{[K^\beta]} \frac{h_n}{n} \text{ exists } \mu - a.e. \tag{24}$$

Let us write $h_n = h'_n + h''_n$ where $h'_n = h_n \cdot \chi_{\{\omega: |h_n| \leq n^\alpha\}}$. Since $\|h_n\|_p \leq C$ we get $\mu(\text{supp } h''_n) \leq n^{-\alpha p}$ so $\sum_n \mu(\text{supp } h''_n) < \infty$. This easily implies

$$\sum_{n=1}^{\infty} \frac{h''_n}{n} \text{ converges } \mu - a.e. \tag{25}$$

From (24) and (25) we infer that

$$\lim_{K \to \infty} \sum_{n=1}^{[K^\beta]} \frac{h'_n}{n} \text{ exists } \mu \text{ a.e.} \tag{26}$$

But for $[K^\beta] \leq N \leq [(K+1)^\beta]$ we have

$$\left| \sum_{[K^\beta]}^{N} \frac{h'_n}{n} \right| \leq ((K+1)^\beta - K^\beta) \frac{(K+1)^{\alpha\beta}}{K^\beta} \leq CK^{\alpha\beta - 1}. \tag{27}$$

The choice of α and β with (26) and (27) gives that $\lim_K \sum_{n=1}^{K} \left(\frac{h'_n}{n}\right)$ exists μ-a.e. and this together with (25) yields (23). ∎

Remark: There is an alternative argument for (23). From the proof of Proposition 28 we infer that $\left\| \sum_{n=N}^{M} \frac{\ln(n+1)h_n}{n} \right\|^s \leq C \sum_{n=N}^{M} n^{-s} \ln(n+1)^s$ so the series $\sum_{n=2}^{\infty} \frac{\ln(n+1)}{n} h_n$ converges unconditionally in $L_p(\Omega, \mu)$, so it unconditionally converges in measure. Corollary III.H.25. shows that (23) holds.

Notes and remarks.

The $L_p(\mu)$-spaces are among the most important and widely used spaces in analysis. It is probably useless to trace back their first appearance in the literature but already in Banach [1932] they are the prime examples of Banach spaces. *Proposition 3* is, as very often in this book, only a sample result. The same holds for spaces $W_p^s(M)$ for $s \geq 0, 1 < p < \infty$, and M a sufficiently regular set in $\mathbb{R}^n$ or a differentiable manifold. It also holds for spaces analogously defined by different sets of derivatives. The reader should consult Pełczyński-Senator [1986] for generalizations.

Proposition 5 and *Theorem 6* are taken from Pełczyński [1960]. Much work has been done on complemented subspaces of $L_p[0,1]$, $1 < p < \infty$. Many local and global characterizations have been given. It was shown in Bourgain-Rosenthal-Schechtman [1981] that there are uncountably many non-isomorphic such subspaces. We will not discuss this subject in our book. The interested reader should consult the above mentioned paper and references quoted there. For $p = 1$ the situation is different. All known complemented subspaces of $L_1[0,1]$ are isomorphic either to ℓ_1 or to $L_1[0,1]$. It is unknown if this is true for all complemented subspaces of $L_1[0,1]$.

Theorem 8 is a special case of results proved by Shields-Williams [1971]. Our proof of this result follows the presentation of Forelli-Rudin [1976] where these results are extended to the unit ball in $\mathbb{C}^n$. The same approach is given in Axler [1988]. *Proposition 9* is well known. Parts (a) and (b) are almost obvious and (c) is usually called Schur's lemma. Actually Schur proved only a very special case of it and the result evolved gradually. *Theorem 11* is taken from Lindenstrauss-Pełczyński [1971]. It is also true that $B_p(\mathbb{D})$ is isomorphic to ℓ_p for $0 < p < 1$ (see Kalton-Trautman [1982]) but the proof has to be different since *Theorem 8* is clearly false for $p < 1$. Those results lead naturally to the following problem: Find a system of functions which is a basis in $B_p(\mathbb{D})$ equivalent to the unit vector basis in ℓ_p. Wojtaszczyk [1984] has shown that for $p \le 1$ analytic versions of spline systems analogous to the Franklin system have this property. In particular the Bočkariov basis for A constructed in III.E.16 and 17 is also a basis in $B_p(\mathbb{D})$, $\frac{1}{2} < p \le 1$, and after suitable normalization is equivalent to the unit vector basis in ℓ_p. It is unknown how those systems behave for $p > 1$. In the case $p > 1$ different bases have been constructed in Matelievič-Pavlovič [1984].

Proposition 7 and *Corollary 16* address special cases of the following question: for what p, q is the space $L_p[0,1]$ or ℓ_p isomorphic to a subspace of $L_q[0,1]$? Under the name of linear dimension this was studied already in Banach [1932] and Banach-Mazur [1933]. Today the full answer is known. It is summarized in the Table 1.

From the results given in this section the interested reader can easily deduce all these facts except the case $2 < p < q < \infty$, which is due to Kadec-Pełczyński [1962] (see Exercises 3 and 4). The answers are the same if we replace ℓ_p by $L_p[0,1]$. This follows from *Proposition 2* and the following

Proposition. *Let X be a Banach space finitely representable in ℓ_p, $1 \le p \le \infty$. Then there exists a measure μ such that X is isomorphic to a subspace of $L_p(\mu)$.*

Table 1.

$\ell_p \backslash L_q$	$q = 1$	$1 < q < 2$	$q = 2$	$2 < q < \infty$	$q = \infty$
$p = 1$	Yes	No	No	No	Yes
$1 < p < 2$	Yes	If $p \geq q$ yes if $p < q$ no	No	No	Yes
$p = 2$	Yes	Yes	Yes	Yes	Yes
$2 < p < \infty$	No	No	No	If $p = q$ yes $p \neq q$ no	Yes
$p = \infty$	No	No	No	No	Yes

The proof can be found in Lindenstrauss-Pełczyński [1968]; it is basically a compactness argument.

Corollary 16 in the context of Banach spaces was observed by Kadec [1958], but all the probabilistic background as given in *Theorem 14* and *Proposition 15* was known much earlier. It can be found in Levy [1925] and our proof of *Theorem 14* follows Bochner [1937]. We presented the existence of stable laws in detail not only to show the very useful *Corollary 16* but also because they are important in Banach space theory (e.g. the notion of p-stable type explained in Notes and remarks to III.H). As remarked already the class of p-stable variables investigated in probability theory contains many more variables than we discuss in this book. Note that our p-stable variables are symmetric (this follows immediately from (7)). The general treatment can be found in many books on probability theory, e.g. Feller [1971] or Lukacs [1970].

The notions of type and cotype were in the air in the early 70's. Type 2 under the name of 'subquadratic Rademacher average' appeared in Dubinsky-Pełczyński-Rosenthal [1972] and the general notion of type and cotype was introduced by Maurey in the Séminaire Maurey-Schwartz 72/73 and Hoffmann-Jørgensen in the Aarhus University 72/73 preprint 'Sums of independent Banach space valued random variables'. It was the French group around L. Schwartz, B. Maurey and G. Pisier who showed the importance of the concept in Banach space theory and in operator theory. We will discuss some important applications of the notions of type and cotype in III.F, III.H and III.I. Let us also note that it is not known in general if a space with the Orlicz property has cotype 2. Our *Proposition 24* (which is due to Figiel-Pisier [1974]) represents all that is known about this problem. We will use it in III.I.

Theorem 18 (Kahane's inequality), which is basic to the theory of type and cotype, can be found in Kahane [1985]. It was first published in the first edition of this book which appeared in 1968. *Theorem 25* for the trigonometric system was shown in Carleman [1918]. Actually the following was shown in Kahane-Katznelson-de Leeuw [1977].

Theorem.   *Given a sequence of numbers* $(a_n)_{n=-\infty}^{+\infty}, a_n \geq 0$ *and* $\sum_{n=-\infty}^{+\infty} a_n^2 < \infty$ *there exists* $f \in C(\mathbb{T})$ *such that* $|\hat{f}(n)| \geq a_n$ *for* $n = 0, \pm 1, \pm 2, \dots$.

As was shown by Kislyakov [1981] one can even get the above f with uniformly convergent Fourier series.

Some generalizations of *Theorem 25* are presented in Wojtaszczyk [1988]. A classical but more complicated proof of this theorem and more precise results can be found in Olevskiĭ [1975], p. 77 and Chapter III sec. 4. It is interesting that our proof of this Theorem is only a small modification of the arguments in Orlicz [1933]. *Theorem 27* was proved in Banach-Saks [1930]. In this formulation it clearly fails for L_1 and L_∞. On the other hand if we consider only weakly convergent sequences the theorem still holds in L_1. This was shown by Szlenk [1965]. *Theorem 29* (also for $p = 1$) was proved by Berkes [1986]. It extends earlier results (without permutations being allowed) of Komlós [1967] and Aldous [1977]. Our proof is a modification of the arguments given in Lyons [1985].

Exercises

1.   Let $\{f_\gamma\}_{\gamma \in \Gamma} \subset L_p(\mu)$, $1 \leq p \leq \infty$, be a subset such that $|f_\gamma| \leq g$ for some $g \in L_p(\mu)$ and all $\gamma \in \Gamma$. Show that there exists $f \in L_p(\mu)$, denoted $f = \sup\{f_\gamma\}_{\gamma \in \Gamma}$ such that $f_\gamma \leq f$ for all $\gamma \in \Gamma$ and if f_1 is such that $f_1 \neq f$ and $f_1 \leq f$ then there exists $\gamma_1 \in \Gamma$ such that we do not have $f_{\gamma_1} \leq f_1$. All the inequalities between functions are understood to be μ-a.e.

Warning. Note that elements of $L_p(\mu)$ are really classes of functions equal μ-a.e., so if Γ is uncountable we have trouble with pointwise supremum.

2. Suppose $X \subset L_p(\mu)$ is a closed subspace such that for some $q < p$ we have $\|x\|_q \geq c\|x\|_p$ for all $x \in X$. Show that $\|x\|_s \geq C_s\|x\|_p$ for all s, $0 < s < p$ and $x \in X$.

3. Suppose $X \subset L_p(\mu)$, $1 < p < \infty$ and $X \sim \ell_p$. Show that there exists $Y \subset X$ such that $Y \sim \ell_p$ and Y is complemented in $L_p(\mu)$.

4. Suppose $p \geq 2$. Show that every infinite dimensional subspace $X \subset L_p[0,1]$ contains an infinite dimensional subspace Y such that Y is complemented in $L_p[0,1]$ and either $Y \sim \ell_p$ or $Y \sim \ell_2$.

5. Let $P: L_p(\mu) \to L_p(\mu)$, $1 \leq p \leq \infty$, be a projection of norm 1 with $\dim Im\, P = n$. Show that $Im\, P \cong \ell_p^n$.

6. Show that $\left(\sum_{n=1}^{\infty} \ell_2^n\right)_p \sim \ell_p$, $1 < p < \infty$.

7. Construct an unconditionally convergent series $\sum_{n=1}^{\infty} x_n$ in ℓ_p, $1 \leq p \leq 2$ such that $\sum_{n=1}^{\infty} \|x_n\|^{2-\varepsilon} = \infty$ for every $\varepsilon > 0$.

8. A family of projections $\mathcal{P} = \{P_j\}_{j \in J}$ on a Banach space X is called a boolean algebra of projections if

    (a) for $j_1, j_2 \in J$ we have $P_{j_1} P_{j_2} = P_{j_2} P_{j_1} \in \mathcal{P}$,

    (b) if $j_1, j_2 \in J$ and $P_{j_1} P_{j_2} = 0$ then $P_{j_1} + P_{j_2} \in \mathcal{P}$.

    A boolean algebra of projections is bounded if $\sup\{\|P_j\| : j \in J\} < \infty$. Assume that X is a subspace of $L_p(\Omega, \mu)$, $1 \leq p < \infty$. Show that if we are given two commuting, bounded boolean algebras of projections $\mathcal{P} = \{P_j\}_{j \in J}$ and $\mathcal{Q} = \{Q_s\}_{s \in S}$ (i.e. we assume $P_j Q_s = Q_s P_j$ for all $s \in S, j \in J$), then the boolean algebra of projections generated by $\mathcal{P} \cup \mathcal{Q}$ is bounded.

9. Show that the sequence $(f_n)_{n=1}^{\infty} = \left(z^{2^n} \|z^{2^n}\|^{-1}\right)_{n=1}^{\infty} \subset B_p(\mathbb{D})$, $0 < p < \infty, n = 1, 2, \ldots$ is a basic sequence equivalent to the unit vector basis in ℓ_p. Show also that $\left(t^{2^n} \|t^{2^n}\|_p^{-1}\right)_{n=1}^{\infty} \subset L_p[0,1]$, $0 < p < \infty$, $n = 1, 2, \ldots$ is equivalent to the unit vector basis in ℓ_p.

10. For f defined on $\mathbb{D}$ its Bergman projection is defined as

$$P(f)(z) = \frac{1}{\pi} \int_{\mathbb{D}} \frac{f(\omega)}{(1 - z\bar{\omega})^2} d\nu(\omega).$$

Show that

    (a) P is the orthogonal projection from $L_2(\mathbb{D}, \nu)$ onto $B_2(\mathbb{D})$,

    (b) P is a bounded projection from $L_p(\mathbb{D}, \nu)$ onto $B_p(\mathbb{D})$ for $1 < p < \infty$,

(c) P is not bounded on $L_1(\mathbb{D}, \nu)$.

11. Show that if g is an analytic function on $\mathbb{D}$ such that the operator
$Tg(f) = f \cdot g$ maps $B_p(\mathbb{D})$ into $B_q(\mathbb{D})$ for some p, q, $0 < p < q < \infty$,
then $g = 0$.

12. Let $i: W_p^1(\mathbb{T}^n) \to \left(\sum_{i=1}^{n+1} L_p(\mathbb{T}^n)\right)_p$ be the natural embedding, and
let P be the orthogonal projection from $\left(\sum_{i=1}^{n+1} L_2(\mathbb{T}^n)\right)_2$ onto
$i(W_2^1(\mathbb{T}^n))$. Show that P is continuous on $\left(\sum_{i=1}^{n+1} L_p(\mathbb{T}^n)\right)_p$, $1 <$
$p < \infty$ and of weak type (1-1).

13. Show that the operator $T_\alpha f(x) = \int_0^x (x - y)^{-\alpha} f(y) dy$, $0 < \alpha < 1$,
is bounded on $L_p[0, 1]$ for $1 \le p \le \infty$.

14. Show that the operator $Tf(x) = x^{-1} \int_0^x f(t) dt$ is a bounded operator on $L_p(0, \infty)$ for $1 < p < \infty$.

15. Show that every linear operator from ℓ_p into ℓ_q, $0 < q < p < \infty$,
is compact and also every operator from c_0 into ℓ_q, $0 < q < \infty$, is
compact and that the same is true for $T: X \to Y$ where X is any
subspace of ℓ_p and Y any subspace of ℓ_q.

16. Let $(\varphi_n)_{n=1}^\infty$ be a complete orthonormal system in $L_2[0, 1]$.

(a) Show that for every $U \subset [0, 1]$ with $|U| > 0$ there exists $f \in$
$L_\infty[0, 1]$, supp $f \subset U$ such that $\sum_{n=1}^\infty |\langle f, \varphi_n \rangle|^p = \infty$ for all
$p < 2$.

(b) Show that there exists an $f \in C[0, 1]$ such that f is linear
on every interval of $[0, 1] \backslash \Delta$ (where Δ is a Cantor set) and
$\sum_{n=1}^\infty |\langle f, \varphi_n \rangle|^p = \infty$ for all $p < 2$.

(c) Show that there exists a set $U \subset [0, 1]$ such that for all $p < 2$.
$\sum_{n=1}^\infty |\langle \chi_U, \varphi_n \rangle|^p = \infty$

(d) Suppose that $(\psi_n)_{n=1}^\infty$ is an orthonormal system in $L_2[0, 1]$ (not
necessarily complete) such that $\inf_n \int |\psi_n| > 0$. Show that
there exists an $f \in C[0, 1]$ such that $\sum_{n=1}^\infty |\langle f, \psi_n \rangle|^p = \infty$ for
all $p < 2$.

(e) Find an example of a non-complete orthonormal system
$(\psi_n)_{n=1}^\infty$ in $L_2[0, 1]$ such that $\sum_{n=1}^\infty |\langle f, \psi_n \rangle| < \infty$ for all
$f \in C[0, 1]$.

17. Suppose that X is a Banach space of type p and that $Y \subset X$. Show
that the quotient space X/Y has type p.

18. Suppose that $(X_n)_{n=1}^\infty$ is a sequence of Banach spaces such that $T_p(X_n) \le C$ for $n = 1, 2, \ldots$ and some p, $1 \le p \le 2$. Show that $\left(\sum_{n=1}^\infty X_n\right)_q$ has type $\min(p, q)$. If $C_p(X_n) \le C$ for $n = 1, 2, \ldots$ and some p, $2 \le p \le \infty$ then $\left(\sum_{n=1}^\infty X_n\right)_q$ has cotype $\max(p, q)$.

19. (a) Find a sequence $(f_n)_{n=1}^\infty \subset C[0, 1]$ such that $\|f_n\| = 1$, $f_n \xrightarrow{\omega} 0$ as $n \to \infty$ but for every subsequence $(f_{n_k})_{k=1}^\infty$ we have $\limsup_{N \to \infty} N^{-1} \left\| \sum_{k=1}^N f_{n_k} \right\|_\infty > 0$.

   (b) Show that a sequence like in (a) can be found in a reflexive space.

III.B. Projection Constants

We discuss projection and extension constants of Banach spaces. The Lewis estimate for the norm of a projection onto a finite dimensional subspace of $L_p(\mu)$ is given as well as the Kadec-Snobar estimate for the projection constant of an n-dimensional Banach space. We compute the projection constant of the space of trigonometric polynomials (Lozinski-Kharshiladze theorem) and apply it to show the divergence of the general interpolating processes in $C(\mathbb{T})$ and to estimates of degrees of polynomial bases for $C(\mathbb{T})$. We also show that spaces of n-homogenous polynomials on $\mathbb{B}_d$ have projection constants bounded uniformly in n. This is applied to some questions in function theory, in particular a construction of a non-constant inner function in $\mathbb{B}_d$ is presented. We conclude this chapter with the dual presentation of extension of operators; we introduce the notion of the projective tensor product and present some basic observations about it.

1. In this paragraph we discuss quantitative notions connected with projections and extensions of linear operators. We say that a Banach space X is injective if for every Banach space Y and every subspace Z of Y and every operator $T\colon Z \to X$ there exists an extension $\widetilde{T}\colon Y \to X$. We define the extension constant $e(X)$ by

$$e(X) = \inf\{c\colon \text{ for every } Y \supset Z \text{ and } T\colon Z \to X$$
$$\text{there exists } \widetilde{T} : Y \to X : \widetilde{T}|Z = T \text{ and } \|\widetilde{T}\| \leq c\|T\|\}.$$

Every injective space X has $e(X) < \infty$, because if not then take $Y_n \supset Z_n$ and $T_n\colon Z_n \to X$ with $\|T_n\| = n^{-2}$ such that every $\widetilde{T}_n\colon Y_n \to X$ such that $\widetilde{T}_n|Z_n = T_n$ has $\|\widetilde{T}\| \geq n$. Put $Y = \left(\sum_{n=1}^{\infty} Y_n\right)_2 \supset Z = \left(\sum_{n=1}^{\infty} Z_n\right)_2$ and define $T\colon Z \to X$ by $T((z_n)) = (T_n(z_n))$. One sees that T is a continuous operator from Z into X without a continuous extension.

2. The Hahn-Banach theorem applied coordinatewise gives $e(\ell_\infty) = 1$. The following generalizes this observation.

Theorem. *For every measure μ we have $e(L_\infty(\mu)) = 1$.*

Proof. Let Ω be a space on which μ is a measure. Let $\Gamma(\Omega)$ denote the set of all partitions $\omega = (\omega_j)$ of Ω into sets of finite, positive measure.

Note that $\Gamma(\Omega)$ is a directed set, if we put $\omega < \tilde{\omega}$ if and only if for every $\omega_j \in \omega$ there exist $\tilde{\omega}_i \in \tilde{\omega}$ such that $\omega_j \subset \tilde{\omega}_i$. Given a pair $Y \supset Z$ and $T\colon Z \to L_\infty(\mu)$ we define $T_\omega\colon Z \to L_\infty(\mu)$ by

$$T_\omega(z) = \sum_j \Big[\frac{1}{\mu(\omega_j)} \int_{\omega_j} T(z)d\mu\Big]\chi_{\omega_j}.$$

Since the mapping $z \mapsto \mu(\omega_j)^{-1} \int_{\omega_j} T(z)d\mu$ defines a linear functional on Z of norm not greater than $\|T\|$ we extend each such functional to $y^*_{\omega_j} \in Y^*$ with $\|y^*_{\omega_j}\| \le \|T\|$. We define $\tilde{T}_\omega\colon Y \to L_\infty(\mu)$ by $\tilde{T}_\omega(y) = \sum_j y^*_{\omega_j}(y)\chi_{\omega_j}$. Clearly, for every $\omega \in \Gamma(\Omega)$, $\tilde{T}_\omega$ is an extension of T_ω with $\|\tilde{T}_\omega\| \le \|T\|$. Let B_t be the closed ball in $L_\infty(\mu)$ with centre 0 and radius t, equipped with the $\sigma(L_\infty(\mu), L_1(\mu))$-topology. By II.A.9 it is a compact set. Put $B = \prod_{y \in Y} B_{\|T\|\,\|y\|}$. Then B is a compact set with the usual product topology. We consider each $\tilde{T}_\omega$ as an element of B; $\tilde{T}_\omega$ at the coordinate y has the value $\tilde{T}_\omega(y)$. Let $\tilde{T}$ be the limit point of the set $\{\tilde{T}_\omega\}_{\omega \in \Gamma(\Omega)} \subset B$. One easily checks that $\tilde{T}$ can be interpreted as a linear operator from Y into $L_\infty(\mu)$ with $\|\tilde{T}\| \le \|T\|$. Also, since for every $z \in Z$, $T_\omega(z) \to T(z)$ along the directed set $\Gamma(\Omega)$ in $\sigma(L_\infty(\mu), L_1(\mu))$-topology we get $\tilde{T}|Z = T$. ∎

3.   Now we will define some other constants. For a pair (X, Y) consisting of a Banach space Y and a subspace X of Y we define

$$\tilde{e}(X, Y) = \inf\{c : \text{for every operator } T\colon X \to Z \text{ there}$$
$$\text{exists an extension } \tilde{T}\colon Y \to Z : \|\tilde{T}\| \le c\|T\|\},$$
$$\lambda(X, Y) = \inf\{\|P\| : P \text{ is a projection from } Y \text{ onto } X\}.$$

For a fixed space X we define

$$\tilde{e}(X) = \sup\{\tilde{e}(X_1, Y) : X_1 \text{ is a subspace of } Y \text{ isometric to } X\},$$
$$\lambda(X) = \sup\{\lambda(X_1, Y) : X_1 \text{ is a subspace of } Y \text{ isometric to } X\},$$
$$\gamma_\infty(X) = \inf\{\|\alpha\| \cdot \|\beta\| : X \xrightarrow{\alpha} L_\infty(\mu) \xrightarrow{\beta} X \text{ with}$$
$$\mu \text{ an arbitrary measure and } \beta\alpha = id_X\}.$$

The constant $\lambda(X, Y)$ is called the relative projection constant while $\lambda(X)$ is called the projection constant.

4 Lemma.   *For every Y and its subspace X we have $\tilde{e}(X, Y) = \lambda(X, Y)$.*

Proof: Since a projection from Y onto X is an extension of $id\colon X \longrightarrow X$ to an operator from Y into X we see that $\lambda(X,Y) \leq \tilde{e}(X,Y)$. On the other hand every projection $P\colon Y \xrightarrow{\mathrm{onto}} X$ gives rise to the extension $\tilde{T} = TP$ with $\|T\| \leq \|P\| \cdot \|T\|$. This shows $\tilde{e}(X,Y) \leq \lambda(X,Y)$. ∎

5 Theorem. *Let X be a Banach space and let X_1 be any subspace of $L_\infty(\mu)$ isometric to X. Then*

$$e(X) = \tilde{e}(X) = \lambda(X) = \gamma_\infty(X) = \tilde{e}(X_1, L_\infty(\mu)) = \lambda(X_1, L_\infty(\mu)). \quad (1)$$

Moreover if X is finite dimensional and X_1 is a subspace of some $C(K)$-space isometric to X then $\lambda(X) = \lambda(X_1, C(K))$.

Proof: We infer from Lemma 4 that $\tilde{e}(X) = \lambda(X)$ and $\tilde{e}(X_1, L_\infty(\mu)) = \lambda(X_1, L_\infty(\mu))$. Obviously $\gamma_\infty(X) \leq \lambda(X_1, L_\infty(\mu)) \leq \lambda(X)$. If we have $X \xrightarrow{\alpha} L_\infty \xrightarrow{\beta} X$ with $\beta\alpha = id_X$ and $\|\alpha\| \cdot \|\beta\| \leq \gamma_\infty(X) + \varepsilon$ and are given $Y \supset Z$ and $T\colon Z \longrightarrow X$ then Theorem 2 gives an operator $S\colon Y \longrightarrow L_\infty$ with $S|Z = \alpha T$ and $\|S\| = \|\alpha T\|$. Since for $z \in Z$ we have $\beta S(x) = \beta\alpha T(z) = T(z)$ the operator βS is the extension of T with $\|\beta S\| \leq \|\alpha\| \cdot \|\beta\| \, \|T\|$, so $e(X) \leq \gamma_\infty(X)$. If X is a subspace of Y we can extend $id\colon X \longrightarrow X$ to a projection $P\colon Y \longrightarrow X$ with $\|P\| \leq e(X)$, so $\lambda(X) \leq \cdot e(X)$. This gives (1). In order to see the 'moreover' part note that from the definition $\lambda(X_1, C(K)) \leq \lambda(X)$. Conversely if P is a projection from $C(K)$ onto X_1 then $P^{**}\colon C(K)^{**} \xrightarrow{\mathrm{onto}} X_1$ is a projection of the same norm, so $\lambda(X_1, C(K)) \geq \lambda(X_1, C(K))^{**})$. Since $C(K)^{**} \cong L_\infty(\mu)$ for some measure μ, (1) gives the claim. ∎

6 Corollary. *For any two Banach spaces X and Y we have $\lambda(X) \leq \lambda(Y) \cdot d(X,Y)$.*

Proof: From the definitions $\gamma_\infty(X) \leq \gamma_\infty(Y) \cdot d(X,Y)$ so the claim follows from (1). ∎

7. Before we proceed to more concrete questions, we want to address the general question how big $\lambda(X, L_p)$, $1 \leq p \leq \infty$, can be for n-dimensional spaces X. The answer is given in Theorem 10 below.

We start with

Proposition. *Let X be an n-dimensional subspace of the space $L_p(\mu)$, $1 < p < \infty$. Then there exist a basis $(h_j)_{j=1}^n \subset X$ and a system of functions $(g_j)_{j=1}^n \subset L_{p'}(\mu)$ such that*

(a) $\left\| \left(\sum_{j=1}^{n} |h_j|^2 \right)^{\frac{1}{2}} \right\|_p \leq n^{\frac{1}{p}},$

(b) $\left\| \left(\sum_{j=1}^{n} |g_j|^2 \right)^{\frac{1}{2}} \right\|_{p'} \leq n^{\frac{1}{p}},$

(c) $\int g_j h_k d\mu = \delta_{jk}, \quad j, k = 1, 2, \ldots, n,$

(d) If $H = \left(\sum_{j=1}^{n} |h_j|^2 \right)^{\frac{1}{2}}$ then $g_j = h_j \cdot H^{p-2}$ (we take $\frac{0}{0} = 0$).

Proof: The basis $(h_j)_{j=1}^n$ is found as a solution of the following extremal problem:

> given $\phi_1, \ldots, \phi_n$, linearly independent functionals
> on X, we find $h_1, \ldots, h_n \in X$ giving   (2)
> $$\max\{\det[\phi_j(h_k)]_{j,k=1}^n \colon \ \|(\sum_{j=1}^{n} |h_j|^2)^{\frac{1}{2}}\|_p \leq n^{\frac{1}{p}}\}.$$

On the n-fold direct sum of $L_p(\mu)$ we define the norm $\||(f_1, \ldots, f_n)\||_p = \left\| \left(\sum_{j=1}^{n} |f_j|^2 \right)^{\frac{1}{2}} \right\|_p$. This direct sum with this particular norm will be denoted ΣL_p. One easily checks that $(\Sigma L_p)^* = \Sigma L_p^* = \Sigma L_{p'}$. We will also consider X_n, the n-fold direct sum of X with the norm inherited from ΣL_p.

Now let us return to the extremal problem (2). Since X is finite dimensional, the solution $(h_1, \ldots, h_n)$ clearly exists. Note also that if $(h_1, \ldots, h_n)$ gives the maximum for one set $\phi_1, \ldots, \phi_n$ it gives the maximum for any other set of n linearly independent functionals on X. Let us fix $\psi_1, \ldots, \psi_n \in X^*$ such that $\psi_j(h_k) = \delta_{kj}$. We consider $\psi = (\psi_1, \ldots, \psi_n)$ as a functional on X_n and claim that

$$\||\psi\|| = n^{\frac{1}{p'}}.$$

Since $\psi(h_1, \ldots, h_n) = n$ we get from (2) that $\||\psi\|| \geq n^{\frac{1}{p'}}$. On the other hand given any $(x_1, \ldots, x_n) \in X_n$ with $\||(x_1, \ldots, x_n)\||_p = 1$ and any $t \in \mathbb{R}$ we have

$$\det(I + t[\psi_j(x_k)]_{j,k=1}^n) = \det[\psi_j(h_k + tx_k)]_{j,k=1}^n$$
$$\leq \||(h_k + tx_k)_{k=1}^n\||_p^n \ n^{-\frac{n}{p}} \det[\psi_j(h_k)]_{j,k=1}^n$$
$$\leq (n^{\frac{1}{p}} + t)^n \ n^{-\frac{n}{p}} = (1 + n^{-\frac{1}{p}}t)^n.$$

Since both sides are polynomials in t of degree n and both sides are equal for $t = 0$ we have

$$\frac{d}{dt} \det(I + t[\psi_j(x_k)]_{j,k=1}^n)\big|_{t=0} \leq \frac{d}{dt}(1 + n^{-\frac{1}{p}}t)^n\big|_{t=0},$$

that is

$$\sum_{j=1}^{n} \psi_j(x_j) \leq nn^{-\frac{1}{p}} = n^{\frac{1}{p'}}$$

so $|||\psi||| \leq n^{\frac{1}{p'}}$. Now we use the Hahn-Banach theorem and extend ψ to an element $(g_1, \ldots, g_n) \in \Sigma L_{p'}$ with $|||(g_1, \ldots, g_n)|||_{p'} \leq n^{\frac{1}{p'}}$.

Since

$$n = \psi(h_1, \ldots, h_n) = \sum_{j=1}^{n} \int h_j g_j d\mu \leq \int \left(\sum_{j=1}^{n} |h_j|^2\right)^{\frac{1}{2}} \left(\sum_{j=1}^{n} |g_j|^2\right)^{\frac{1}{2}} d\mu$$

$$\leq |||(h_j)_{j=1}^{n}|||_p \cdot |||(g_j)_{j=1}^{n}|||_{p'} = n$$

both inequalities are in fact equalities so the conditions for equality in the Hölder inequality (I.B.3) give (d). ∎

8 Lemma. *With the notation of Proposition 7 we define a Hilbertian norm $N(f)$ as $N(f) = (\int |f|^2 w d\mu)^{\frac{1}{2}}$ where $w = H^{p-2}$. The following inequalities hold:*

(a) *if $1 < p < 2$, then $\|f\|_p \leq n^{\frac{1}{p} - \frac{1}{2}} N(f)$ and for $x \in X$ we have $N(x) \leq \|x\|_p$;*

(b) *if $2 \leq p < \infty$, then $N(f) \leq n^{\frac{1}{2} - \frac{1}{p}} \|f\|_p$ and for $x \in X$ we have $\|x\|_p \leq N(x)$.*

Proof: It follows directly from Proposition 7(d) that $(h_j)_{j=1}^{n}$ is an orthonormal basis in the norm $N(\cdot)$. For $1 < p < 2$ and $x = \sum \alpha_j h_j \in X$ we have

$$N(x)^2 = \int |x|^p |x|^{2-p} w d\mu \leq \|x\|_p^p \cdot \left\| \left(\sum_{j=1}^{n} \alpha_j h_j\right)^{2-p} w \right\|_{\infty}$$

$$\leq \|x\|_p^p \left(\sum_{j=1}^{n} |\alpha_j|^2\right)^{\frac{2-p}{2}} \|H^{2-p} \cdot w\|_{\infty} = \|x\|_p^p N(x)^{2-p}$$

so $N(x) \leq \|x\|_p$.

For $1 < p < 2$ and arbitrary f we have

$$\|f\|_p^p = \int |f|^p w^{\frac{p}{2}} w^{-\frac{p}{2}} d\mu \leq \left(\int |f|^2 w d\mu\right)^{\frac{p}{2}} \cdot \left(\int w^{(\frac{-p}{2})(\frac{2}{(2-p)})} d\mu\right)^{\frac{(2-p)}{2}}$$

$$= N(f)^p \left(\int H^p d\mu\right)^{\frac{(2-p)}{2}} = N(f)^p |||(h_j)_{j=1}^{n}|||_p^{\frac{(2-p)}{2}}$$

$$= N(f)^p n^{\frac{(2-p)}{2}}$$

so $\|f\|_p \le n^{\frac{1}{p}-\frac{1}{2}} N(f)$.

This proves part (a). For $p \ge 2$ and arbitrary f we have

$$N(f)^2 \le \left(\int |f|^p d\mu\right)^{\frac{2}{p}} \left(\int w^{\frac{p}{(p-2)}} d\mu\right)^{\frac{(p-2)}{p}}$$

$$= \|f\|_p^2 \left(\int H^p d\mu\right)^{\frac{(p-2)}{p}}$$

$$= \|f\|_p^2 \, n^{\frac{(p-2)}{p}}$$

so $N(f) \le n^{\frac{1}{2}-\frac{1}{p}} \|f\|_p$. For $p \ge 2$ and $x = \sum_{j=1}^n \alpha_j h_j$ we have

$$\|x\|_p^p = \int |x|^2 \left|\sum_{j=1}^n \alpha_j h_j\right|^{p-2} d\mu \le \int |x|^2 \left(\sum |\alpha_j|^2\right)^{\frac{(p-2)}{2}} H^{p-2} d\mu$$

$$= N(x)^{p-2} \int |x|^2 w d\mu = N(x)^p$$

so $\|x\|_p \le N(x)$. This gives (b). $\blacksquare$

9 Corollary. *Let X be an n-dimensional subspace of $L_p(\mu), 1 \le p \le \infty$. Then $d(X, \ell_2^n) \le n^{|\frac{1}{2}-\frac{1}{p}|}$. In particular for any n-dimensional space X we have $d(X, \ell_2^n) \le \sqrt{n}$.*

Proof: For $1 < p < \infty$ the corollary follows directly from Lemma 8(a) and (b). Let X be any n-dimensional Banach space. By the Banach-Mazur theorem II.B.4 X is isometric to a subspace $\widetilde{X}$ of $C[0,1]$. Let X_p denote $\widetilde{X}$ with the norm from $L_p[0,1]$. One easily sees that $d(X_p, X) \longrightarrow 1$ as $p \to \infty$, so

$$d(X, \ell_2^n) = d(\widetilde{X}, \ell_2^n) \le n^{\frac{1}{2}-\frac{1}{p}} d(\widetilde{X}, X_p) \longrightarrow n^{\frac{1}{2}} \text{ as } p \to \infty. \quad \blacksquare$$

The estimates given in this corollary are in general best possible; see II.B.8.

10 Theorem. *Let X be an n-dimensional subspace of $L_p(\mu)$, $1 < p < \infty$. Then there exists a projection P onto X with $\|P\| \le n^{|\frac{1}{2}-\frac{1}{p}|}$. If Y is any Banach space and X is an n-dimensional subspace of Y, then there exists a projection $P: Y \xrightarrow{\text{onto}} X$ with $\|P\| \le \sqrt{n}$. In particular for any Banach space X we have $\lambda(X) \le \sqrt{\dim X}$.*

Proof: For $1 < p < \infty$ we take as P the orthogonal projection with respect to the Hilbertian norm $N(\cdot)$ considered in Lemma 8. For $2 \le p < \infty$ Lemma 8 gives

$$\|Pf\|_p \le N(Pf) \le N(f) \le n^{\frac{1}{2}-\frac{1}{p}}\|f\|_p.$$

Given any Banach space Y we can assume that Y is a subspace of some $L_\infty(\mu)$. Let X_s be the space X with the norm $L_s(\mu), s \ge 2$. Let P_s be a projection from $L_s(\mu)$ onto X_s with $\|P_s\| \le n^{\frac{1}{2}-\frac{1}{s}}$. Let us fix any basis $(x_j)_{j=1}^n$ in X and write $P_s(f) = \sum_{j=1}^n \varphi_j^s(f)x_j$. If $\varphi_j, j = 1, 2, \ldots, n$, is any cluster point of φ_j^s as $s \to \infty$ in L_∞^*, then $P(f) = \sum_{j=1}^n \varphi_j(f)x_j$ is a projection from $L_\infty(\mu)$, and so from Y, onto X with $\|P\| \le \sqrt{n}$.

To show that the orthogonal projection P works also for $1 < p < 2$ requires some additional duality arguments. The adjoint projection $P^*\colon L_{p'} \to L_{p'}$ is given by $P^*(f) = \sum_{j=1}^n \int fh_j d\mu\; g_j$. This is a projection orthogonal in the scalar product $\langle \varphi, \psi \rangle = \int \varphi\bar\psi w^{-1}d\mu$, and $(g_j)_{j=1}^n$ is an orthonormal basis with respect to this product. Analogously as in Lemma 8(b) we get

$$\left(\int \left|\sum_{j=1}^n \alpha_j g_j\right|^{p'} d\mu\right)^{\frac{1}{p'}} \le \left(\int \left|\sum_{j=1}^n \alpha_j g_j\right|^2 w^{-1}d\mu\right)^{\frac{1}{2}}.$$

Thus for any $f \in L_{p'}$ we have

$$\|P^*f\|_{p'} \le \left(\int |f|^2 w^{-1}d\mu\right)^{\frac{1}{2}} \le \|f\|_{p'}\, n^{\frac{1}{2}-\frac{1}{p}}.$$

So $\|P\| = \|P^*\| \le n^{\frac{1}{2}-\frac{1}{p}}$. ∎

11 Corollary. Let $X \subset Y$ be a subspace of codimension n. Then $\lambda(X,Y) \le \sqrt{n} + 1$.

Proof: Let P be a projection from Y^* onto $X^\perp$ with $\|P\| \le \sqrt{n}$. We write $P(y^*) = \sum_{j=1}^n y_j^{**}(y^*)x_j^\perp$ with $x_j^\perp \in X^\perp$ and $y_j^{**}(x_k^\perp) = \delta_{j,k}$. Using the principle of local reflexivity, II.E.14, for span $(y_j^{**})_{j=1}^n$ we find $(y_j)_{j=1}^n \subset Y$ such that $x_j^\perp(y_k) = \delta_{j,k}$ and such that $\widetilde{P}$ defined as $\widetilde{P}(y) = \sum_{j=1}^n x_j^\perp(y)y_j$ is a projection with norm $\le \sqrt{n} + \varepsilon$. Obviously $\ker \widetilde{P} = X$ so $I - \widetilde{P}$ is a projection onto X with $\|I - \widetilde{P}\| \le \sqrt{n} + 1 + \varepsilon$. Since ε was arbitrary we have the claim. ∎

12 Corollary. *Suppose Z and V are finite dimensional subspaces of X with $Z \subset V$, $\dim Z = m$, $\dim V = n$. Then*

$$\frac{\lambda(Z,X)}{\lambda(V,X)} \leq \sqrt{n-m}+1, \tag{3}$$

$$\frac{\lambda(V,X)+1}{\lambda(Z,X)+1} \leq \sqrt{n-m}+1. \tag{4}$$

Proof: Clearly $\lambda(Z,X) \leq \lambda(Z,V) \cdot \lambda(V,X)$ and Corollary 11 gives $\lambda(Z,V) \leq \sqrt{n-m}+1$. This yields (3). Given a projection R from X onto Z and Q a projection from $\ker R$ onto $\ker R \cap V$ we put $P = R + Q(I - R)$. One checks that P is a projection from X onto V and $\|P\| \leq \|R\| + \|Q\|(\|R\| + 1)$. Thus

$$\lambda(V,X) + 1 \leq (\lambda(Z,X) + 1)(\lambda(\ker R \cap V,\ \ker R) + 1).$$

Since $\dim(\ker R \cap V) = n - m$ Theorem 10 gives (4).   ∎

13.   The following theorem is used quite often for computation of projection constants of concrete spaces.

Theorem. *Let X be a Banach space and Y a complemented subspace. Let G be a compact group and let $g \mapsto T_g$ be a representation of G into $L(X)$ such that*

(a) *$T_g(x)$ is a continuous function of g, for every $x \in X$,*

(b) *$T_g(Y) \subset Y$ for all $g \in G$.*
Then for every $\varepsilon > 0$ there exists a projection $P\colon X \longrightarrow Y$ such that $T_g P\, T_{g^{-1}} = P$ for all $g \in G$ and $\|P\| \leq (\lambda(Y,X) + \varepsilon)\ \sup_{g \in G} \|T_g\|^2$.

Proof: Observe first that the Banach-Steinhaus theorem I.A.7 implies $\sup_{g \in G} \|T_g\| < \infty$. Let us fix a projection Q from X onto Y such that $\|Q\| \leq \lambda(Y,X) + \varepsilon$ and for $x \in X$ define

$$P(x) = \int_G T_g Q T_{g^{-1}}(x)dg \tag{5}$$

where dg denotes the normalized Haar measure on G. Condition (a) ensures that there is no problem with the definition of the integral. For every $x \in X$ we see that $T_g Q T_{g^{-1}}(x) \in Y$. Also for $y \in Y$ we see that $T_g Q T_{g^{-1}}(y) = y$ so P is a projection onto Y. For $h \in G$

$$T_h P T_{h^{-1}} = \int_G T_h T_g Q T_{g^{-1}} T_{h^{-1}} dg = \int_G T_{hg} Q T_{(hg)^{-1}} dg = P.$$

Also

$$\|Px\| \le \int_G \|T_g\| \, \|Q\| \, \|T_{g^{-1}}\| \, \|x\| dg$$
$$\le \|Q\| \, \|x\| \sup \|T_g\|^2. \qquad \blacksquare$$

In many concrete application T_g are isometries and the invariant projection is unique. Then the theorem implies that the norm of the invariant projection equals $\lambda(Y, X)$.

14.   Let us recall that

$$\mathbf{S}_d = \left\{ (z_1, \dots, z_d) \in \mathbf{C}^d : \sum_{j=1}^d |z_j|^2 = 1 \right\}$$

and that σ is the normalized rotation invariant measure on $\mathbf{S}_d$. By $W_n(\mathbf{S}_d)$ we will denote the space of all homogenous polynomials of degree n on $\mathbf{C}^d$. Equipped with the norm $\|p\|_\infty = \sup_{\zeta \in \mathbf{S}_d} |p(\zeta)|$ it will be denoted by $W_n^\infty(\mathbf{S}_d)$ and equipped with the norm

$$\|p\|_2 = \left(\int_{\mathbf{S}_d} |p(\zeta)|^2 d\sigma(\zeta) \right)^{1/2}$$

it will be denoted by $W_n^2(\mathbf{S}_d)$. Clearly every polynomial $p(z_1, \dots, z_d)$ can be uniquely written as $p = \sum_{k=0}^s p_k$ with $p_k \in W_k(\mathbf{S}_d)$. The compact group of all unitary operators on $\mathbf{C}^d$ is denoted by $U(d)$. It acts on $C(\mathbf{S}_d)$ by the formula $U(d) : g \mapsto T_g$ where $T_g(f) = f \circ g$. Obviously T_g is an isometry on $C(\mathbf{S}_d)$.

15 Theorem. $\lambda(W_n^\infty(\mathbf{S}_d)) = \frac{\Gamma(d+n)\Gamma(1+\frac{n}{2})}{\Gamma(1+n)\Gamma(d+\frac{n}{2})}.$

Proof: Let us consider the operator $P_{d,n}$ from $C(\mathbf{S}_d)$ into $W_n^\infty(\mathbf{S}_d)$ defined by

$$P_{d,n}(f)(\zeta) = \frac{(d-1+n)!}{(d-1)!n!} \int_{\mathbf{S}_d} f(z)\langle \zeta, z \rangle^n d\sigma(z).$$

The binomial expansion of the Cauchy kernel

$$\frac{1}{(1 - \langle \zeta, z \rangle)^d} = \sum_{n=0}^\infty \frac{(d-1+n)!}{(d-1)!n!} \langle \zeta, z \rangle^n,$$

together with the Cauchy formula I.B.21 and the fact that homogenous polynomials of different degrees are orthogonal, show that $P_{d,n}$ is actually a projection onto $W_n^\infty(\mathbf{S}_d)$. The operator $P_{d,n}$ commutes with the action T_g of the unitary group $U(d)$ and Theorem 12.3.8 of Rudin [1980], i.e. the fact that $U(d)$ acts irreducibly on $W_n(\mathbf{S}_d)$, shows that it is the unique such projection. Theorems 5 and 13 give

$$\lambda(W_n^\infty(\mathbf{S}_d)) = \|P_{d,n}\| = \frac{(d-1+n)!}{(d-1)!n!}\int_{\mathbf{S}_d}|\langle z,\zeta\rangle|^n d\sigma(\zeta). \qquad (6)$$

In order to estimate this integral we use the formula (see Rudin [1980] 1.4.5)

$$\int_{\mathbf{S}_d} f(\langle z,\zeta\rangle)d\sigma(\zeta) = \frac{d-1}{\pi}\int_{\mathbf{D}} f(z)(1-|z|^2)^{d-2}d\nu(z).$$

which reduces it to

$$\frac{d-1}{\pi}\int_{\mathbf{D}}(1-|z|^2)^{d-2}|z|^n d\nu(z).$$

Passing to polar coordinates and substituting $|z|^2 = r$ we get $(d-1)\int_0^1(1-r)^{d-2}r^{\frac{n}{2}}dr$. Known values of the Euler integrals of the first kind yield

$$\int_{\mathbf{S}_d}|\langle z,\zeta\rangle|^n d\sigma(\zeta) = \frac{\Gamma(d)\Gamma(1+\frac{n}{2})}{\Gamma(d+\frac{n}{2})}. \qquad (7)$$

Substituting (7) into (6) we get the desired estimate.   ■

Remark: The integrals computed in (7) are clearly well known. An alternative argument can be found e.g. in Garling [1970].

We want to note two special cases of Theorem 15. For $n = 1$ one easily checks that $W_1^\infty(\mathbf{S}_d) \cong \ell_2^d$. So using the properties of the Γ function we get

16 Corollary. $\sqrt{n} \geq \lambda(\ell_2^n) = \Gamma(1.5)\frac{\Gamma(n+1)}{\Gamma(n+\frac{1}{2})} \geq \frac{1}{2}\sqrt{\pi}\sqrt{n}$.

An easy computation also yields

17 Corollary. $\lambda(W_n^\infty(\mathbf{S}_d)) \leq 2^{d-1}$.

18. Now we will present some applications to function theory on the unit ball of $\mathbb{C}^d$. Our main goal is to present the construction of a non-constant inner function in $\mathbb{B}_d$. A function $f(z) \in H_\infty(\mathbb{B}_d)$ is called inner if $|f(z)| = 1$ σ-a.e. on $\mathbf{S}_d$ (by $f(z)$ for $z \in \mathbf{S}_d$ we mean the radial boundary value; see I.B.21). Clearly inner functions are characterized by $1 = \|f\|_2 = \|f\|_\infty$ so the following proposition can be considered as a 'first approximation'.

Proposition. *For every $d \geq 1$ there exist polynomials $p_n \in W_n(\mathbf{S}_d)$, $n = 1, 2, \ldots$, such that*

$$\|p_n\|_\infty \leq 1 \text{ and } \|p_n\|_2 \geq 2^{-d}\sqrt{\pi}. \tag{8}$$

Proof: Let I denote the identity map from $W_n^\infty(\mathbf{S}_d)$ into $W_n^2(\mathbf{S}_d)$. We have to show $\|I\| \geq 2^{-d}\sqrt{\pi}$. From Corollary 6 we get

$$\begin{aligned}
\lambda(W_n^2(\mathbf{S}_d)) &\leq \lambda(W_n^\infty(\mathbf{S}_d)) \cdot d(W_n^\infty(\mathbf{S}_d), W_n^2(\mathbf{S}_d)) \\
&\leq \lambda(W_n^\infty(\mathbf{S}_d)) \cdot \|I\| \cdot \|I^{-1}\|.
\end{aligned} \tag{9}$$

The space $W_n^2(\mathbf{S}_d)$ is obviously a Hilbert space. Since $P_{d,n}$ is an orthogonal projection from $L_2(\mathbf{S}_d, \sigma)$ onto $W_n^2(\mathbf{S}_d)$ the dimension of $W_n^2(\mathbf{S}_d)$ equals the square of the Hilbert-Schmidt norm of $P_{d,n}$ (III.G.12 and 13) so

$$\dim W_n^2(\mathbf{S}_d) = \left[\frac{(d-1+n)!}{(d-1)!n!}\right]^2 \int_{\mathbf{S}_d} |\langle z, \zeta\rangle|^{2n} d\sigma(\zeta).$$

From (7) we infer that $\dim W_n^2(\mathbf{S}_d) = \frac{(d-1+n)!}{(d-1)!n!}$. The same conclusion can also be reached counting monomials. Thus Corollaries 16 and 17 applied to (9) give

$$\|I\| \geq \sqrt{\pi} \cdot 2^{-d}\sqrt{\frac{(d+n-1)!}{(d-1)!n!}}\|I^{-1}\|^{-1}. \tag{10}$$

In order to compute $\|I^{-1}\|$ take $p \in W_n(\mathbf{S}_d)$ such that $\|p\|_2 = 1$ and $\|p\|_\infty = \|I^{-1}\|$. We can assume $p(1, 0, \ldots, 0) = \|p\|_\infty$. Let us average p over the subgroup of $U(d)$ fixing $(1, 0, \ldots, 0)$. For the averaged polynomial p_0 we have $p_0(1, 0, \ldots, 0) = \|I^{-1}\|$ and $\|p_0\|_2 \leq \|p\|_2$, so $\|p_0\| = \|p\|_2 = 1$. But one easily sees that $p_0(z_1, \ldots, z_d) = cz_1^n$. The normalization gives $c = \left(\int_{\mathbf{S}_d} |z_1|^2 d\sigma(z)\right)^{-\frac{1}{2}}$. Since $c = \|I^{-1}\|$, using (7) and substituting into (10) we get the claim. ∎

19. Our next result is a technical device to construct the inner function.

Proposition. For every $\varphi \in C(\mathbf{S}_d), \varphi > 0$ there exists a natural number $N = N(\varphi)$ such that for every $n > N$ there exists a polynomial $R = R_{\varphi,n}(z_1, \ldots, z_d)$ such that

(a) $|R| < \varphi$ on $\mathbf{S}_d$,

(b) $\int_{\mathbf{S}_d} |R|^2 d\sigma > 4^{-d} \int_{\mathbf{S}_d} \varphi^2 d\sigma$,

(c) $R = \sum_{k=n}^{n+N} R_k$ where $R_k \in W_k(\mathbf{S}_d)$.

Proof: It is enough to show the Proposition for φ polynomial in $z_1, \ldots, z_d$ and $\bar{z}_1, \ldots, \bar{z}_d$ since such polynomials are dense in $C(\mathbf{S}_d)$. For arbitrary $f \in C(\mathbf{S}_d)$ we have

$$\int_{U(d)} \int_{\mathbf{S}_d} \varphi^2(\zeta) |f(g(\zeta))|^2 d\sigma(\zeta) dg = \int_{\mathbf{S}_d} \varphi^2(\zeta) \int_{U(d)} |f(g(\zeta))|^2 dg d\sigma(\zeta)$$
$$= \int_{\mathbf{S}_d} \varphi^2(\zeta) d\sigma(\zeta) \cdot \int_{\mathbf{S}_d} |f(\zeta)|^2 d\sigma(\zeta).$$

Applying this for $f = \tilde{p}_n$, where the $\tilde{p}_n$ satisfy (8) and using the fact that $U(d)$ is connected we infer that there are polynomials $p_n = \tilde{p}_n \circ g_n \in W_n(\mathbf{S}_d), n = 1, 2, \ldots$ such that

$$\|p_n\|_\infty = 1, \ \|p_n\|_2 \geq 2^{-d}\sqrt{\pi} \text{ and} \qquad (11)$$
$$\int_{\mathbf{S}_d} |\varphi(\zeta) p_n(\zeta)|^2 d\sigma(\zeta) \geq 4^{-d}\pi \int_{\mathbf{S}_d} \varphi^2 d\sigma.$$

The desired polynomial R will be equal to $\frac{2}{3}\mathcal{C}(\varphi \cdot p_n)$ where $\mathcal{C}$ is the Cauchy projection (I.B.21) and n is big enough. Let $\alpha, \beta, \gamma, \delta$ denote d-tuples of non-negative integers. Then

$$\mathcal{C}(z^\beta \bar{z}^\gamma z^\delta) = \sum_\alpha \int_{\mathbf{S}_d} z^{\beta+\delta} \bar{z}^\gamma \overline{z^\alpha} d\sigma(z) \cdot \frac{z^\alpha}{\|z^\alpha\|_2}.$$

One sees that $\int_{\mathbf{S}_d} z^{\beta+\delta} \bar{z}^{\gamma+\alpha} d\sigma(z) = 0$ if $\beta + \delta \neq \gamma + \alpha$. From this we infer that if

$$\varphi = \sum_{\substack{\beta,\gamma \\ |\beta|,|\gamma| \leq K}} a_{\beta\gamma} z^\beta \bar{z}^\gamma \text{ then } \mathcal{C}(\varphi \cdot p_k) = \sum_{\substack{\alpha \\ k-K \leq |\alpha| \leq k+K}} b_\alpha z^\alpha.$$

This shows that for n big enough $\frac{2}{3}\mathcal{C}(\varphi \cdot p_n)$ satisfies (c) with $N(\varphi) = 2K$. That $\frac{2}{3}\mathcal{C}(\varphi \cdot p_n)$ for large n also satisfies (a) and (b) follows from

$$\lim_{n \to \infty} \|\varphi p_n - \mathcal{C}(\varphi p_n)\|_{C(\mathbf{S}_d)} = 0. \qquad (12)$$

This in turn is an immediate consequence of the fact that p_n tends weakly to zero in $H_p(\mathbb{B}_d), 1 \leq p < \infty$ and the following lemma which will be also used later.

20 Lemma. *Let $\varphi \in C(\mathbf{S}_d)$ satisfy the Lipschitz condition. Then $T_\varphi(f) = \varphi \cdot f - C(\varphi f)$ is a compact operator from $H_q(\mathbb{B}_d)$ into $C(\mathbf{S}_d)$ for $q > 2d$.*

Proof: Using the Cauchy formula we can write

$$T_\varphi(f)(z) = \int_{\mathbf{S}_d} \frac{\varphi(z) - \varphi(\zeta)}{(1 - \langle z, \zeta \rangle)^d} f(\zeta) d\sigma(\zeta) = \int_{\mathbf{S}_d} \Gamma_z(\zeta) f(\zeta) d\sigma(\zeta).$$

Since $|z - \zeta| \leq C\sqrt{|1 - \langle z, \zeta \rangle|}$ for $\zeta \in \mathbf{S}_d$ and $z \in \mathbb{B}_d \cup \mathbf{S}_d$, we have

$$|\Gamma_z(\zeta)| \leq C \frac{|z - \zeta|}{|1 - \langle z, \zeta \rangle|^d} \leq C \frac{1}{|1 - \langle z, \zeta \rangle|^{d-1/2}}$$

so $|\Gamma_z(\zeta)|^q$ is a uniformly integrable family of functions for $q < \frac{2d}{(2d-1)}$ (see III.C.11 for definitions). Since for $(z_n)_{n=1}^\infty \subset \mathbf{S}_d \cup \mathbb{B}_d$ such that $z_n \longrightarrow z_0$ we have $\Gamma_{z_n} \longrightarrow \Gamma_{z_0}$ pointwise σ-a.e., the uniform integrability and the Egorov theorem imply that $z \mapsto \Gamma_z$ is a continuous map from $\mathbf{S}_d \cup \mathbb{B}_d$ into L_q, for $q < \frac{2d}{(2d-1)}$. This yields the compactness of T_φ. ∎

21. Now we are ready to prove

Theorem. *There exists a non-constant inner function in $\mathbb{B}_d$, $d = 1, 2, 3, \ldots$.*

Proof: Inductively using Proposition 19 we construct a sequence $(R^n)_{n=1}^\infty$ of polynomials such that

(a) $R^n(0) = 0, \quad n = 1, 2, \ldots,$

(b) $\int_{\mathbf{S}_d} R^n \overline{R^k} d\sigma = 0$ if $n \neq k$,

(c) $|R^{n+1}| < 1 - |\sum_{j=1}^n R^j|$ on $\mathbf{S}_d$,

(d) $\int_{\mathbf{S}_d} |R^{n+1}|^2 d\sigma > 4^{-d} \int_{\mathbf{S}_d} (1 - |\sum_{j=1}^n R^j|)^2 d\sigma.$
The series $\sum_{n=1}^\infty R^n$ converges in $H_2(\mathbb{B}_d)$ since it is an orthogonal series (see (b)) and its partial sums are uniformly bounded by 1 (see (c)). This implies that $\int_{\mathbf{S}_d} |R^n|^2 d\sigma \to 0$ as $n \to \infty$ so by (d) $|\sum_{j=1}^n R^j| \longrightarrow 1$ σ-a.e., so $\sum_{n=1}^\infty R^n$ is an inner function. Condition (a) gives that it is non-constant. ∎

22.   Now we will discuss the projection constant of the space T_n^∞ of trigonometric polynomials of degree at most n equipped with the sup-norm. When naturally embedded into $C(\mathbb{T})$, T_n^∞ is a rotation invariant subspace and the unique rotation invariant projection is the Dirichlet projection $\mathcal{D}_n$. Its norm equals (I.B.15) $4\pi^{-2}\log(n+1) + o(1)$. Using Theorem 13 we obtain

Theorem. (Lozinski-Kharshiladze). $\lambda(T_n^\infty) = \left(\frac{4}{\pi^2}\right)\log(n+1) + o(1)$. ∎

23.   This theorem implies some classical results about the divergence of trigonometric interpolation. For example we have

Theorem. *For every system of points $x_0^n, x_1^n, \ldots, x_{2n}^n \in \mathbb{T}, n = 1, 2, \ldots$ there exists a function $f \in C(\mathbb{T})$ such that the sequence of interpolating polynomials $I_n(f)$, i.e. polynomials in T_n^∞ such that $I_n(f)(x_j^n) = f(x_j^n), j = 0, \ldots, 2n$, is not uniformly bounded.*

Proof: Observe that I_n is a projection from $C(\mathbb{T})$ onto T_n^∞, so $\|I_n\| \geq \lambda(T_n^\infty)$. Theorem 22 and the uniform boundedness principle (see I.A.7) give the claim. ∎

24.   We also have

Theorem. *Suppose $(f_n)_{n=1}^\infty$ is a Schauder basis for $C(\mathbb{T})$ such that $f_n \in T_{k(n)}^\infty$ with $k(n)$ an increasing sequence of integers. Then $k(n) \geq \frac{n}{2} + c\log^2 n$ for some $c > 0$.*

Proof: Let $F_n = \text{span} (f_k)_{k=1}^n$. Since $(f_n)_{n=1}^\infty$ is a basis for $C(\mathbb{T})$, II.B.6 and Theorem 5 show that $\lambda(F_n) \leq C$ for some constant $C, n = 1, 2, \ldots$. Using (4) we obtain

$$\frac{\lambda(T_{k(n)}^\infty, C(\mathbb{T})) + 1}{\lambda(F_n, C(\mathbb{T})) + 1} \leq \sqrt{2k(n) + 1 - n} + 1.$$

Thus Theorem 22 yields $\log n \leq \log 2k(n) + 1 \leq C\sqrt{2k(n) + 1 - n}$ so

$$k(n) \geq \frac{n}{2} + c\log^2 n. \qquad \blacksquare$$

25.   We would like to conclude this chapter with a more abstract treatment of the extension of operators. This is connected with the concept of tensor product. Actually we will consider only one tensor product. We assume that the reader is familiar with the algebraic notion of tensor product.

Definition. *If X and Y are Banach spaces then their projective tensor product $X \widehat{\otimes} Y$ is a completion of the algebraic tensor product $X \otimes Y$ with respect to the norm*

$$\left\| \sum_{j=1}^{N} x_j \otimes y_j \right\|_{\wedge} = \inf\left\{ \sum_j \|x_j'\| \, \|y_j'\| : \sum_{j=1}^{N} x_j \otimes y_j = \sum_j x_j' \otimes y_j' \right\}. \quad (13)$$

26 Lemma. *The dual space of $X \widehat{\otimes} Y$ can be isometrically identified with $L(X, Y^*)$.*

Proof: The duality which we use is basically the trace duality (compare with III.F.16). Explicitly it is given as

$$\langle T, \sum_j x_j \otimes y_j \rangle = \sum_j T(x_j)(y_j)$$

for $T \colon X \to Y^*$ and $\sum_j x_j \otimes y_j \in X \otimes Y$. One checks that this definition is independent of the representation of the tensor. Since

$$\left| \langle T, \sum_j x_j \otimes y_j \rangle \right| \leq \sum_j \left| T(x_j)(y_j) \right| \leq \|T\| \cdot \sum_j \|x_j\| \cdot \|y_j\|$$

we infer from (13) that every $T \in L(X, Y^*)$ induces a linear functional on $X \widehat{\otimes} Y$ of norm $\leq \|T\|$. Considering the elementary tensors we get

$$\|T\| = \sup\{|(Tx)(y)| : \|y\| = \|x\| = 1\}$$
$$= \sup\{|\langle T, x \otimes y \rangle| : \|x\| \cdot \|y\| = 1\}$$
$$\leq \sup\left\{ \left| \langle T, \sum_j x_j \otimes y_j \rangle \right| : \left\| \sum_j x_j \otimes y_j \right\|_{\wedge} \leq 1 \right\}.$$

This shows that each $T \in L(X, Y^*)$ gives a functional in $(X \widehat{\otimes} Y)^*$ of norm $\|T\|$ so $L(X, Y^*)$ is isometrically a subspace of $(X \widehat{\otimes} Y)^*$. Now suppose that we are given $\varphi \in (X \widehat{\otimes} Y)^*$. Note that for a fixed $x \in X$, $\varphi_x(y) = \varphi(x \otimes y)$ is a linear functional on Y with norm $\leq \|\varphi\|$. One easily checks that the map $x \mapsto \varphi_x$ is a linear operator from X into Y^*. This shows that $L(X, Y^*) = (X \widehat{\otimes} Y)^*$. ∎

27 Corollary. *Suppose $Y \subset X$ and Z^* are given. The following conditions are equivalent:*

(a) *every operator $T: Y \to Z^*$ extends to an operator $\widetilde{T}: X \to Z^*$;*

(b) *the map $r: L(X, Z^*) \to L(Y, Z^*)$ given by $r(T) = T|Y$ is onto;*

(c) *the identity map $i: Y \widehat{\otimes} Z \to X \widehat{\otimes} Z$ is an isomorphic embedding.*

Proof: This is a routine duality (see I.A.13 and 14) and the observation that $i^* = r$. ∎

28. Now we will formulate the proposition which is the dual version of Theorem 2. Before we do so we will define the space $L_1(\Omega, \mu, Y)$ of Bochner integrable Y-valued functions (Y is a Banach space). First we consider the step functions of the form $f(t) = \sum_{j=1}^{n} y_j \chi_{A_j}(t)$ where $y_j \in Y, j = 1, \ldots, N$ and $(A_j)_{j=1}^{N}$ are disjoint subsets of Ω with $\mu(A_j) < \infty$ for $j = 1, 2, \ldots, N$. For such f we put

$$\|f\| = \sum_{j=1}^{N} \|y_j\| \mu(A_j) = \int_{\Omega} \|f(t)\| d\mu(t).$$

A Y-valued function $f(t)$ is Bochner integrable if there exists a sequence $(f_n)_{n=1}^{\infty}$ of step functions, as above, such that $\int_{\Omega} \|f(t) - f_n(t)\| d\mu(t) \to 0$ as $n \to \infty$.

The space of all Y-valued, Bochner integrable functions (when we identify functions equal μ-a.e.) is denoted by $L_1(\Omega, \mu, Y)$. One easily checks that it is a Banach space.

This description makes the following proposition easy to believe.

Proposition. *If (Ω, μ) is a σ-finite measure space and Y is a Banach space then*

$$Y \widehat{\otimes} L_1(\mu) = L_1(\Omega, \mu, Y).$$

Proof: We define a map $\varphi: Y \otimes L_1(\mu) \to L_1(\Omega, \mu, Y)$ by the formula $\varphi(\sum_{j=1}^{N} y_j \otimes f_j) = \sum_{j=1}^{N} f_j(t) \cdot y_j$. One notes that this map is well defined and

$$\left\| \varphi \left(\sum_{j=1}^{N} y_j \otimes f_j \right) \right\| \leq \sum_{j=1}^{N} \|f_j(t) \cdot y_j\| = \sum_{j=1}^{N} \|f_j\| \cdot \|y_j\|$$

so $\|\varphi\| \leq 1$. On the other hand for every step function $f = \sum_{j=1}^{N} y_j \cdot \chi_{A_j}$ we see that

$$\varphi\left(\sum_{j=1}^{N} y_j \otimes \chi_{A_j}\right) = f \text{ and } \|f\| = \sum_{j=1}^{N} \|y_j\| \, \|\chi_{A_j}\| \geq \left\|\sum_{j=1}^{N} y_j \otimes \chi_{A_j}\right\|_\wedge.$$

This shows that φ is an isometry onto. ∎

Notes and remarks.
The question when a linear operator can be extended from a subspace to the whole space is quite natural and important. It was asked in Banach [1932], remark to Chapter IV, and the first example showing that it is not always possible was given by Banach and Mazur [1933] in the form of an uncomplemented subspace of $C[0,1]$. Thus the connection between extensions of operators and projections was noted very early. *Theorem 2* is due to Kantorovič [1935] with the proof that directly mimics the proof of the Hahn-Banach theorem. Some extensions in terms of Banach lattice theory were given by Akilov [1947] and the following theorem was proved by Kelley [1952] building on earlier work of Nachbin [1950] and Goodner [1950].

Theorem. $e(X) = 1$ *if and only if X is isometric to $C(K)$ for some extremaly disconnected compact space K.*

The notion of the projection constant and its properties summarized in *Theorem 5* evolved from those works in the 50's (except for the constant γ_∞ which is the child of the theory of operator ideals and appeared later).

The beautiful estimate $\lambda(X) \leq \sqrt{\dim X}$ was proved by Kadec-Snobar [1971]. The original proof used the important result of John [1948], a special case of which is our *Corollary 9* for $p = \infty$. *Theorem 10* and *Corollary 9* were proved by Lewis [1978]. Our presentation follows Lorentz–Tomczak-Jaegerman [1984]. *Corollary 11* was observed by Garling-Gordon [1971]. The Kadec-Snobar theorem stimulated further research in the local theory of Banach spaces, which is summarized in Tomczak-Jaegerman [1989]. The Kadec-Snobar theorem is also quite useful in various questions of approximation theory. It may be of some interest that there exist finite dimensional spaces Y_n, $\dim Y_n = n$ such that $\lambda(Y_n) \geq \sqrt{n} - \frac{2}{\sqrt{n}}$. Also an improvement in the Kadec-Snobar estimate is possible. There exists $c > 0$ such that for any finite dimensional Banach space X we have $\lambda(X) \leq \sqrt{n} - \frac{c}{\sqrt{n}}$. All this is beyond the scope of this book (see König [1985], König-Lewis [P] and Tomczak-Jaegerman [1989]). It should be pointed out that methods used in those more sophisticated studies use extensively various Banach ideal norms, in particular those discussed in III.F.

Corollary 12 and Theorem 24 are taken from Kadec [1974]. Theorem 24 for $k(n) = n$ and Theorem 23 are classical results of Faber [1914]. The estimate from below for $k(n)$ in Theorem 24 was the best known for some time. Only recently Privalov [1987] has shown that $k(n) \geq (1+\varepsilon)n$ (where $\varepsilon > 0$ depends on the basis). On the other hand Bočkarev [1985] has constructed a basis $(f_n)_{n=1}^{\infty}$ for $C(\mathbb{T})$ such that $f_n \in T_{4n}$.

Theorem 13 was proved by Rudin [1962] but its main idea, the averaging of projections, was already used in Faber [1914] in the case of the circle group and interpolating projections. Theorem 15, Corollary 17 and Proposition 18 are taken from Ryll-Wojtaszczyk [1983]; Corollary 16 however is quite old. With basically the same proof the upper bound was given in Grünbaum [1960] and this bound was shown to be exact by Rutowitz [1965]. The non-constant inner function in $\mathbb{B}_d, d \geq 2$ was constructed independently in Aleksandrov [1982] and Løw [1982]. This was an unexpected solution to the long standing problem. The proof presented here is due to Aleksandrov [1984]. Further results stemming from the solution of the inner function problem are discussed in detail in Rudin [1986] and Aleksandrov [1987]. Proposition 18 seems to be quite useful in function theory. Besides many applications presented in Rudin [1986] some are presented in Ryll-Wojtaszczyk [1983] and Wojtaszczyk [1982]. Alexander [1982] used Proposition 18 to construct a function $f \in A(\mathbb{B}_d)$ such that $f^{-1}(0)$ has infinite $(2d - 2)$-dimensional volume. Also, the papers Ullrich [1988] and [P] should be consulted for further applications and improvements.

Theorem 22 is due to S.M. Lozinski and F.I. Kharshiladze but the first published proof we are aware of is in Natanson [1949]. It is however one in the long line of averaging arguments starting from Faber [1914] or even earlier.

The general theory of tensor products of Banach spaces was created in the 1940's by R. Schatten in a series of papers. The presentation of this work is contained in Schatten [1950] which even today seems to be the most complete exposition. Later Grothendieck [1955] and [1956] generalized the concept to more general linear topological spaces and applied it to Banach space theory in the most profound way. What we have presented here is mostly folklore and can be found in the works mentioned earlier.

The reader should be informed however that there are many important norms on $X \otimes Y$ besides the projective tensor norm. We have only scratched the surface.

Exercises

1. Show that ℓ_∞ is isomorphic to $L_\infty[0,1]$.

2. Find a one-complemented subspace in $A(\mathbb{T}^2)$ isometric to T_n^∞.

3. Let $W_n = span\{1, t, t^2, \ldots, t^n\} \subset C[0,1]$. Show that $\lambda(W_n) \geq c \log n$ for some positive c.

4. Show the analogue of Theorem 24 for $L_1(\mathbb{T})$ and $A(\mathbb{T})$.

5. (a) Show that if $2 \leq p \leq \infty$ then $cn^{\frac{1}{p}} \leq \lambda(\ell_p^n) \leq n^{\frac{1}{p}}$ for some $c > 0$.

    (b) Show that ℓ_p, $1 < p \leq 2$, is not isomorphic to a complemented subspace of any $L_1(\mu)$.

6. Estimate $\lambda(\ell_1^n)$.

7. Show that, if X is separable and $Y \subset X$ is isomorphic to c_0, then Y is complemented in X.

8. (a) Show that ℓ_∞/c_0 contains a subspace isometric to $c_0(\Gamma)$, where Γ has continuum cardinality.

    (b) Show that $c_0 \subset \ell_\infty$ is not complemented.

    (c) Show that if $X \subset \ell_\infty$ and $X \sim c_0$ then X is not complemented in ℓ_∞.

9. Suppose that $p(z)$ is a polynomial of degree N on $\mathbb{C}^d, d \geq 1$ such that $\|p\|_{H_\infty(\mathbb{B}_d)} < 1$. Show that there exists an inner function φ on $\mathbb{B}_d$ such that if $\varphi = \sum_{n=0}^\infty \varphi_n$ with φ_n a polynomial homogeneous of degree n, then $p = \sum_{n=1}^N \varphi_n$. Show also that for every $r < 1$ and every $\varepsilon > 0$ this inner function φ can be constructed in such a way that $\sup\{|\varphi(z) - p(z)|: |z| < r\} < \varepsilon$.

10. For $d > 1$ let $V \subset \mathbb{B}_d$ be the set $\{(z_1, \ldots, x_d): |z_j| \leq d^{-\frac{1}{2}},\ j = 1, \ldots, d\}$. Show that there exists an inner function φ on $\mathbb{B}_d$ such that $|\varphi(z)| < \frac{1}{2}$ for $z \in V$.

11. If $f(z)$ is holomorphic on $\mathbb{B}_d$, then its radial derivative $Rf(z)$ is defined as $Rf(z) = \sum_{j=1}^d \left(z_j \frac{\partial f(z)}{\partial z_j} \right)$. Show that if the d-th radial derivative of f, $R^d f$, is in $H_1(\mathbb{B}_d)$ then $f \in A(\mathbb{B}_d)$.

12. Show that the operator $T \colon A(\mathbf{B}_d) \longrightarrow \ell_2$ defined as $T(f) = \left(\int_{\mathbf{S}_d} f(\zeta) \overline{p_{2^n}(\zeta)} d\sigma(\zeta) \right)_{n=1}^{\infty}$, where (p_k) are the polynomials constructed in Proposition 18, is onto.

13. A function $f(z)$ holomorphic in $\mathbf{D}$ is called a Bloch function if

$$\sup_{z \in \mathbf{D}} |f'(z)| \cdot (1 - |z|^2) < \infty.$$

   (a) Show that $\sum_{n=1}^{\infty} a_n z^{2^n}$ is a Bloch function if and only if $(a_n)_{n=1}^{\infty} \in \ell_\infty$.

   (b) Find a Bloch function which does not have radial limits a.e. on the circle $\mathbf{T}$.

14. For a function analytic in $\mathbf{B}_d$ and $\zeta \in \mathbf{S}_d$ we define a slice function $f_\zeta(z)$ analytic in $\mathbf{D}$ by $f_\zeta(z) = f(z \cdot \zeta)$. We say that a function f analytic in $\mathbf{B}_d$ is Bloch if

$$\sup_{\zeta \in \mathbf{S}_d} \sup_{z \in \mathbf{D}} |f'_\zeta(z)|(1 - |z|^2) < \infty.$$

   Find a Bloch function in $\mathbf{B}_d$ which does not have radial limits σ-a.e. on $\mathbf{S}_d$.

III.C. $L_1(\mu)$-Spaces

This chapter discusses some topics connected with $L_1(\mu)$-spaces. We start with the general notion of semi-embedding and investigate semi-embeddings of $L_1(\mu)$ into various Banach spaces. This is applied to the class $M_0(\mathbb{T})$ of all measures such that $\hat{\mu}(n) \to 0$ as $n \to \infty$. We prove the classical Menchoff theorem that there are singular such measures and a theorem of Lyons characterising zero sets for the class $M_0(\mathbb{T})$. Next we describe relatively weakly compact sets in ℓ_1 (Schur's theorem) and in general $L_1(\mu)$-spaces for a probability measure μ (the Dunford-Pettis theorem). We also discuss the connection between type and finite representability of ℓ_1. Some characterizations of reflexive subspaces of L_1 are given. We conclude with some results connected with the classical result of Nevanlinna about cosets $F + H_\infty(\mathbb{T}) \subset L_\infty(\mathbb{T})$.

1. We have already seen that properties of L_1 spaces differ from properties of L_p-spaces for $1 < p < \infty$. In particular, L_1 being non-reflexive, its unit ball is not weakly compact. Actually more is true.

Proposition. *If μ is a non-atomic measure and $T\colon L_1(\mu) \to X$ is a 1-1, weakly compact linear operator then $T(B_{L_1(\mu)})$ is not norm-closed.*

Proof: Since $B_{L_1(\mu)}$ has no extreme points and T is 1-1 $T(B_{L_1(\mu)})$ also has no extreme points. The Krein-Milman theorem I.A.22 implies that $T(B_{L_1(\mu)})$ is not weakly compact, thus it cannot be norm-closed.∎

2. Let us introduce the following

Definition. *A 1-1 linear operator $T\colon X \to Y$ is called a semi-embedding if $T(B_X)$ is closed in Y.*

Clearly every isomorphic embedding is a semi-embedding. Also if X is reflexive then every 1-1 map $T\colon X \to Y$ is a semi-embedding. Proposition 1 clearly says that there is no semi-embedding from $L_1[0,1]$ into any reflexive space.

3. The following is a general, topological observation showing that $T^{-1}|T(B_X)$ has at least one point of continuity. We formulate it for Banach spaces only, because this is the way we will use it.

Proposition.   *Let X and Y be separable Banach spaces and let
$T: X \to Y$ be a semi-embedding. Then there exists an $x \in X$, $\|x\| = 1$
such that if $(x_n)_{n=1}^{\infty} \subset X, \|x_n\| \leq 1, n = 1, 2, \ldots$ and $Tx_n \to Tx$ then
$x_n \to x$.*

Proof: Since T is a semi-embedding, the image of every closed ball
in X is closed in Y. Let us fix $\varepsilon > 0$ and a relatively open set $V \subset
T(B_X)$. Let us fix a sequence $(v_j)_{j=1}^{\infty} \subset B_X$ such that we can write
$B_X = \bigcup_{j=1}^{\infty} B(v_j, \varepsilon) \cap B_X$. Applying the Baire category theorem to
the covering of V by closed sets $T\big(B(v_j, \varepsilon) \cap B_X\big)$ we get the following
statement:

> for every $\varepsilon > 0$ and for every relatively open set $V \subset T(B_X)$
> there exists a non-empty relatively open set $U \subset V$ such   (1)
> that diam $T^{-1}(U) < \varepsilon$.

Applying (1) inductively we find a sequence of relatively open sets
$(U_n)_{n=1}^{\infty}$ in $T(B_X)$ such that $U_n \supset \overline{U}_{n+1} \supset U_{n+1}$ and diam $T^{-1}(U_n) < \frac{1}{n}$
and $0 \notin U_1$. Clearly $\bigcap_{n=1}^{\infty} T^{-1}U_n$ consists of exactly one point x_0. The
desired x equals $\frac{x_0}{\|x_0\|}$.   ■

4.   Applying Proposition 3 for $X = L_1[0, 1]$ we get

Lemma.   *If $T: L_1[0, 1] \to X$ is a semi-embedding, then there exists
an $f \in L_1[0, 1]$ with $\|f\| = 1$ and a number $\delta > 0$ such that for all
real-valued functions φ with $\int_0^1 \varphi(t)dt = 0$ and $|\varphi| = 1$, μ-a.e. we have
$\|T(\varphi f)\| \geq \delta$.*

Proof: From Proposition 3 we infer that there exists an $f \in
L_1[0, 1]$, $\|f\| = 1$ and $\delta > 0$ such that $\|f - g\| < \frac{1}{2}$ whenever $\|g\| \leq 1$
and $\|Tf - Tg\| < \delta$. Given φ as above we take $\psi = \varphi$ or $\psi = -\varphi$ so
that we have $\int(1 + \psi)|f|d\mu \leq 1$. For $g = (1 + \psi)f$ we have $\|g\| \leq 1$ and
$\|g - f\| = \|\psi f\| = 1$, thus $\delta < \|Tf - Tg\|$. But $\|T(\varphi f)\| = \|Tf - Tg\|$
so the claim follows.   ■

5.   Now we are ready to prove

Theorem.   *Let μ be an atom-free, separable measure. There is no
semi-embedding from $L_1(\mu)$ into c_0.*

Proof: We see from III.A.1 that it is enough to consider $L_1[0, 1]$ only.
Assume to the contrary that $T: L_1[0, 1] \to c_0$ is a semi-embedding, and

$\|T\| = 1$. Take $f \in L_1[0,1]$ and $\delta > 0$ as in Lemma 4. Fix an integer $N > 2/\delta$, and let $A_1, \ldots, A_N$ be disjoint sets such that $\int_{A_j} |f| = \frac{1}{N}$. For each n, $n = 1, 2, \ldots, N$ let r_j^n denote a sequence of Rademacher-like functions on A_n, i.e.

$$|r_j^n| = \chi_{A_n}, \quad n = 1, \ldots, N, \; j = 1, 2, \ldots, \tag{2}$$

$$\int r_j^n = 0, \quad n = 1, \ldots, N, \; j = 1, 2, \ldots, \tag{3}$$

$$r_j^n f \xrightarrow{\omega} 0 \text{ as } j \to \infty, \text{ for every } n, \; n = 1, 2, \ldots, N. \tag{4}$$

Using the standard 'gliding hump' argument (cf. proof of II.B.17) condition (4) yields numbers $j(n)$, $n = 1, \ldots, N$, such that

$$\left\| \sum_{n=1}^{N} T(r_{j(n)}^n f) \right\| \leq 1.5 \max_n \|T(r_{j(n)}^n f)\| \leq 1.5 \cdot N^{-1} < \delta.$$

But (2), (3) and Lemma 4 give

$$\left\| \sum_{n=1}^{N} T(r_{j(n)}^n f) \right\| = \left\| T\left(\left(\sum_{n=1}^{N} r_{j(n)}^n \right) f \right) \right\| \geq \delta$$

which is impossible. This contradiction proves the theorem.   ■

6.   The spirit of the above results is that it is rather difficult to put a weaker, linear topology on the unit ball of $L_1[0,1]$ and make it compact or even complete. The intuition is that somehow we always have to add some singular measures in the completion. The following classical theorem of Menchoff is only an example of this.

Theorem. (Menchoff).   *Let G be a compact, infinite, metrizable abelian group with dual group Γ. Let $M_0(G) \subset M(G)$ denote the set of all measures ν with $\hat{\nu}(\gamma) \in c_0(\Gamma)$. Then M_0 is a non-separable closed band in $M(G)$.*

Proof: Since the Fourier transform $^\wedge: M(G) \to \ell_\infty(\Gamma)$ is continous and since $M_0 = (^{\wedge-1})(c_0(\Gamma))$, we see that M_0 is a closed linear subspace. Also for $\nu \in M_0$ and $p = \sum_{\gamma \in A} a_\gamma \gamma$ with finite $A \subset \Gamma$ one easily checks that $p \cdot \nu \in M_0$. Since M_0 is closed this gives that M_0 is a band. If M_0 is separable, then there exists a positive measure $\mu \in M(G)$ such that $M_0 = L_1(\mu)$. One easily checks that μ has no atoms. By Theorem 5 $(B_{M_0})^\wedge$ is not closed in $c_0(\Gamma)$, i.e. there exist $\alpha(\gamma) \in c_0(\Gamma) \backslash (B_{M_0})^\wedge$

and $\mu_n \in B_{M_0}$ such that $\hat{\mu}_n \to \alpha$ in $c_0(\Gamma)$. Let μ be an ω^*-cluster point of $\{\mu_n\}_{n=1}^{\infty}$. Since $\|\mu\| \leq 1$ and $\hat{\mu} = \alpha$ we infer that $\mu \in B_{M_0}$. This contradicts the choice of α.   ■

Note that the group structure plays almost no role in the above argument.

7.   Given a class of measures it is natural to seek its zero sets, i.e. sets of measure zero with respect to every measure in the class. We will discuss zero sets for $M_0(\mathbb{T})$. This is a small part of the classical branch of the theory of Fourier series or harmonic analysis on more general groups. In order to proceed we need some definitions. We say that a sequence $(x_n)_{n=1}^{\infty} \subset \mathbb{T}$ has an asymptotic distribution if and only if $N^{-1} \sum_{n=1}^{N} \delta_{x_n}$ converges $\sigma(M(\mathbb{T}), C(\mathbb{T}))$ to some measure $\nu \in M(\mathbb{T})$. Let us recall that δ_x denote the Dirac measure concentrated at the point x. Since for every $m \in \mathbb{Z}$ we have $\hat{\nu}(m) = \lim_{N \to \infty} \left(N^{-1} \sum_{n=1}^{N} \delta_{x_n}\right)^{\wedge}(m) = \lim_{N \to \infty} \left(N^{-1} \sum_{n=1}^{N} \exp(-imx_n)\right)$ we see that the sequence $(x_n)_{n=1}^{\infty} \subset \mathbb{T}$ has an asymptotic distribution if and only if $\lim_{N \to \infty} \left(\frac{1}{N} \sum_{n=1}^{N} \exp(-imx_n)\right)$ exists for every $m \in \mathbb{Z}$. A Borel subset $E \subset \mathbb{T}$ is called a Weyl set if there exists an increasing sequence of integers $(n_k)_{k=1}^{\infty}$ such that for every $x \in E$ the sequence $(n_k \cdot x)_{k=1}^{\infty}$ has an asymptotic distribution with the corresponding measure ν_x different from the normalized Lebesgue measure.

8.   The following theorem characterizes $M_0(\mathbb{T})$.

Theorem.   The measure μ is in $M_0(\mathbb{T})$ if and only if $\mu(E) = 0$ for every Weyl set $E \subset \mathbb{T}$.

Proof: Throughout the whole proof it is sufficient to consider only positive measures (see Theorem 6).

    $\Rightarrow$. Let μ be a non-zero measure in $M_0(\mathbb{T})$ and let E be a Weyl set with the corresponding sequence $(n_k)_{k=1}^{\infty}$. For $m \in \mathbb{Z}, m \neq 0$ we put

$$c_m(t) = \begin{cases} \lim_{N \to \infty} \dfrac{1}{N} \sum_{k=1}^{N} \exp(-imn_k t), & t \in E, \\ 0, & t \notin E. \end{cases} \tag{5}$$

For a Borel subset $F \subset E$ we have

$$\int_F c_m(t)d\mu(t) = \int_F \lim_{N \to \infty} \frac{1}{N} \sum_{k=1}^{N} \exp(-imn_k t)d\mu(t)$$

$$\tag{6}$$

$$= \lim_{N \to \infty} \frac{1}{N} \sum_{k=1}^{N} (\widehat{\mu|F})(mn_k).$$

Since $M_0(\mathbb{T})$ is a band (Theorem 6) $\mu|F \in M_0(\mathbb{T})$ so (6) gives $\int_F c_m(t)d\mu(t) = 0$. Since F was an arbitrary Borel subset of E we infer that $c_m(t) = 0, \mu$-a.e. for $m \neq 0$. But E is a Weyl set, thus for every $t \in E$ there is an $m \neq 0$ such that $c_m(t) \neq 0$. This shows that $\mu(E) = 0$.

$\Leftarrow$. Let us take $\mu \notin M_0(\mathbb{T})$ and let us fix a sequence of integers $n_k \longrightarrow \infty$ as $k \to \infty$ (or $n_k \longrightarrow -\infty$ as $k \to \infty$) such that $\hat{\mu}(n_k) \longrightarrow \alpha \neq 0$ as $k \to \infty$. Applying Theorem III.A.29 to the family of sequences $\{\exp(-imn_k t)\}_{k=1}^{\infty}$ in $L_2(\mathbb{T}, d\mu)$ we get a further subsequence $(n'_k)_{k=1}^{\infty}$ such that $f_N^m(t) = N^{-1} \sum_{k=1}^{N} \exp(-imn'_k t)$ converges μ-a.e for each $m \in \mathbb{Z}$. Let us put

$$E = \{t \in \mathbb{T} : \lim_{N \to \infty} f_N^m(t) \text{ exists for each } m \in \mathbb{Z}$$

$$\text{and } \lim_{N \to \infty} f_N^1(t) \text{ is not zero}\}.$$

This is a Weyl set. We have

$$\int_E \lim_{N \to \infty} f_N d\mu = \int_{\mathbb{T}} \lim_{N \to \infty} f_N d\mu = \lim_{N \to \infty} \int_{\mathbb{T}} f_N d\mu$$

$$= \lim_{N \to \infty} \frac{1}{N} \sum_{k=1}^{N} \hat{\mu}(n'_k) = \alpha \neq 0.$$

Thus $\mu(E) > 0$. ∎

9. Our goal now is to characterize weakly compact sets in $L_1(\mu)$-spaces. This characterization has many further applications (see III.C.19 or III.H.10 among others) and generalizations (see e.g. III.D.31). It also nicely connects the general functional analytic notions with measure-theoretical concepts. We start with the distinctive special case of the space ℓ_1.

Theorem. (Schur). *For a bounded subset $H \subset \ell_1$ the following conditions are equivalent:*

(a)   *H is relatively compact;*

(b)   *H is relatively weakly compact;*

(c)   *there is no sequence $(a_n)_{n=1}^\infty \subset H$ which is a basic sequence equivalent to the unit vector basis of ℓ_1.*

The proof clearly follows from the following.

10 Lemma.   *If $H \subset \ell_1$ is a bounded subset, not relatively compact, then there exists a basic sequence $(a_n)_{n=1}^\infty \subset H$ equivalent to the unit vector basis of ℓ_1.*

Proof: We find $\{b_n\}_{n=1}^\infty \subset H$ such that $\|b_n\| \le C, n = 1, 2, \ldots$ for some C and $\|b_n - b_m\| \ge \delta$, for $m \ne n$ and $\delta > 0$. A standard diagonal procedure gives a subsequence $\{b_{n_j}\}_{j=1}^\infty$ such that $b_{n_j}(k) \longrightarrow b(k)$ as $j \to \infty$, for every $k = 1, 2, \ldots$. Clearly $b \in \ell_1$ and $\|b_{n_j} - b\| \ge \delta, j = 1, 2, \ldots$. II.B.17 gives a further subsequence (call it also b_{n_j}) such that $(b_{n_j} - b)_{j=1}^\infty$ is equivalent to a block-basic sequence, thus to the unit vector basis in ℓ_1. Let $Y = \mathrm{span}\{(b_{n_j} - b)\}_{j=1}^\infty$. Omitting if necessary a finite number of j's we can assume that $b \notin Y$. Then

$$\left\| \sum \alpha_j b_{n_j} \right\| = \left\| \sum \alpha_j (b_{n_j} - b) + \left(\sum \alpha_j \right) b \right\| \ge K \left\| \sum \alpha_j (b_{n_j} - b) \right\|$$
$$\ge K \sum |\alpha_j|,$$

thus $\{b_{n_j}\}_{j=1}^\infty$ is the desired sequence.   ■

11.   Now we will discuss relatively weakly compact sets in $L_1(\mu)$ for a general probability measure μ. Our main tool will be the notion of uniform integrability.

Definition.   *A subset $H \subset L_1(\mu)$ is called uniformly integrable if for every $\varepsilon > 0$ there exists an $\eta > 0$ such that*

$$\sup \left\{ \int_A |f| d\mu \: : \: \mu(A) \le \eta, \; f \in H \right\} \le \varepsilon. \tag{7}$$

If μ is an atom-free probability measure then every uniformly integrable set in $L_1(\mu)$ is norm-bounded. This follows from (7) and the observation that for every $f \in L_1(\mu)$ there exists a set A with $\mu(A) = \frac{1}{n}$ such that $\int_A |f| d\mu \ge n^{-1} \int |f| d\mu$. In the other direction let us observe that every one-element set, and thus every finite set, is uniformly integrable. To see this put $A_n = \{t : |f(t)| > n\}$. Since $\mu(A_n) \to 0$, the Lebesgue dominated convergence theorem gives $\int_{A_n} |f|d\mu \longrightarrow 0$ as $n \to \infty$.

12.　　The next theorem gives the promised characterization of relatively weakly compact sets in $L_1(\mu)$. This is the main result of this chapter. It says that basically there is only one reason for a bounded set not to be relatively weakly compact. In this sense it is similar to Proposition II.D.5 and also to Theorem 9. Note also the equivalence of finite and infinite conditions. This will be investigated later.

Theorem.　　*Let μ be a probability measure and let H be a bounded subset of $L_1(\mu)$. The following conditions are equivalent:*

(a)　*H is not relatively weakly compact in $L_1(\mu)$;*

(b)　*H is not uniformly integrable;*

(c)　*there exists an $\varepsilon > 0$ and a sequence of disjoint sets $(A_n)_{n=1}^{\infty}$ such that*
$$\sup\left\{\int_{A_n} |f|d\mu : f \in H\right\} \geq \varepsilon; \qquad n = 1, 2, \ldots$$

(d)　*·there exists a basic sequence $(f_n)_{n=1}^{\infty} \subset H$ equivalent to the unit vector basis in ℓ_1;*

(e)　*there exists an $\varepsilon > 0$ such that for every integer N there exist N disjoint sets $A_1, \ldots, A_N$ such that*
$$\sup\left\{\int_{A_n} |f|d\mu : f \in H\right\} \geq \varepsilon, \quad n = 1, 2 \ldots, N;$$

(f)　*there exists a constant K such that for every integer N there exist $f_1, \ldots, f_N \subset H$, K-equivalent to the unit vector basis in ℓ_1^N.*

Proof:　　The proof will consist of the following implications:

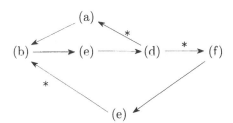

with the implications marked $*$ being obvious.

(a)$\to$(b). Suppose $H \subset L_1(\mu)$ is uniformly integrable, thus bounded, i.e. for $f \in H$ we have $\|f\| \leq M$. Given an integer n we write every function $f \in H$ as $f = f^n + f_n = f \cdot \chi_{\{|f| \geq n\}} + f \cdot \chi_{\{|f| < n\}}$. The set $\{f_n : f \in H\} \subset n \cdot B_{L_\infty(\mu)}$ is clearly weakly relatively compact. Since $\mu\{|f| \geq n\} \leq \frac{M}{n}$, uniform integrability gives $\sup\{\|f^n\|_1 : f \in H\} \longrightarrow 0$ as $n \to \infty$. From Lemma II.C.7 we infer that H is relatively weakly compact.

(b)$\to$(c). The condition (b) means explicitly that there exists a positive ε such that for every $\eta > 0$ there exist a set A_η and a function $f_\eta \in H$ such that $\mu(A_\eta) < \eta$ and $\int_{A_\eta} |f_\eta| d\mu > \varepsilon$. Let us fix positive numbers $\delta_{j,k}, k = 2, 3, \ldots, \ j = 1, 2, \ldots, k-1$, such that $\sum_{j,k} \delta_{j,k} < \frac{\varepsilon}{2}$. Repeatedly using the observation that a finite set of functions is uniformly integrable we find sets $(B_k)_{k=1}^\infty$ and functions $(f_k)_{k=1}^\infty$ in H such that

$$\int_{B_k} |f_k| \geq \varepsilon, \qquad k = 1, 2, 3 \ldots,$$

$$\int_{B_k} |f_j| \leq \delta_{j,k}, \qquad j = 1, 2, \ldots, k-1.$$

We put $A_k = B_k \backslash \bigcup_{j>k} B_j$. Clearly the $(A_k)_{k=1}^\infty$ are disjoint and

$$\int_{A_k} |f_k| d\mu \geq \int_{B_k} |f_k| d\mu - \sum_{j>k} \int_{B_j} |f_k| d\mu \geq \frac{\varepsilon}{2}.$$

(c)$\to$(d). Fix disjoint sets $(A_n)_{n=1}^\infty$ and functions $f_n \in H$, $n = 1, 2, \ldots$ such that $\int_{A_n} |f_n| d\mu \geq \varepsilon > 0$. Put $h_n = (\int_{A_n} |f_n| d\mu)^{-1} f_n \cdot \chi_{A_n}$ and $\varphi_n = \mathrm{sgn} f_n \cdot \chi_{A_n}$. Clearly $Y = \mathrm{span}\{h_n\}_{n=1}^\infty$ is isometric to ℓ_1 and $P(f) = \sum_{n=1}^\infty \int f \bar\varphi_n d\mu \cdot h_n$ is a projection from $L_1(\mu)$ onto Y. One easily sees that $P(\{f_n\}_{n=1}^\infty)$ is not relatively compact in norm. From Theorem 9 we get a subsequence $\{f_{n_j}\}_{j=1}^\infty$ such that $\{P(f_{n_j})\}_{j=1}^\infty$ is equivalent to the unit vector basis in ℓ_1, but this implies that $\{f_{n_j}\}_{j=1}^\infty$ itself is equivalent to the unit vector basis in ℓ_1.

(f)$\to$(e). We can assume $\|f_j\| \leq 1$, $j = 1, \ldots, N$ and thus $K^{-1} \sum_{j=1}^N |\alpha_j| \leq \int |\sum_{j=1}^N \alpha_j f_j|$ for all sequences of scalars $(\alpha_j)_{j=1}^N$. Let $r_j(t)$ be, as usual, the Rademacher functions. We have

$$K^{-1} \sum_{j=1}^N |\alpha_j| \leq \int_0^1 \int \left| \sum_{j=1}^N \alpha_j r_j(t) f_j \right| d\mu dt$$

$$\leq \int \left(\int_0^1 \left| \sum_{j=1}^N \alpha_j r_j(t) f_j \right|^2 dt \right)^{\frac{1}{2}} d\mu$$

$$= \int \left(\sum_{j=1}^{N} |\alpha_j f_j|^2 \right)^{\frac{1}{2}} d\mu$$

$$\leq \int (\max_j |\alpha_j f_j|)^{\frac{1}{2}} \cdot \left(\sum_{j=1}^{N} |\alpha_j f_j| \right)^{\frac{1}{2}} d\mu \qquad (8)$$

$$\leq \left(\int \max_j |\alpha_j f_j| d\mu \right)^{\frac{1}{2}} \left(\int \sum_{j=1}^{N} |\alpha_j f_j| d\mu \right)^{\frac{1}{2}}$$

$$\leq \left(\int \max_j |\alpha_j f_j| d\mu \right)^{\frac{1}{2}} \left(\sum_{j=1}^{N} |\alpha_j| \right)^{\frac{1}{2}}.$$

Thus

$$K^{-2} \sum_{j=1}^{N} |\alpha_j| \leq \int \max_j |\alpha_j f_j| d\mu \qquad (9)$$

and in particular

$$K^{-2} N \leq \int \max_j |f_j| d\mu. \qquad (10)$$

Let B_s, $s = 1, 2, \ldots, N$, be disjoint sets such that $(\max_j |f_j|)|B_s = |f_s|$ (some of them can be empty). From (10) we get

$$K^{-2} N \leq \sum_{s=1}^{N} \int_{B_s} |f_s| d\mu.$$

Since $\|f_s\| \leq 1$ we infer that for at least $\frac{N}{(2K^2-1)}$ indices s we have $\int_{B_s} |f_s| d\mu \geq \frac{1}{2K^2}$. ∎

13. Remarks. (a) If we keep track of the constants in the proof of (f)$\to$ (e) we get the following statement: If $(f_j)_{j=1}^{N} \subset L_1(\mu)$ are such that

$$K^{-1} \sum_{j=1}^{N} |\alpha_j| \leq \int \left| \sum_{j=1}^{N} \alpha_j f_j \right| d\mu \leq \sum_{j=1}^{N} |\alpha_j| \quad \text{for all scalars} (\alpha_j)_{j=1}^{N}$$

then for every $\delta < 1$ there exists a subset $\Lambda \subset \{1, \ldots, N\}$ and disjoint sets $\{A_j\}_{j\in\Lambda}$ such that

$$\int_{A_j} |f_j| d\mu \geq \delta/K^2 \text{ and } |\Lambda| \geq \varphi_\delta(N)$$

where for every δ, $\varphi_\delta(N) \longrightarrow \infty$ as $N \to \infty$.

(b) If μ is an arbitrary measure on Ω and $H \subset L_1(\Omega, \mu)$ is weakly relatively compact then there exists a set $\Omega_1 \subset \Omega$ of σ-finite measure μ such that all functions from H are supported on Ω_1. Thus when dealing with relatively weakly compact sets in $L_1(\mu)$ we can restrict our attention to σ-finite μ. But this case, as we know (II.B.2(c)), easily reduces to the probability measure μ.

14 Corollary. (Steinhaus). *The space $L_1(\mu)$, μ arbitrary, is weakly sequentially complete, i.e., weakly Cauchy sequences are weakly convergent.*

Proof: Since every sequence is supported on a σ-finite set it is enough to consider a probability measure μ (Remark 13(b)). A weakly Cauchy sequence which is not weakly convergent is not relatively weakly compact, so from Theorem 12 it has a subsequence equivalent to the unit vector basis in ℓ_1, thus not weakly Cauchy. This contradiction proves the corollary.   ■

15.   We have seen in Theorem 12 the interesting interplay between global notions like weak compactness and the local concept of finite representability of ℓ_1. We want now to discuss the finite representability of ℓ_1 in a general Banach space X. We start with an interesting lemma about finite dimensional isomorphs of ℓ_1^N.

Lemma.   *Let X be an N-dimensional Banach space with $d(X, \ell_1^N) = \alpha$. Then there exists a subspace $X_1 \subset X$ with $\dim X_1 = [\sqrt{N}]$ and $d(X_1, \ell_1^{[\sqrt{N}]}) \le \sqrt{\alpha}$.*

Proof: Let us fix a basis $(x_j)_{j=1}^N$ in X such that for every sequence of scalars $(\alpha_j)_{j=1}^N$ we have

$$\frac{1}{\alpha} \sum_{j=1}^N |\alpha_j| \le \left\| \sum_{j=1}^N \alpha_j x_j \right\| \le \sum_{j=1}^N |\alpha_j|.$$

Let us also fix $[\sqrt{N}]$ disjoint subsets $A_s \subset \{1, 2, \ldots, N\}$ each with cardinality $[\sqrt{N}]$. For each $s = 1, 2, \ldots, [\sqrt{N}]$ we define

$$d_s = \inf \left\{ \left\| \sum_{j \in A_s} \alpha_j x_j \right\| : \sum_{j \in A_s} |\alpha_j| = 1 \right\}.$$

If for some s we have $d_s \geq \frac{1}{\sqrt{\alpha}}$ than $X_1 = \mathrm{span}\{x_j : j \in A_s\}$ is a good choice. On the other hand if for all s we have $d_s < \frac{1}{\sqrt{\alpha}}$ then we fix $y_s = \sum_{j \in A_s} \alpha_j x_j$ such that $\|y_s\| < \frac{1}{\sqrt{\alpha}}$ and $\sum_{j \in A_s} |\alpha_j| = 1$ for $s = 1, 2, \ldots, [\sqrt{N}]$. For every sequence of scalars $(\beta_s)_{s=1}^{[\sqrt{N}]}$ we have

$$\frac{1}{\alpha} \sum_{s=1}^{[\sqrt{N}]} |\beta_s| \leq \left\| \sum_{s=1}^{[\sqrt{N}]} \beta_s y_s \right\| \leq \frac{1}{\sqrt{\alpha}} \sum_{s=1}^{[\sqrt{N}]} |\beta_s|$$

so for $X_1 = \mathrm{span}\{y_s\}_{s=1}^{[\sqrt{N}]}$ we get $d(X_1, \ell_1^{[\sqrt{N}]}) \leq \sqrt{\alpha}$. ∎

16 Theorem. *Let X be an infinite dimensional Banach space. The following conditions are equivalent:*

(a) *X does not have type p for any $p > 1$;*

(b) *for every $n = 1, 2, \ldots$ and every $\varepsilon > 0$ there exist norm-one vectors $x_1, \ldots, x_n$ in X such that*

$$\min_{\varepsilon_i = \pm 1} \left\| \sum_{i=1}^{n} \varepsilon_i x_i \right\| \geq n - \varepsilon;$$

(c) *ℓ_1 is finitely representable in X;*

(d) *for every $n = 1, 2, \ldots$ and every $\varepsilon > 0$ there exists a subspace $X_{n,\varepsilon} \subset X$ with $d(X_{n,\varepsilon}, \ell_1^n) \leq 1 + \varepsilon$.*

This is a remarkable theorem. We will use it in III.I to study some questions about the disc algebra. Note that it contains the passage from a probabilistic context of the definition of type to the purely deterministic situation described in (c) and (d). It is also nice because it tells us that certain abstract things (like (a)) happen only due to the presence of a very 'concrete' subspaces, namely ℓ_1^n's.

For the proof of this theorem we introduce constants $\gamma_n(X)$, $n = 1, 2, \ldots$, defined by the formula

$$\gamma_n(X) = \inf \left\{ \gamma : \left(\int \left\| \sum_{i=1}^{n} r_i(t) x_i \right\|^2 dt \right)^{\frac{1}{2}} \right.$$

$$\leq \gamma \sqrt{n} \left(\sum_{i=1}^{n} \|x_i\|^2 \right)^{\frac{1}{2}} \tag{11}$$

$$\left. \text{for all } (x_i)_{i=1}^{n} \subset X \right\}.$$

Note that $\gamma_n(X) \leq 1$ for $n = 1, 2, \ldots$. The following lemma really explains some consequences of condition (a) of Theorem 16.

17 Lemma.   (a) *The constants $\gamma_n(X)$ are submultiplicative, i.e.* $\gamma_{n \cdot k}(X) \leq \gamma_n(X) \cdot \gamma_k(X)$ *for all $n, k \in \mathbb{N}$.*

(b) *If $\gamma_n(X) < 1$ for some n then X has type p for some $p > 1$.*

Proof: (a) Fix integers n and k, a number $\varepsilon > 0$ and a sequence $(x_j)_{j=1}^{n \cdot k}$ such that

$$\left(\int_0^1 \left\| \sum_{j=1}^{n \cdot k} r_j(t) x_j \right\|^2 dt \right)^{\frac{1}{2}} \geq (\gamma_{n \cdot k}(X) - \varepsilon) \sqrt{n \cdot k} \left(\sum_{j=1}^{n \cdot k} \|x_j\|^2 \right)^{\frac{1}{2}}. \quad (12)$$

For $s = 0, \ldots, (k-1)$ define

$$\phi_s(\theta) = \sum_{j=s \cdot n + 1}^{(s+1)n} r_j(\theta) x_j.$$

For every θ we have

$$\int \left\| \sum_{s=0}^{k-1} r_s(t) \phi_s(\theta) \right\|^2 dt \leq \gamma_k^2(X) k \sum_{j=0}^{k-1} \|\phi_j(\theta)\|^2,$$

and integrating over θ we get

$$\int \left\| \sum_{j=1}^{n \cdot k} r_j(t) x_j \right\|^2 dt = \int \int \left\| \sum_{s=0}^{k-1} r_s(t) \phi_s(\theta) \right\|^2 dt d\theta$$

$$\leq \gamma_k^2(X) k \sum_{s=0}^{k-1} \int \|\phi_s(\theta)\|^2 d\theta \quad\quad\quad (13)$$

$$\leq \gamma_k^2(X) k \sum_{s=0}^{k-1} \gamma_n^2(X) n \sum_{j=s \cdot n+1}^{(s+1)n} \|x_j\|^2$$

$$= \gamma_k^2(X) \cdot \gamma_n^2(X) \cdot k \cdot n \sum_{j=1}^{k \cdot n} \|x_j\|^2.$$

Since ε was arbitrary, comparing (12) and (13) we get $\gamma_{n \cdot k}(X) \leq \gamma_k(X) \gamma_n(X)$.

    (b) Fix $q > 1$ such that $\gamma_n(X) = n^{-1/q'}$ where $\frac{1}{q} + \frac{1}{q'} = 1$. Observe also that it follows directly from (11) that $(\gamma_n(X)\sqrt{n})_{n=1}^{\infty}$ is an increasing sequence. Take $1 < p < q$. For an arbitrary finite sequence $(x_j)_{j=1}^{n^k}$ with $\sum_{j=1}^{n^k} \|x_j\|^p = 1$ (we can put in some additional zeros to have the right length) we define sets of indices $A_s = \{j : n^{-(s+1)/p} \leq \|x_j\| \leq n^{-s/p}\}$. Clearly $|A_s| \leq n^{s+1}$. We have

$$
\left(\int \left\| \sum_{j=1}^{n^k} r_j(t) x_j \right\|^2 dt \right)^{\frac{1}{2}} \leq \sum_{s=0}^{\infty} \left(\int \left\| \sum_{j \in A_s} r_j(t) x_j \right\|^2 dt \right)^{\frac{1}{2}}
$$

$$
\leq \sum_{s=0}^{\infty} \gamma_{|A_s|}(X) \sqrt{|A_s|} \left(\sum_{j \in A_s} \|x_j\|^2 \right)^{\frac{1}{2}}
$$

$$
\leq \sum_{j=0}^{\infty} \gamma_{n^{s+1}}(X) \sqrt{n^{s+1}} (n^{s+1} \cdot n^{-2s/p})^{\frac{1}{2}}
$$

$$
\leq \sum_{s=0}^{\infty} (n^{-\frac{1}{q'}})^{s+1} n^{s+1} n^{-s/p}
$$

$$
\leq n^{1-\frac{1}{q'}} \sum_{s=0}^{\infty} n^{s(1-\frac{1}{q'}-\frac{1}{p})} < \infty.
$$

This shows that X is of type p for every $p < q$. ∎

Proof of Theorem 16. From Lemma 15 we see that (c)$\Leftrightarrow$(d). Also obviously both (c) and (d) imply (a). (a)$\Rightarrow$(b). From Lemma 17 we see that $\gamma_n(X) = 1$ for $n = 1,2,3\ldots$. This means that for every $n = 1,2,\ldots$ and every $\varepsilon > 0$ there are vectors $x_1, \ldots, x_n$ in X such that

$$
(1-\varepsilon)\sqrt{n} \left(\sum_{j=1}^{n} \|x_j\|^2 \right)^{\frac{1}{2}} \leq \left(\int \left\| \sum_{j=1}^{n} r_j(t) x_j \right\|^2 dt \right)^{\frac{1}{2}} \leq \sum_{j=1}^{n} \|x_j\|. \quad (14)
$$

If ε is very small (depending on n) we see from (14) that $\|x_j\|$, $j = 1, 2, \ldots, n$, has to be practically constant so X has the property

for every n and every $\varepsilon > 0$ there exist vectors
$x_1, \ldots, x_n \in X$ with $\|x_j\| = 1$ for $j = 1, \ldots, n$ such that $\quad (15)$
$(1-\varepsilon)n \leq (\int \| \sum_{j=1}^{n} r_j(t) x_j \|^2)^{\frac{1}{2}}$.

Since obviously

$$
\left(\int \left\| \sum_{j=1}^{n} r_j(t) x_j \right\|^2 dt \right)^{\frac{1}{2}}
$$

$$
\leq \left(2^{-n} \min_{\varepsilon_j = \pm 1} \left\| \sum_{j=1}^{n} \varepsilon_j x_j \right\|^2 + (1 - 2^{-n}) \left(\sum_{j=1}^{n} \|x_j\| \right)^2 \right)^{\frac{1}{2}}
$$

we infer from (15) (make ε very small) that X satisfies (b).

(b)$\Rightarrow$(c). For each sequence $\eta = (\varepsilon_j)_{j=1}^n$ with $\varepsilon_j = \pm 1$ there exists a functional $x_\eta^* \in X^*$ such that $\|x_\eta^*\| = 1$ and $\sum_{j=1}^n x_\eta^*(\varepsilon_j x_j) > n - \varepsilon$. Since $\|x_j\| = 1$, an elementary computation shows that for every η and j, we have $|x_\eta^*(\varepsilon_j x_j) - 1| < \sqrt{3\varepsilon}$. Using this, for any sequence of numbers $(\alpha_j)_{j=1}^n$ with $\sum_{j=1}^n |\alpha_j| = 1$ we obtain

$$\left\| \sum_{j=1}^n \alpha_j x_j \right\| \geq \max_\eta \left| \sum_{j=1}^n \alpha_j x_\eta^*(x_j) \right|$$

$$\geq \max_\eta \left| \sum_{j=1}^n \alpha_j \varepsilon_j \right| - \sqrt{3\varepsilon} \geq \frac{1}{2} - \sqrt{3\varepsilon}.$$

So $d(\mathrm{span}(x_j)_{j=1}^n, \ell_1^n) \leq (\frac{1}{2} - \sqrt{3\varepsilon})^{-1}$.   ∎

18.   As an application of our previous considerations we have the following useful

Corollary. *Let X be a closed subspace of $L_1(\mu)$. The following conditions are equivalent:*

(a) *X is reflexive;*

(b) *X has type p for some $p > 1$;*

(c) *X does not contain a subspace isomorphic to ℓ_1;*

(d) *ℓ_1 is not finitely representable in X.*

Proof: Since each of the conditions holds for X if and only if it holds for every separable subspace of X we can assume that μ is a probability measure (see III.A.2). Now the corollary immediately follows from Theorem 12 and Theorem 16 (see also II.A.14).   ∎

19.   We wish to conclude this section with the proof of the following.

Theorem. *Suppose $F_0 \in L_\infty(\mathbb{T})$ is such that $\inf\{\|F_0 + h\|_\infty : h \in H_\infty(\mathbb{T})\} < 1$. Then there exists an $F \in F_0 + H_\infty(\mathbb{T})$ and $h \in H_1(\mathbb{T}), h \neq 0$ such that*

$$F \cdot h = |h| \qquad a.e. \ on \ \mathbb{T}. \qquad (16)$$

The proof of this theorem is a nice application of Theorem 12. Moreover the following lemma is relevant to some questions which will be discussed in Chapter III.E.

20 Lemma. *If $\{E_n\}$ is a sequence of measurable subsets of $\mathbb{T}$ such that $|E_n| \to 0$, then there is a sequence $\{g_n\}$ of functions in H_∞ such that*

(a) $\operatorname{supess}\{|g_n(t)| : t \in E_n\} \longrightarrow 0$ *as* $n \to \infty$,

(b) $g_n(0) = (2\pi)^{-1} \int_{\mathbb{T}} g_n(t)dt \longrightarrow 1$ *as* $n \to \infty$,

(c) $|g_n| + |1 - g_n| \leq 1 + \varepsilon_n$ *where* $\lim \varepsilon_n = 0$.

Proof: Fix numbers A_n such that $A_n \longrightarrow \infty$ as $n \to \infty$ and $A_n|E_n| \longrightarrow 0$ as $n \to \infty$. Let f_n be the Poisson integral of $A_n\chi_{E_n} + iA_n\tilde{\chi}_{E_n}$ ($\tilde{\chi}_{E_n}$ is the harmonic conjugate of χ_{E_n}, I.B.22). Clearly f_n is an analytic function on $\mathbb{D}$ taking values in the right half plane. Since the map $z \mapsto \frac{1}{(1+z)}$ maps the right half plane onto the disc

$$\left| w - \frac{1}{2} \right| < \frac{1}{2} \tag{17}$$

we get that $h_n(z) = \frac{1}{(1+f_n(z))}$ maps $\mathbb{D}$ into the disc given by (17). Since $f_n(0) = A_n|E_n|$ we get $h_n(0) \longrightarrow 1$ as $n \to \infty$. Also

$$\operatorname{supess}\{|h_n(t)| : t \in E_n\} \leq \operatorname{supess}\{\frac{1}{(1 + Ref_n(t))} : t \in E_n\}$$

$$= \frac{1}{(1 + A_n)} \longrightarrow 0 \text{ as } n \to \infty.$$

Now observe that the map $z \to z^\delta$ compresses the disc (17) into the ellipse $|w| + |1 - w| \leq 1 + \varepsilon(\delta)$ where $\varepsilon(\delta) \longrightarrow 0$ as $\delta \to 0$. All the above yields that for some sequence $\delta_n \to 0$ slowly enough the functions $g_n = h_n^{\delta_n}$ satisfy (a), (b) and (c). ∎

Proof of Theorem 19. Put

$$a = \sup\left\{ \frac{1}{2\pi} \left| \int_0^{2\pi} f(e^{i\theta})d\theta \right| : f \in F_0 + H_\infty, \|f\| \leq 1 \right\}. \tag{18}$$

Since the unit ball in H_∞ is $\sigma(L_\infty, L_1)$-compact this supremum is attained at some $F \in F_0 + H_\infty$. Clearly $1 \geq \operatorname{dist}(F, H_\infty^0) = \inf\{\|F - h\| : h \in H_\infty, h(0) = 0\}$. If $\|F - h\|_\infty < 1$ for some $h \in H_\infty^0$ we see that for small ε's the function $F - h + \varepsilon$ is an admissible f in (18) giving a larger mean than F. This shows that $\operatorname{dist}(F, H_\infty^0) = 1$. Since $H_1^* = L_\infty/H_\infty^0$ we get by duality

$$\sup\left\{ \frac{1}{2\pi} \left| \int_0^{2\pi} F(e^{i\theta})h(e^{i\theta})d\theta \right| : h \in H_1, \|h\| \leq 1 \right\} = 1.$$

Fix a sequence $(h_n)_{n=1}^{\infty}$ in H_1 with $\|h_n\| \leq 1$ and

$$\frac{1}{2\pi} \int_0^{2\pi} F(e^{i\theta}) h_n(e^{i\theta}) d\theta \longrightarrow 1. \qquad (19)$$

If $(h_n)_{n=1}^{\infty}$ has a weakly convergent subsequence $(h_{n_k})_{k=1}^{\infty}$ we put $h = \omega\text{-}\lim h_{n_k} \in H_1(\mathbb{T})$. Now (19) gives

$$\frac{1}{2\pi} \int_0^{2\pi} F(e^{i\theta}) h(e^{i\theta}) d\theta = 1,$$

so in particular $h \neq 0$. Since $\|F\|_{\infty} \leq 1$ and $\|h\|_1 \leq 1$ we get (16).

We complete the proof by showing that the assumption that $(h_n)_{n=1}^{\infty}$ has no weakly convergent subsequence leads to a contradiction. If $(h_n)_{n=1}^{\infty}$ has no weakly convergent subsequence Theorem II.C.3 (Eberlein-Šmulian) and Theorem 12 give sets $(E_n) \subset \mathbb{T}$ with $|E_n| \to 0$ such that

$$\left| \frac{1}{2\pi} \int_{E_n} h_n(e^{i\theta}) d\theta \right| > \beta > 0 \qquad (20)$$

at least for a subsequence of (h_n). Now let g_n and ε_n be given by Lemma 20 and put

$$H_n = \frac{g_n h_n}{1 + \varepsilon_n} \quad \text{and} \quad K_n = \frac{(1 - g_n) h_n}{1 + \varepsilon_n}.$$

We infer from Lemma 20(c) that $\|H_n\|_1 + \|K_n\|_1 \leq \|h_n\| \leq 1$. Also since $\varepsilon_n \to 0$, (19) gives

$$1 = \lim_{n \to \infty} \frac{1}{2\pi} \left| \int_0^{2\pi} F \frac{h_n}{1 + \varepsilon_n} \right|$$

$$\leq \lim_{n \to \infty} \left(\frac{1}{2\pi} \left| \int F H_n \right| + \frac{1}{2\pi} \left| \int F K_n \right| \right) \qquad (21)$$

$$\leq \lim_{n \to \infty} (\|H_n\|_1 + \|K_n\|_1) \leq 1.$$

Thus the limit in the middle exists and equals 1. From (20) and Lemma 20 (a) we get $\lim_n \|K_n\|_1 \geq \beta > 0$ so (21) yields

$$\frac{1}{2\pi} \left| \int F \frac{K_n}{\|K_n\|_1} \right| \longrightarrow 1. \qquad (22)$$

Note that Lemma 20 (b) gives $K_n(0) \longrightarrow 0$ as $n \to \infty$. The duality relation $(H_1^0)^* = L_{\infty}/H_{\infty}$ and (22) give $\text{dist}(F, H_{\infty}) \geq 1$. But

$\text{dist}(F, H_\infty) = \text{dist}(F_0, H_\infty) < 1$. This contradiction shows that $(h_n)_{n=1}^\infty$ actually has a weakly convergent subsequence. ∎

21 Corollary. Let $(z_j)_{j=1}^\infty \subset \mathbb{D}$ be such that $\sum_j (1 - |z_j|) < \infty$. For every $f \in H_\infty$ with $\|f\|_\infty < 1$ there exists an inner function φ with $\varphi(z_j) = f(z_j)$, $j = 1, 2, 3, \ldots$.

Proof: Let B be the Blaschke product with zeros $(z_j)_{j=1}^\infty$. We define $F_0 \in L_\infty(\mathbb{T})$ by $F_0 = \overline{B}f$ and apply Theorem 19. We obtain a unimodular $F = F_0 + g$ for some $g \in H_\infty$. Since on the circle $\mathbb{T}$ we have $BF = f + Bg$, we see that BF is a boundary value of an analytic function. This function is clearly inner and satisfies $BF(z_j) = f(z_j)$ for $j = 1, 2, \ldots$. Thus BF is the desired φ. ∎

Notes and remarks.
The fact that there is no weaker topology on $L_1(\mu)$, μ-atom-free, with some compactness properties is well established. Our *Proposition 1* only states the easiest fact of this type. A more detailed study of semi-embeddings is in Bourgain-Rosenthal [1983]. This paper contains our *Theorem 5* and the proof of *Theorem 6*. The first proof of *Theorem 6* was given in Menchoff [1916]. Much more detailed information on supports of measures in $M_0(G)$ can be found in Varopoulos [1966]. The following fact was shown in Pigno-Saeki [1973].

Theorem A. If μ is a measure in $M(G)$ such that $\mu * M_0(G) \subset L_1(G)$ then $\mu \in L_1(G)$.

Theorem 8 is taken from Lyons [1985]. It completes a long line of investigations in the theory of Fourier series and solves problems going back to Rajchman in the 20's. The non-trivial implication (b)⇒(a) in *Theorem 9* is due to Schur [1921] in the language of summability methods. Banach spaces where this implication holds are nowadays said to satisfy the Schur property. The fundamental *Theorem 12* is usually called the Dunford-Pettis theorem. They established the equivalence between (a) and (b) and successfully used it in their papers Dunford [1939] and Dunford-Pettis [1940]. The relevance of condition (d) was realized in Kadec-Pełczyński [1962]. This theorem is by now classical and various versions of it with many different applications are presented in Dunford-Schwartz [1958], Diestel-Uhl [1977], Kopp [1984]. This last book shows the fundamental importance of this theorem in probability theory. *Corollary 14* is a classical theorem of Steinhaus [1919]. We will discuss important generalizations of these facts in the next chapter.

Lemma 15 is a well known finite dimensional version of a result of James [1964] (see also Exercise 9). The important *Theorem 16* was proved by Pisier [1974]. We basically reproduce his proof here with the changes necessary to obtain the result for complex spaces as well. This theorem was the beginning of the study of connections between type and cotype on one side and geometry on the other side. By now the subject has grown enormously. A presentation of this is contained in the beautiful monograph Milman-Schechtman [1986]. Let us only note the direct generalization of *Theorem 16* proved by Maurey-Pisier [1976].

Theorem B. *Let X be an infinite dimensional Banach space and let*

$$p_X = \sup\{p\colon X \text{ has type } p\},$$
$$q_X = \inf\{q\colon X \text{ has cotype } q\}.$$

Then ℓ_{p_X} and ℓ_{q_X} are finitely representable in X.

The connection between the reflexivity of subspaces of $L_1(\mu)$ and conditions like (b) in *Corollary 18* was recognized in the important paper Rosenthal [1973].

Theorem 19 and *Lemma 20* is due to Garnett [1977] (see also Garnett [1981]). The first version of *Lemma 20*, with a more complicated proof, was discovered by Havin [1973]. Various variants, usually referred to as the Havin lemma, are known. The main idea is to show that on small sets there are analytic functions almost peaking on them. Theorem III.I.9 presents a very elaborate version of this idea. *Corollary 21* is an easy special case of a classical theorem of Nevanlinna. For more details on such matters the reader should consult Garnett [1981], in particular chapter IV.4.

Exercises

1. Suppose that $T\colon c_0 \to X$ is a semi-embedding. Show that T is an embedding.

2. Suppose X is a Banach space and X^* is separable. Show that X^* does not contain a subspace isomorphic to $L_1[0,1]$.

3. Construct a sequence which does not have the asymptotic distribution.

4. Suppose that μ is an arbitrary measure on Ω and that $H \subset L_1(\mu)$ is a relatively weakly compact subset. Show that there exists a subset $V \subset \Omega$ of σ-finite measure such that for every $f \in H$ we have supp$f \subset V$.

5. Suppose (Ω, μ) is a probability measure space and K is a compact space. Show that, if $T: L_1(\mu) \to M(K)$ is a continuous linear operator, then there exists $\nu \in M(K), \nu \geq 0$ such that $T(L_1(\mu)) \subset L_1(\nu)$.

6. Show that $H \subset L_1(\mu)$ is relatively norm-compact if and only if it is relatively weakly compact and relatively compact with respect to convergence in measure.

7. Find a weakly compact set $H \subset L_1[0,1]$ such that there is no function $\varphi \geq 1, \varphi$ finite almost everywhere, such that $\{f : f \cdot \varphi \in H\} \subset L_p[0,1]$ for some $p > 1$.

8. Suppose that $(f_n)_{n=1}^{\infty}$ is a sequence in $H_1(\mathbb{T})$ and that $f_n \xrightarrow{w} f$ and $\|f_n\| \to \|f\|$ as $n \to \infty$ for some $f \in H_1(\mathbb{T})$. Show that $\|f_n - f\| \to 0$ as $n \to \infty$.

9. Suppose X is isomorphic to ℓ_1. Show that for every $\varepsilon > 0$ there exists an infinite dimensional subspace $X_1 \subset X$ with $d(X_1, \ell_1) \leq 1 + \varepsilon$.

10. Show that if ℓ_1 is not finitely representable in X, then ℓ_1 is not finitely representable in X^*.

11. Suppose that the sets (E_n) in Lemma 20 are closed. Show that the functions (g_n) can be chosen in the disc algebra A.

12. Show that if $H \subset L_1[0,1]$ is not uniformly integrable then there exists a sequence $(\varphi_n)_{n=1}^{\infty} \subset C[0,1]$ such that $\sum_{n=1}^{\infty} |\varphi_n(t)| \leq 1$ and $\limsup_{n \to \infty} \sup_{h \in H} |\int_0^1 \varphi_n(t)h(t)dt| > 0$.

13. Suppose that $T: X \to \ell_1$ is a non-compact operator. Show that there exists a complemented subspace $X_1 \subset X$ such that $X_1 \sim \ell_1$ and $T|X_1$ is an isomorphic embedding.

III.D. $C(K)$-Spaces

We start this chapter with the general notion of an M-ideal and show that for every element one can find a best approximation to it in any M-ideal. We discuss the space $H_\infty + C$ and show that every function $f \in L_\infty(\mathbb{T})$ has a best approximation in $H_\infty + C$. We prove the linear extension theorem of Michael and Pełczyński and the Milutin theorem that all spaces $C(K)$ for K compact, metric and uncountable are isomorphic. We present the construction of the periodic Franklin system and prove its basic properties. We investigate its behaviour in $L_p(\mathbb{T})$, $C(\mathbb{T})$ and $Lip_\alpha(\mathbb{T})$. We show that $Lip_\alpha(\mathbb{T}) \sim \ell_\infty$. Then we investigate weakly compact sets in duals of $C^k(\mathbb{T}^s)$ and $A(\mathbb{B}_d)$ and show that they have properties similar to weakly compact sets in $L_1(\mu)$. The unifying framework for this study is provided by the concept of a rich subspace of $C(K, E)$. We also introduce and study the concepts of the Dunford-Pettis property and the Pełczyński property.

1. Our subject in this chapter is the spaces of the form $C(K)$ where K is a compact space. This includes also spaces $L_\infty(\mu)$ since they can be realized as $C(K)$ (I.B.10). This fact is a standard result in the theory of Banach algebras. It shows the importance of the multiplicative structure existing in $C(K)$-spaces. Given a closed subset $S \subset K$ put $C(K; S) = \{f \in C(K): f|S = 0\}$. This is clearly a closed subspace and actually an ideal in the algebra $C(K)$.

Proposition. *Every closed ideal in $C(K)$ is of the form $C(K; S)$ for some closed $S \subset K$.*

Proof: Given a non-trivial closed ideal $I \subset C(K)$ we define $S_I = \bigcap_{f \in I} f^{-1}(0)$. Since every $f \in I$ is non-invertible we get $f^{-1}(0) \neq \emptyset$. Given f_1 and f_2 in I we see that $|f_1|^2 + |f_2|^2 = f_1 \bar{f_1} + f_2 \bar{f_2} \in I$ which gives that the family $\{f^{-1}(0)\}_{f \in I}$ is a family with the finite intersection property so its intersection S_I is not empty. Thus $C(K; S_I)$ is a proper ideal containing I. Note that if $f \in I$ then $\bar{f} \in I$. To see this let $g \in C(K)$ be such that $g = \bar{f}/f$ on the set $\{k \in K: |f| \geq \varepsilon\}$ and $\|g\| \leq 1$. Such a g exists by the Tietze extension theorem. Then $g \cdot f \in I$ and $\|g \cdot f - \bar{f}\|_\infty \leq 2\varepsilon$. Since ε was arbitrary and I was closed we get $\bar{f} \in I$. Note also that for any two points $k_1 \neq k_2$ in $K \backslash S_I$ there is an

$f \in I$ with $f(k_1) \neq f(k_2) \neq 0$. Thus the Stone-Weierstrass theorem I.B.12 yields that $I = C(K; S_I)$.   ∎

2.   The notion of an ideal is not a linear concept but the following concept generalizes the above considerations.

Definition. *A subspace M of a Banach space X is called an M-ideal if there exists a projection E from X^* onto $M^\perp = \{x^* \in X^* : x^*|M = 0\}$ such that for every $x^* \in X^*$ we have*

$$\|x^*\| = \|Ex^*\| + \|x^* - Ex^*\|.$$

One checks that every subspace of $C(K)$ of the form $C(K; S)$ is an M-ideal. The projection E of a measure μ is given by $E(\mu) = \mu|S$. Other examples of M-ideals are given in Exercises 1 and 2 and also in Theorem 8.

3 Proposition.   *Let M be an M-ideal in X and let open balls $B(x_1, r_1) = B_1$ and $B(x_2, r_2) = B_2$ be such that $B_1 \cap B_2 \neq \emptyset, B_1 \cap M \neq \emptyset$ and $B_2 \cap M \neq \emptyset$. Then $B_1 \cap B_2 \cap M \neq \emptyset$.*

Proof: Consider $B_1 \times B_2 \subset X \oplus X$ and let $\Delta = \{(m, m) : m \in M\} \subset X \oplus X$. If the conclusion does not hold then $(B_1 \times B_2) \cap \Delta = \emptyset$. Since B_1 and B_2 are open we can diminish r_1 and r_2 a bit in such a way that the assumptions are still satisfied and there exists $\varphi = (\varphi_1, \varphi_2) \in X^* \oplus X^*$ such that

$$0 = \varphi(\Delta) < \varepsilon \leq Re\{\varphi_1(b_1) + \varphi_2(b_2) : b_1 \in B_1, \ b_2 \in B_2\}$$

for some positive ε. Passing to biduals we get from Goldstine's theorem II.A.13 that $\Delta^{**} = \{(m^{**}, m^{**}) : m^{**} \in M^{**}\}$ is disjoint from $B_1^{**} \times B_2^{**} = B(x_1, r_1) \times B(x_2, r_2)$ where the balls are in X^{**} this time. This translates back to the statement that there are two open balls B_1^{**}, B_2^{**} in X^{**} such that $B_1^{**} \cap M^{**} \neq \emptyset, B_2^{**} \cap M^{**} \neq \emptyset, B_1^{**} \cap B_2^{**} \neq \emptyset$ and $B_1^{**} \cap B_2^{**} \cap M^{**} = \emptyset$. But $X^{**} = (M^{**} \oplus Z)_\infty$ (since M is an M-ideal) so this is clearly impossible.   ∎

4.   Using this proposition we will obtain the following useful

Theorem.   *If M is an M-ideal in X, then the quotient map $q: X \rightarrow X/M$ maps the closed unit ball in X onto the closed unit*

ball in X/M, or equivalently, for every $x \in X$ the set $P_M(x) = \{m \in M : \|x - m\| = \text{dist}(x, M)\} \neq \emptyset$.

Let us note that the open mapping theorem yields that every quotient map maps the open unit ball onto the open unit ball. For the closed unit ball this is generally false. As an example take map $T: \ell_1 \to \ell_2$ defined by $T(e_n) = x_n$ where $(x_n)_{n=1}^{\infty}$ is a sequence dense in the open unit ball of ℓ_2. Clearly T maps the closed unit ball of ℓ_1 onto the open unit ball of ℓ_2.

Proof of the theorem.   Fix $x \in X \backslash M$ and let $d = \text{dist}(x, M)$. Given $\varepsilon > 0$ there exists $m_1 \in M$ such that $\|x - m_1\| < d + \varepsilon$. The open balls $B(m_1, \frac{3}{4}\varepsilon)$ and $B(x, d + \frac{1}{2}\varepsilon)$ satisfy the assumptions of Proposition 3 so there exists $m_2 \in M$ such that $\|m_1 - m_2\| \leq \frac{3}{4}\varepsilon$ and $\|x - m_2\| < d + \frac{1}{2}\varepsilon$. Now the balls $B(m_2, (\frac{3}{4})^2\varepsilon)$ and $B(x, d + (\frac{1}{2})^2\varepsilon)$ satisfy the assumptions of Proposition 3, so we get $m_3 \in M$ such that $\|m_2 - m_3\| < (\frac{3}{4})^2\varepsilon$ and $\|x - m_3\| < d + (\frac{1}{2})^3\varepsilon$. Continuing in this manner we find a sequence $(m_k)_{k=1}^{\infty}$ in M such that $\|m_k - m_{k+1}\| < (\frac{3}{4})^k\varepsilon$ and $\|x - m_{k+1}\| < d + (\frac{1}{2})^k\varepsilon$. This sequence clearly converges to a limit $m \in M$ and $\|x - m\| \leq d$.   ∎

Remark.   Actually the above argument gives the following statement.

> For every $m_1 \varepsilon M$ with $\|m_1 - x\| < d + \varepsilon$ there exists
> $m \in P_M(x)$ such that $\|m - m_1\| \leq 3\varepsilon$.   (1)

5.   Using (1) we obtain the following improvement of Theorem 4.

Proposition.   If M is an M-ideal in X, then for every $x \in X \backslash M$ the set $P_M(x)$ algebraically spans M.

Proof: For every $m \in M$ with $\|m\| < d = \text{dist}(x, M)$ we will show that $m = z - v$ for some $z, v \in P_M(x)$. Since $P_M(x + m) = m + P_M(x)$ we have to show that $P_M(x) \cap P_M(x + m) \neq \emptyset$. Assume to the contrary that $P_M(x) \cap P_M(x + m) = \emptyset$. Then

$$\text{dist}(P_M(x), P_M(x + 2m)) \geq \|m\|. \qquad (2)$$

From Proposition 3 we get that for every $\varepsilon > 0$ there exists a point

$$m_\varepsilon \in B(x, d + \varepsilon) \cap B(x + 2m, d + \varepsilon) \cap M.$$

From (1) we get $m_1 \in P_M(x)$ such that $\|m_1 - m_\varepsilon\| \leq 3\varepsilon$ and $m_2 \in P_M(x+2m)$ such that $\|m_2 - m_\varepsilon\| \leq 3\varepsilon$, so $\mathrm{dist}(P_M(x), P_M(x+2m)) \leq 6\varepsilon$. This contradicts (2) if we choose $\varepsilon < \frac{1}{10}\|m\|$.   ∎

This proposition shows that the best approximation by elements of an M-ideal is always possible and in a very non-unique way. A concrete application of this fact will be given in Corollary 9.

6.   Let us introduce now the space $H_\infty + C$ which has many important applications in operator theory and function theory. To be more precise $H_\infty + C$ is the subspace of $L_\infty(\mathbb{T})$ algebraically spanned by $C(\mathbb{T})$ and boundary values of H_∞.

Our study of this space is based on the following general

Lemma.   *Suppose Y and Z are closed subspaces of a Banach space X and suppose that there is a family Φ of uniformly bounded operators from X into X such that*

(a)   *every Λ in Φ maps X into Y,*

(b)   *every Λ in Φ maps Z into Z,*

(c)   *for every $y \in Y$ and $\varepsilon > 0$ there exists $\Lambda \in \Phi$ such that $\|y - \Lambda y\| < \varepsilon$.*

Then the algebraic sum $Y + Z$ is closed.

Proof: Let $x \in \overline{Y + Z}$. We can find a sequence $x_n \in Y + Z$ such that $\sum_{n=1}^{\infty} x_n = x$ and $\|x_n\| \leq 2^{-n}$ for $n > 1$. Every x_n, $n = 1, 2, \ldots$ can be written as $x_n = y_n + z_n$ with $y_n \in Y$ and $z_n \in Z$. Using (c) we can find Λ_n, $n = 1, 2, \ldots$, in Φ such that $\|y_n - \Lambda_n y_n\| < 2^{-n}$. Let us put $\tilde{y}_n = y_n - \Lambda_n y_n + \Lambda_n x_n$ and $\tilde{z}_n = z_n - \Lambda_n z_n$. Properties (a) and (b) show that $\tilde{y}_n \in Y$ and $\tilde{z}_n \in Z$. Since Φ is uniformly bounded we see that $\|\tilde{y}_n\| \leq C2^{-n}$. Since $x_n = \tilde{y}_n + \tilde{z}_n$ we have $\|\tilde{z}_n\| \leq C2^{-n}$. This gives $x = \sum_{n=1}^{\infty} x_n = \sum_{n=1}^{\infty} \tilde{y}_n + \sum_{n=1}^{\infty} \tilde{z}_n$. But Y and Z are complete so x is really in $Y + Z$.   ∎

7 Corollary.   *The algebraic sum $H_\infty + C$ is closed. It is also a Banach algebra.*

Proof: We apply Lemma 6 with $X = L_\infty(\mathbb{T})$, $Y = C(\mathbb{T})$, $Z = H_\infty(\mathbb{T})$ and Φ the family of Fejér operators. This shows that $H_\infty + C$ is closed.

In order to show that $H_\infty + C$ is a Banach algebra it is enough to show that $h \cdot f \in H_\infty + C$ for $h \in H_\infty$ and $f \in C(\mathbb{T})$. Since $H_\infty + C$ is closed it is enough to consider only trigonometric polynomials f, but then

$$h \cdot f = \sum_{n=0}^{\infty} a_n e^{in\theta} \cdot \sum_{-N}^{N} b_n e^{in\theta}$$

$$= \left(\sum_{n=0}^{N} a_n e^{in\theta} \right) \cdot \left(\sum_{-N}^{N} b_n e^{in\theta} \right)$$

$$+ \left(\sum_{N+1}^{\infty} a_n e^{in\theta} \right) \left(\sum_{-N}^{N} b_n e^{in\theta} \right).$$

The first summand is in $C(\mathbb{T})$ and the second is in H_∞, so $H_\infty + C$ is an algebra.   ∎

8 Theorem.   *The quotient space* $(H_\infty + C)/H_\infty$ *is an* M*-ideal in* L_∞/H_∞.

Proof: Let us identify $L_\infty(\mathbb{T})$ with $C(\mathcal{M})$ where $\mathcal{M} = \mathcal{M}(L_\infty(\mathbb{T}))$ is the space of all non-zero linear and multiplicative functionals on $L_\infty(\mathbb{T})$ (I.B.10). Then $(L_\infty/H_\infty)^* = H_\infty^\perp = \{\mu \in M(\mathcal{M}): \int f d\mu = 0$ for all $f \in H_\infty\}$. The annihilator of $(H_\infty + C)/H_\infty$ in $(L_\infty/H_\infty)^*$ can be identified with

$$(H_\infty + C)^\perp = \{\mu \in M(\mathcal{M}) : \int f d\mu = 0 \text{ for all } f \in H_\infty + C\}.$$

Let m denote the Lebesgue measure considered as a measure on $\mathcal{M}$. From the generalized F.-M. Riesz theorem (see Garnett [1981] V.4.4 or Gamelin [1969] II.7.9) we get

$$H_\infty^\perp = (L_1(m) \cap H_\infty^\perp) \oplus_1 (M_{\text{sing}} \cap H_\infty^\perp) \tag{3}$$

where M_{sing} denotes the space of all measures on $\mathcal{M}$ singular with respect to m. Suppose that $\mu \in (M_{\text{sing}} \cap H_\infty^\perp)$. Let $\alpha = \int e^{-i\theta} d\mu$. Clearly $e^{-i\theta}\mu - \alpha m \in H_\infty^\perp$ and from (3) we get $e^{-i\theta}\mu - \alpha m = fm + \nu$. But both ν and $e^{-i\theta}\mu$ are singular with respect to m so $-\alpha = f$ and since $f \in H_\infty^\perp$ we get $\alpha = 0$. This means that $e^{-i\theta}\mu \in H_\infty^\perp$. Repeating this we get that $\int f d\mu = 0$ for $f \in \bigcup_{n=1}^{\infty} e^{-in\theta} \cdot H_\infty$ which is dense in $H_\infty + C$, so $\mu \in (H_\infty + C)^\perp$. Since no $f \in L_1(m) \cap H_\infty^\perp$ can annihilate $H_\infty + C$ we get that $(H_\infty + C)^\perp = M_{\text{sing}} \cap H_\infty^\perp \subset H_\infty^\perp = (L_\infty/H_\infty)^*$. From (3) we see that $(H_\infty + C)/H_\infty$ is an M-ideal in L_∞/H_∞.   ∎

9 Corollary.   *For every $f \in L_\infty$ there exists $h \in H_\infty + C$ such that*
$\|f - h\|_\infty = \inf\{\|f - g\| : g \in H_\infty + C\}$.

Proof: From Theorems 4 and 8 and the definition of the quotient
norm we infer that given $f \in L_\infty$ there exists a $g \in H_\infty + C$ such
that $\operatorname{dist}(f - g, H_\infty) = \operatorname{dist}(f, H_\infty + C)$. Since balls from H_∞ are ω^*-
compact in L_∞ we obtain that there exists an $h_1 \in H_\infty$ such that
$\|f - g - h_1\| = \operatorname{dist}(f - g, H_\infty) = \operatorname{dist}(f, H_\infty + C)$. The function $h = g + h_1$
is in $H_\infty + C$, so it is the desired best approximant.

10.   Now we want to address the problem of linear extensions. If K
is a compact space and $S \subset K$ a closed subset then for every $f \in C(S)$
there exists a $g \in C(K)$ such that $\|g\| = \|f\|$ and $g|S = f$. This is easy
to see directly but can also be derived from Theorem 4 since the map
$g \rightarrow g|S$ is a quotient map from $C(K)$ onto $C(S) = C(K)/C(K; S)$.
The question is when there exists a linear map $u: C(S) \rightarrow C(K)$ such
that $u(f)|S = f$. In some cases such a map obviously exists (see II.B.4)
and in some cases it does not exist (see Exercise III.B. 8). Because of
applications to the disc algebra (III.E.3) we will study this question in
some generality.

   Let T be a topological space and let S be a closed subset. Suppose
we are given linear subspaces $E \subset C(S)$ and $H \subset C(T)$. We say that
$u: E \rightarrow H$ is a linear extension operator if $u(f)|S = f$ for all $f \in E$.
Clearly, in order to be able to talk about such operators we need $H|S \supset$
E. Actually we will always assume that $H|S = E$. We will denote the set
of such extension operators $\Lambda(E, H)$. If a linear extension $u \in \Lambda(E, H)$
exists, then the operator $P(h) = h - u(h|S)$ defines a projection in H
with $\operatorname{Im} P = \{h \in H : h|S = 0\}$ and $\ker P = u(E) \simeq E$.

11 Definition.   *The pair (E, H) as above has the bounded extension*
property if there exists a constant C such that for every $\varepsilon > 0$ and every
open set $W \supset S$ and for every $f \in E$ there exists $g \in H$ such that
$\|g\| \leq C\|f\|$, $g|S = f$ and $|g(t)| \leq \varepsilon\|f\|$ for $t \in T\backslash W$.

   Let us start with two easy observations.

12 Lemma.   *If $G \subset C(S)$ is a finite dimensional subspace and $u \in$*
$\Lambda(G, C(T))$ then for every $\varepsilon > 0$ there exists an open set $W \supset S$ such
that
$$|u(g)(t)| \leq (1 + \varepsilon)\|g\| \quad \text{for} \quad g \in G \text{ and } t \in W.$$

Proof: Since the closed unit ball of G is compact and u is continuous one checks that $\varphi(t) = \sup\{|u(g)(t)| : g \in G, \|g\| \le 1\}$ is a continuous function. Since $\varphi(t) \le 1$ for $t \in S$ the lemma follows.   ∎

13 Lemma.   *If (E, H) has the bounded extension property and $G \subset E$ is a finite dimensional subspace, then there exists a constant C such that for every open set $W \supset S$ and every $\varepsilon > 0$ there exists $u \in \Lambda(G, H)$ with $\|u\| < C$ and $|u(g)(t)| \le \varepsilon\|g\|$ for $t \in T\backslash W$.*

Proof: We choose an algebraic basis in G and extend each function separately using Definition 11. This yields the desired operator u.   ∎

14 Proposition.   *Let $F \subset G$ be finite dimensional subspaces of E. Assume that π: $G \xrightarrow{\text{onto}} F$ is a projection with $\|\pi\| \le 1$. Given $\varepsilon > 0$ and $u \in \Lambda(F, H)$ there exists $\tilde{u} \in \Lambda(G, H)$ such that $\tilde{u}|F = u$ and $\|\tilde{u}\| \le \max(1, \|u\|) + \varepsilon$.*

Remark.   The mysterious expression $\max(1, \|u\|)$ is justified by the fact that we allow $F = \{0\}$ and $u = 0$.

Proof: Let us put $F_1 = \ker \pi$. We start with an arbitrary open set $W_1 \supset S$ and from Lemma 13 we get $v_1 \in \Lambda(F_1, H)$ with $\|v_1\| \le C$ and $|v_1(f)(t)| \le \frac{1}{2}\varepsilon\|f\|$ for $t \in T\backslash W_1$. We define $u_1 \in \Lambda(G, H)$ as $u_1(g) = u(\pi(g)) + v_1(g - \pi(g))$.

Now Lemma 12 gives an open set W_2, $S \subset W_2 \subset W_1$ such that $|u_1(g)(t)| \le (1 + \frac{\varepsilon}{4})\|g\|$ for $g \in G$ and $t \in W_2$.

Using Lemma 13 we get $v_2 \in \Lambda(F_1, H)$ such that $\|v_2\| \le C$ and $|v_2(f)(t)| \le \frac{1}{4}\varepsilon\|f\|$ for $t \in T\backslash W_2$. We define $u_2 \in \Lambda(G, H)$ as $u_2(g) = u(\pi g) + v_2(g - \pi(g))$.

Repeating this procedure N times we obtain a decreasing sequence of open sets $W_1 \supset W_2 \supset \cdots \supset W_{N+1} \supset S$ and a sequence of extensions $u_j = u \circ \pi + v_j \circ (id - \pi) \in \Lambda(G, H)$, $j = 1, 2, \ldots, N$, such that $u_j|F = u$, $j = 1, 2, \ldots, N$ and

$$|u_j(g)(t)| \le \begin{cases} \left(1 + \frac{\varepsilon}{4}\right)\|g\| & \text{for} & t \in W_{j+1}, \\ \left(\|u\| + \frac{\varepsilon}{2}\right)\|g\| & \text{for} & t \in T\backslash W_j, \\ (\|u\| + 2C)\|g\| & \text{otherwise.} \end{cases} \quad (4)$$

The desired extension $\tilde{u}$ is defined as $\tilde{u} = N^{-1}\sum_{j=1}^{N} u_j$. Obviously $\tilde{u}|F = u$. From (4) we see that for any given $t \in T$, we have

$|u_j(g)(t)| > \max(1, \|u\|) + \frac{\varepsilon}{2}$ for at most one index j, so we obtain that $\|u\| \leq \max(1, \|u\|) + \varepsilon$, provided N was big enough.   ∎

Remark.   The same argument gives for every $e \in E$ and every $\varepsilon > 0$ and every open set $W \supset S$ an extension $h \in H$ such that $\|h\| \leq 2\|e\|$ and $|h(t)| \leq \varepsilon$ for $t \in T\backslash W$.

15 Corollary.   *If (E, H) has the bounded extension property and E is a separable π_1-space then there exists a linear extension operator $u \colon E \to H$.*

Proof: In E we have an increasing sequence of finite dimensional subspaces E_n and a sequence of projections $\pi_n \colon E \overset{\text{onto}}{\longrightarrow} E_n$ with $\|\pi_n\| = 1$. Using Proposition 14 with ε_n such that $\Sigma\varepsilon_n < \infty$ we get a sequence of extensions $u_n \in \Lambda(E_n, H)$ with $u_n|E_k = u_k$ for $k < n$ and $\sup_n \|u_n\| < \infty$. Since $\overline{\cup E_n} = E$ we infer that $u(f) = \lim_{n \to \infty} u_n(f)$ extends to a well defined linear extension operator on E.   ∎

16.   Being a π_1-space is a rather restrictive condition. It is difficult however to modify the above proof using the weaker approximation condition. Nevertheless the following theorem is true, albeit with a rather roundabout proof.

Theorem.   *Let T be a topological space and let $S \subset T$ be a closed subset. Let $E \subset C(S)$ and $H \subset C(T)$ be closed linear subspaces and let (E, H) have the bounded extension property. Assume that E is separable and has the bounded approximation property. Then there exists a linear extension operator $u \colon E \to H$.*

Proof: We will deduce Theorem 16 from Corollary 15 applied to a properly enlarged space. Let $\omega = \{1, 2, 3, \ldots\} \cup \{\infty\}$ be the one-point compactification of the natural numbers, and let $\widetilde{S} = S \times \omega \subset \widetilde{T} = T \times \omega$. Let $T_n \colon E \to E$, $n = 1, 2, \ldots$, be a sequence of finite dimensional operators with $T_n(e) \to e$ for $e \in E$. Denote $E_n = \text{span} \bigcup_{k=1}^{n} T_k(E)$. We define $\widetilde{E} \subset C(\widetilde{S})$ by

$$\widetilde{E} = \{(f_\gamma)_{\gamma \in \omega} : f_\infty \in E, f_n \in E_n, \ n = 1, 2, \ldots,$$
$$\text{and } f_n \to f_\infty \text{ as } n \to \infty\}. \tag{5}$$

We also define

$$\widetilde{H} = \{(f_\gamma)_{\gamma \in \omega} : f_\gamma \in H \text{ for } \gamma \in \omega$$
$$\text{and } f_n \to f_\infty \text{ as } n \to \infty\}.$$

One easily sees that $\widetilde{E}$ is a closed subspace of $C(\widetilde{S})$ and $\widetilde{H}$ is a closed subspace of $C(\widetilde{T})$.

Claim.   The pair $(\widetilde{E}, \widetilde{H})$ has the bounded extension property.

Proof of the claim.   If $W \subset \widetilde{T}$ is an open set containing $\widetilde{S}$ then there exists an open set $W_\infty \subset T$, such that $W_\infty \supset S$ and $W_\infty \times \omega \supset \widetilde{S}$. Given $(f_\gamma)_{\gamma \in \omega} \in \widetilde{E}$ and $\varepsilon > 0$ we can find $g_\infty \in H$ such that $g_\infty|S = f_\infty$ and $|g_\infty(t)| < \frac{\varepsilon}{2}$ for $t \in T \backslash W_\infty$. Using the remark after Proposition 14 we can find $h_n \in H$ with $\|h_n\| \leq 2\|f_\infty - f_n\|$, $h_n|S = f_\infty - f_n$ and $|h_n(t)| < \frac{\varepsilon}{2}$ for $t \in T \backslash W_\infty$. Since $\|f_\infty - f_n\| \longrightarrow 0$ as $n \to \infty$ we also have $\|h_n\| \longrightarrow 0$ as $n \to \infty$. The desired extension $(g_\gamma)_{\gamma \in \omega}$ is defined by $g_n = g_\infty - h_n$.

Returning to the proof of the theorem let us observe that $\widetilde{E}$ is a π_1-space. To see this we define $\widetilde{E}_n = \{(f_\gamma)_{\gamma \in \omega} \in \widetilde{E}: f_k = f_n$ for $k \geq n\}$. The projections $P_n: \widetilde{E} \overset{\text{onto}}{\longrightarrow} \widetilde{E}_n$ are defined by $P_n((f_\gamma)_{\gamma \in \omega}) = (f_1, f_2, \ldots, f_n, f_n, \ldots, f_n)$. Obviously $\|P_n\| = 1$ and $\cup \widetilde{E}_n$ is dense in $\widetilde{E}$. From Corollary 15 we get a linear extension operator $\tilde{u}: \widetilde{E} \to \widetilde{H}$. We define

$$u: E \to H \quad \text{by} \quad u(f) = \tilde{u}(T_1(f), T_2(f), \ldots, f)|T \times \{\infty\}. \quad \blacksquare$$

17 Corollary. *If $S \subset T$ and S is compact metric and T is normal then there exists a linear extension operator $u: C(S) \to C(T)$.*

Proof: Since S is compact metrizable the space $C(S)$ is separable. The Tietze extension theorem implies that $(C(S), C(T))$ has the bounded extension property. Since $C(S)$ has the bounded approximation property (see II.E.5(c)) the corollary follows. $\blacksquare$

18.   The above corollary exhibits many complemented subspaces of $C(K)$-spaces. One more, different and very important example is provided by the following.

Proposition. (Milutin)   *Let $\mathbb{T}^\mathbb{N}$ denote the countable product of circles. The space $C(\Delta)$ contains a 1-complemented copy of $C(\mathbb{T}^\mathbb{N})$.*

Proof: Let us identify Δ with $\{-1, 1\}^\mathbb{N}$. By $\tilde{\rho}$ we mean the classical Cantor map from Δ onto $\mathbb{T}$, i.e. $\tilde{\rho}((\varepsilon_j)_{j=1}^\infty) = 2\pi \sum_{j=1}^\infty \frac{1}{2}(1 + \varepsilon_j)2^{-j}$. It is an easy and well known fact that there exists a map $\tilde{\tau}: \mathbb{T} \to \Delta$ such that $\tilde{\rho}\tilde{\tau} = id_{\mathbb{T}}$ and $\tilde{\tau}$ is measurable and continuous on $\mathbb{T}\backslash D$ where D is a countable set of dyadic points. Since Δ is homeomorphic to Δ^N, we infer (take products) that there exists a continuous map $\rho\colon \Delta \overset{\text{onto}}{\longrightarrow} \mathbb{T}^N$ and a measurable map $\tau\colon \mathbb{T}^N \to \Delta$ such that $\rho\tau = id_{\mathbb{T}^N}$ and τ is continuous on $\mathbb{T}^N\backslash D_\infty$ where D_∞ has measure 0. Note that both Δ and $\mathbb{T}^N$ have a natural group structure, so we can perform algebraic operations. We define the isometric embedding $I\colon C(\mathbb{T}^N) \longrightarrow C(\Delta \times \Delta) \cong C(\Delta)$ as $I(f)(\alpha,\beta) = f(\rho(\alpha)+\rho(\beta))$. We define a norm-1 map $Q\colon C(\Delta \times \Delta) \longrightarrow C(\mathbb{T}^N)$ by

$$Q(g)(t) = \int_{\mathbb{T}^N} g(\tau(s), \tau(t-s))ds.$$

To see that $Q(g) \in C(\mathbb{T}^N)$ let us take a sequence $t_n \in \mathbb{T}^N, t_n \to t$. Then

$$g(\tau(s), \tau(t_n - s)) \longrightarrow g(\tau(s), \tau(t - s)) \text{ as } n \to \infty$$

for all $s \in \mathbb{T}^N\backslash(\bigcup_{n=1}^\infty (t_n - D_\infty) \cup (t - D_\infty) \cup D_\infty)$, i.e. for almost all $s \in \mathbb{T}^N$. By the Lebesgue dominated convergence theorem

$$Q(g)(t_n) = \int_{\mathbb{T}^N} g(\tau(s), \tau(t_n-s))ds \longrightarrow \int_{\mathbb{T}^N} g(\tau(s), \tau(t-s))ds = Q(g)(t)$$

as $n \to \infty$, so $Q(g)$ is continuous. Since

$$QI(f)(t) = \int_{\mathbb{T}^N} f(\rho(\tau(s)) + \rho(\tau(t - s)))ds = \int_{\mathbb{T}^N} f(t)ds = f(t)$$

the proposition follows.   ∎

This proposition should be compared with Exercise 4, which indicates that some ingenious embedding is needed.

19. The above proposition allows us to prove the following surprising result.

Theorem. (Milutin) *For every compact, metric, uncountable space K, the space $C(K)$ is isomorphic to $C(\Delta)$.*

Proof: As is well known, every uncountable, compact metric space K contains a subset K_1 homeomorphic to Δ (see Kuratowski [1968] or Semadeni [1971]). So by Corollary 17 the space $C(K)$ contains a complemented subspace isomorphic to $C(\Delta)$. It is also elementary and well known (cf. Kuratowski [1968] ch.4§41.VI.) that every compact metric space is homeomorphic to a subset of $\mathbb{T}^{\mathbb{N}}$ so in particular we obtain from Corollary 17 that $C(K)$ is isomorphic to a complemented subspace of $C(\mathbb{T}^{\mathbb{N}})$. This and Proposition 18 yield that $C(K)$ is isomorphic to a complemented subspace of $C(\Delta)$. Since $C(\Delta) \sim (\Sigma C(\Delta))_0$ Theorem II.B.24 gives the claim.   ∎

This theorem in particular implies that every $C(K)$-space for K compact, metric, and uncountable has a Schauder basis. Also, such a $C(K)$-space is isomorphic to $(\Sigma C(K))_0$.

20.   Our aim now is to present in some detail the orthonormal Franklin system. Usually it is constructed on $[0, 1]$. We will present the detailed construction on the circle (i.e. we will construct the periodic Franklin system). This will be useful for some of the future applications, in particular Theorem III.E.17. The reader interested in $[0,1]$ should be able to repeat the construction in this case without any difficulty.

We will identify the circle $\mathbb{T}$ with the interval $[0,1)$. For an integer $n = 2^k + j$, with $k = 0, 1, 2, \ldots$ and $0 \le j < 2^k$ we define $t_n = \frac{(2j+1)}{2^{k+1}}$ and we put $t_0 = 0$. The Franklin system $(f_n)_{n=0}^{\infty}$ is an orthonormal system of real valued, continuous, piecewise linear functions such that f_n has nodes at points t_j, $j = 0, 1, \ldots, n$, for $n = 0, 1, 2, \ldots$. This definition specifies f_n up to the sign. Let $\mathbf{F}_n = \mathrm{span}\{f_j\}_{j \le n}$. Clearly $\mathbf{F}_n$ is the space of all continuous, piecewise linear functions with nodes at $\{t_j\}_{j \le n}$. For a fixed n we will denote by $(s_j)_{j=0}^{n}$ the increasing renumbering of $(t_j)_{j=0}^{n}$, i.e. $0 = s_0 < s_1 < \cdots < s_n = 1$. Let $\mathbb{Z}_{n+1}$ denote the group of integers $0, 1, \ldots, n$ with addition mod $(n + 1)$. The natural group invariant distance $\rho(\cdot, \cdot)$ on $\mathbb{Z}_{n+1}$ is defined as $\rho(k, l) = \min(|k - l|, |n + 1 + l - k|, |n + 1 + k - l|)$, for $k, l = 0, 1, \ldots, n$.

We define also (for fixed n) the 'tent' functions Λ_j, $j \in \mathbb{Z}_{n+1}$ by the conditions $\Lambda_j \in \mathbf{F}_n$, $\Lambda_j(s_j) = 1$ and $\Lambda_j(s_k) = 0$ for $k \ne j$. Let us also note that

$$\frac{1}{2n} \le \mathrm{dist}(s_j, s_{j+1}) \le \frac{2}{n} \quad \text{for} \quad j \in \mathbb{Z}_{n+1}. \tag{6}$$

21.   Our main goal now is to establish the following technical proposition. It describes the behaviour of an individual Franklin function and of the integral kernel of the partial sum projection. This proposition will allow us to investigate the properties of the Franklin-Fourier series $\sum_{n=0}^{\infty} \langle f, f_n \rangle f_n$ for f in various classes of functions.

Proposition.   *There exist constants $C < \infty$ and $q < 1$ such that for every $n = 0, 1, 2 \ldots$, we have*

(a) $|f_n(t)| \le C\sqrt{n+1}\, q^{n\cdot\text{dist}(t,t_n)}$,

(b) $|\sum_{n=0}^{n} f_j(x)f_j(y)| \le C(n+1)q^{n\cdot\text{dist}(x,y)}$.

Before we start the proof we need some preliminary remarks. Let us write

$$K_n(x,y) = \sum_{j=0}^{n} f_j(x)f_j(y) = \sum_{i,j\in\mathbf{Z}_{n+1}} \alpha_{ij}\Lambda_i(x)\Lambda_j(y). \qquad (7)$$

Since $P_n f(x) = \int_{\mathbf{T}} f(y)K_n(x,y)dy$ is an orthogonal projection onto the space $\mathbf{F}_n$ we see that $P_n(\Lambda_j) = \Lambda_j$, for $j \in \mathbf{Z}_{n+1}$. From this we infer that $(\alpha_{ij})_{i,j\in\mathbf{Z}_{n+1}}$ is the inverse matrix to the matrix $\left(\int_{\mathbf{T}} \Lambda_i(x)\Lambda_j(x)dx\right)_{i,j\in\mathbf{Z}_{n+1}}$. The key to the proof of Proposition 21 is the analysis of the matrix $(\alpha_{ij})_{i,j\in\mathbf{Z}_{n+1}}$. This is what we will do now.

22 Lemma. *There exists a constant C (independent of n) such that the matrix*

$$A_n = \left((n+1)\int_{\mathbf{T}} \Lambda_i(x)\Lambda_j(x)dx \right)_{i,j\in\mathbf{Z}_{n+1}} \qquad (8)$$

defines .an operator A_n: $\ell_2(\mathbf{Z}_{n+1}) \to \ell_2(\mathbf{Z}_{n+1})$ with $\|A_n\| \le C$ and $\|A_n^{-1}\| \le C$.

Proof: One checks that for any affine function $f(t)$ on the interval $[a,b]$ we have

$$\frac{1}{6}(b-a)[|f(a)|^2 + |f(b)|^2] \le \int_a^b |f(x)|^2 dx \le \frac{1}{2}(b-a)[|f(a)|^2 + |f(b)|^2]. \qquad (9)$$

Define an operator S: $\ell_2(\mathbf{Z}_{n+1}) \to L_2(\mathbf{T})$ by $S(e_j) = \sqrt{n+1}\Lambda_j$, for $j \in \mathbf{Z}_{n+1}$. From (6) and (9) we obtain

$$\frac{1}{\sqrt{2}}\|S(x)\| \le \|x\| \le 2\sqrt{3}\|S(x)\| \quad \text{for} \quad x \in \ell_2(\mathbf{Z}_{n+1}). \qquad (10)$$

Since $A_n = S^*S$ (10) implies the claim.   ■

23. The important feature of the matrix A given by (8) is that $\int_{\mathbf{T}} \Lambda_i(x)\Lambda_j(x)dx = 0$ if $\rho(i,j) > 1$. We say that the matrix $(a_{ij})_{i,j\in\mathbf{Z}_{n+1}}$ is m-banded if $a_{ij} = 0$ for $\rho(i,j) > m$. One easily sees that the inverse of an m-banded matrix (if it exists at all) need not be banded. The following proposition shows that something remains. The entries of the inverse of an m-banded matrix are exponentially small far away from the diagonal.

Proposition. Let $A = (a_{ij})_{i,j \in \mathbf{Z}_{n+1}}$ be an m-banded invertible matrix with $\|A\| \le 1$ and $\|A^{-1}\| \le C$ where the norm is understood as the operator norm on $\ell_2(\mathbf{Z}_{n+1})$. There exist numbers $K = K(C,m)$ and $q = q(C,m)$, $0 < q < 1$ such that

$$|b_{ij}| \le K q^{\rho(i,j)} \quad \text{where} \quad A^{-1} = (b_{ij})_{i,j \in \mathbf{Z}_{n+1}}. \tag{11}$$

The proof of this proposition uses the following easy approximation fact.

24 Lemma. Let $0 < a < b$ and let $f(x) = \frac{1}{x}$. Then $a_n = \inf\{\|f - p\|_{C[a,b]}: p$ is an algebraic polynomial of degree $l \le n\} \le K q^{n+1}$ for some $K = K(a,b)$ and $q = q(a,b) < 1$.

Proof: Let $c = \frac{(a+b)}{2}$. Then $\frac{c}{x} = \sum_{k=0}^{\infty}(\frac{(c-x)}{c})^k$ and this series converges in $C[a,b]$. We have

$$a_n \le \frac{1}{c} \sum_{k=n+1}^{\infty} \left\| \left(\frac{c-x}{c}\right)^k \right\|_{C[a,b]} = \frac{1}{c} \sum_{k=n+1}^{\infty} \left(\frac{b-a}{b+a}\right)^k$$

$$\le K \left(\frac{b-a}{b+a}\right)^{n-1}. \qquad \blacksquare$$

Proof of Proposition 23. First note that if A_1 is m_1-banded and A_2 is m_2-banded then $A_1 A_2$ is $(m_1 + m_2)$-banded. In particular AA^* is a positive definite $2m$-banded invertible matrix with $(AA^*)^{-1} = (u_{ij})_{i,j \in \mathbf{Z}_{n+1}}$. Obviously $\sigma(AA^*) \subset [C^{-2}, 1]$ so from the spectral mapping theorem and Lemma 24 there exists a sequence of algebraic polynomials $(p_k)_{k=0}^{\infty}$ with $\deg p_k \le k$ such that

$$\|(AA^*)^{-1} - p_k(AA^*)\| \le K q^{k+1}.$$

For a given pair $(i,j) \in \mathbf{Z}_{n+1} \times \mathbf{Z}_{n+1}$ we take the largest integer k such that $2k \cdot m < \rho(i,j)$. Since $p_k(AA^*)$ is $2km$-banded it has a zero entry at the place (i,j) so

$$|u_{ij}| \le \|(AA^*)^{-1} - p_k(AA^*)\| \le K q^{k+1} \le C q_1^{\text{dist}(i,j)}.$$

Since the product of an m_1-banded matrix and a matrix satisfying (11) also satisfies (11) (for different K and $q < 1$) the formula $A^{-1} = A^*(AA^*)^{-1}$ gives the claim. $\blacksquare$

Proof of Proposition 21.   Part (b) follows immediately from Lemma 22, Proposition 23 and (7). Given n let $\Lambda(t) \in \mathbf{F}_{n+1}$ be such that

$$\Lambda(t_{n+1}) = 1 \quad \text{and} \quad \Lambda(t_j) = 0 \quad \text{for} \quad j = 0, 1, \ldots, n.$$

Clearly $f_{n+1} = \|\Lambda - P_n\Lambda\|_2^{-1}(\Lambda - P_n(\Lambda))$. From (7) and Proposition 23 we get $|P_n\Lambda(x)| \leq Cq^{(n+1) \operatorname{dist}(x,t_{n+1})}$ so also $|\Lambda(x) - P_n\Lambda(x)| \leq Cq^{(n+1) \operatorname{dist}(x,t_{n+1})}$ for some $C < \infty$ and $q < 1$. Since

$$\|\Lambda - P_n\Lambda\|_2 \geq \inf_{a,b} \left(\int_{\operatorname{supp}\Lambda} |\Lambda(x) - ax - b|^2 dx \right)^{\frac{1}{2}}$$
$$= C\sqrt{|\operatorname{supp}\Lambda|} \geq C\frac{1}{\sqrt{n+1}}$$

the proposition follows.   ∎

25 Theorem.   The Franklin system $(f_n)_{n=0}^\infty$ is a Schauder basis in the space $C(\mathbb{T})$.

Proof: Clearly $\operatorname{span}\{f_n\}_{n=0}^\infty = C(\mathbb{T})$. In order to estimate the norm of P_n (see Proposition II.B.8) we have from Proposition 21(b)

$$|P_nf(x)| \leq \left| \int_{\mathbb{T}} \sum_{j=0}^n f_j(x)f_j(y)f(y)dy \right|$$
$$\leq \|f\|_\infty \int_{\mathbb{T}} \left| \sum_{j=0}^n f_j(x)f_j(y) \right| dy$$
$$\leq C\|f\|_\infty \cdot (n+1) \int_0^\infty q^{nx} dx$$
$$\leq C\|f\|_\infty.$$ ∎

26 Corollary.   The Franklin system is a Schauder basis in $L_p(\mathbb{T})$ for, $1 \leq p < \infty$.

Proof: The case $p = 1$ follows by duality from Theorem 25, because the Franklin system is orthonormal. It also follows from Theorem 25 that the partial sum projections P_n are uniformly bounded on $L_\infty(\mathbb{T})$. Thus the Riesz-Thorin theorem I.B.6 gives that they are uniformly bounded on $L_p(\mathbb{T})$ for $1 < p < \infty$. This gives the claim; see Proposition II.B.8.
                                                                        ∎

27. Our next application of the Franklin system is to the spaces $Lip_\alpha(\mathbb{T})$, $0 < \alpha < 1$. We have

Theorem. *The space $Lip_\alpha(\mathbb{T})$, $0 < \alpha < 1$, is isomorphic to ℓ_∞. The isomorphism is given explicitly as $f \mapsto \left((n+1)^{\alpha+\frac{1}{2}} \int_\mathbb{T} f(x)f_n(x)dx\right)_{n=0}^\infty$.*

28 Lemma. *There exists a constant C such that for $n = 0, 1, 2, \ldots$, the circle $\mathbb{T}$ can be partitioned into intervals (I_j^n) with $|I_j^n| < \frac{C}{(n+1)}$ and such that $\int_{I_j^n} f_n(t)dt = 0$ for all j.*

Proof: Since f_n is orthogonal to $\mathbf{F}_{n-1}$ **(20)** the function $\varphi(x) = \int_0^x f_n(t)dt$ is orthogonal to all step functions with discontinuities at $t_0, t_1, \ldots, t_{n-1}$. This implies that $\varphi(x)$ has a zero in each interval (s_j, s_{j+1}). The zeros of $\varphi(x)$ give the endpoints of the desired intervals (I_j^n). ∎

Proof of Theorem 27. For a given integer n, let $(I_j)_{j=1}^s$ be the intervals given by Lemma 28. Fix points $u_j \in I_j$, $j = 1, \ldots, s$. For $f \in Lip_\alpha(\mathbb{T})$ we have from Lemma 28 and Proposition 21(a)

$$\left| \int_\mathbb{T} f(t)f_n(t)dt \right| = \left| \sum_{j=1}^s \int_{I_j} f(t)f_n(t)dt \right|$$

$$\leq \sum_{j=1}^s \int_{I_j} |f(t) - f(u_j)| \, |f_n(t)|dt$$

$$\leq \sum_{j=1}^s \max_{t \in I_j} |f(t) - f(u_j)| \int_{I_j} |f_n(t)|dt$$

$$\leq \|f\|_{Lip\alpha} \frac{C}{(n+1)^\alpha} \int_\mathbb{T} |f_n(t)|dt$$

$$\leq C\|f\|_{Lip_\alpha} (n+1)^{-\alpha-\frac{1}{2}}.$$

In order to show the other estimate let us fix a sequence of scalars $(a_n)_{n=0}^\infty$ such that $|a_n| \leq (n+1)^{-\alpha-\frac{1}{2}}$. Let us write

$$f = \sum_{n=0}^\infty a_n f_n = a_0 f_0 + \sum_{k=0}^\infty \sum_{2^k}^{2^{k+1}-1} a_n f_n = a_0 f_0 + \sum_{k=0}^\infty F_k.$$

From Proposition 21(a) we infer

$$|F_k(t)| \leq \sum_{2^k}^{2^{k+1}-1} |a_n| \, |f_n(t)| \leq C2^{-k\alpha}. \tag{12}$$

Each $F_k(t)$ is a continuous, piecewise linear function with nodes at least 2^{-k-1} apart, so

$$|F_k(t_1) - F_k(t_2)| \le C \cdot 2^{-k\alpha} \cdot 2^{k+1} \ \text{dist}(t_1, t_2). \tag{13}$$

For $t_1, t_2 \in \mathbb{T}$ let N be such that $2^{-N-1} < \text{dist}(t_1, t_2) \le 2^{-N}$. Using (12) and (13) we have

$$|f(t_1) - f(t_2)| \le |a_0| \ \text{dist}(t_1, t_2) + \sum_{k=0}^{N} |F_k(t_1) - F_k(t_2)| + 2 \sum_{N+1}^{\infty} \|F_k\|_\infty$$

$$\le C \ \text{dist}(t_1, t_2)^\alpha + C \sum_{k=0}^{N} 2^{-k\alpha} 2^{k+1} \ \text{dist}(t_1, t_2) + C \sum_{k=N+1}^{\infty} 2^{-k\alpha}$$

$$\le C \ \text{dist}(t_1, t_2)^\alpha + C 2^{-N\alpha} \le C \ \text{dist}(t_1, t_2)^\alpha. \qquad \blacksquare$$

29. Now we will introduce a special class of subspaces of $C(S, E)$, the space of continuous E-valued functions on S where E is a finite dimensional Banach space. Before we proceed a few comments about notation are in order. The norm in E will be denoted by $|\cdot|$. Actually, since all finite dimensional spaces of a given dimension are isomorphic different choices of norm in E give isomorphic spaces $C(S, E)$. Any linear functional on $C(S, E)$ can be identified with an E^*-valued measure. Really, we can identify E with $\mathbb{R}^n$ or $\mathbb{C}^n$ using the Auerbach lemma (see II.E.11), so then each functional on $C(S, E)$ can be identified with a sequence $(\mu_1, \dots, \mu_n)$ of measures. For every E^*-valued measure $\mu = (\mu_1, \dots, \mu_n)$ there exists a positive measure $|\mu|$ such that for $f \in C(S, E)$ we have

$$\left| \int_S f d\mu \right| \le \int_S |f| d|\mu|. \tag{14}$$

and $\|\mu\|_{C(S,E)^*} = \| |\mu| \|$. This requires some work, but the measure $\sum_{j=1}^{n} |\mu_j|$ clearly satisfies (14) and also $\| \sum_{j=1}^{n} |\mu_j| \| \le C\|\mu\|_{C(S,E)^*}$. This is all we will need.

Definition. *Let E be a finite dimensional Banach space and let S be a compact space. A subspace $X \subset C(S, E)$ is called rich if there exists a probability measure σ on S such that for every $\varphi \in C(S)$ and every bounded sequence $(x_n)_{n=1}^\infty \subset X$ such that $\int_S |x_n| d\sigma \longrightarrow 0$ as $n \to \infty$ we have $\text{dist}(\varphi \cdot x_n, X) \longrightarrow 0$ as $n \to \infty$.*

As we will see later, rich subspaces of $C(S, E)$ and their duals share with $C(K)$-and $L_1(\mu)$-spaces many properties related to weak compactness. Also some important spaces can be realized as rich subspaces of $C(S, E)$.

30. Examples (a) $C(S, E)$ itself is obviously rich.

(b) Let us consider the space $C^k(\mathbf{T}^s)$, $k \geq 0, s \geq 1$ (see I.B.30). If D is the set of all multiindices $i = (i_1, \ldots, i_s)$ such that i_j is a non-negative integer and $\sum_{j=1}^{s} |i_j| \leq k$ then we can embed $C^k(\mathbf{T}^s)$ into $C(\mathbf{T}^s, \ell_\infty^{|D|})$. The embedding Φ simply assigns to f the collection $(\frac{\partial^{|i|} f}{\partial_{i_1} x_1, \ldots, \partial_{i_s} x_s})_{i \in D} = \Phi(f)$. The subspace $\Phi(C^k(\mathbf{T}^s)) \subset C(\mathbf{T}^s, \ell_\infty^{|D|})$ is rich. To see this let us take a trigonometric polynomial $\varphi \in C(\mathbf{T}^s)$ and a function $f \in C^k(\mathbf{T}^s)$. It is easy to check that $\Phi(\varphi \cdot f) = \varphi \cdot \Phi(f) + G$ where G is the sum of derivatives of f of order $< k$ multiplied by some derivatives of φ. This shows that

$$\|\varphi \Phi(f) - \Phi(\varphi f)\| \leq C_\varphi \cdot \|f\|_{C^{k-1}(\mathbf{T}^s)}.$$

Take now a bounded sequence $(f_n)_{n=1}^\infty \subset C^k(\mathbf{T}^s)$ such that $\int_{\mathbf{T}^s} |\Phi(f_n)| dm \longrightarrow 0$ as $n \to \infty$ where m is the normalized Lebesgue measure on $\mathbf{T}^s$. Since $id: C^k(\mathbf{T}^s) \to C^{k-1}(\mathbf{T}^s)$ is a compact operator (see Exercise 24), we get $\|f_n\|_{C^{k-1}(\mathbf{T}^s)} \longrightarrow 0$ as $n \to \infty$. This shows that for every trigonometric polynomial φ we have

$$\text{dist}(\varphi \cdot \Phi(f_n), \Phi(C^k(\mathbf{T}^s))) \leq \|\varphi \Phi(f_n) - \Phi(\varphi \cdot f_n)\|$$
$$\leq C_\varphi \|f_n\|_{C^{k-1}(\mathbf{T}^s)} \longrightarrow 0.$$

Since the trigonometric polynomials are dense in $C(\mathbf{T}^s)$ the above relation holds for all $\varphi \in C(\mathbf{T}^s)$. This shows that $\Phi(C^k(\mathbf{T}^s))$ is a rich subspace.

(c) Let us consider $A(\mathbf{B}_d) \subset C(\mathbf{S}_d)$ where $\mathbf{B}_d$ is the ball in $\mathbf{C}^d$ and $\mathbf{S}_d$ is the unit sphere. Let $\mathcal{C}$ denote the Cauchy projection (I.B.21). Given φ, a polynomial in $z_1, \ldots, z_d, \bar{z}_1, \ldots, \bar{z}_d$, and a bounded sequence $(f_n)_{n=1}^\infty \subset A(\mathbf{B}_d)$ such that $\int_{\mathbf{S}_d} |f_n| d\sigma \to 0$ (σ is a normalized rotation invariant measure on $\mathbf{S}_d$) we get from III.B.20

$$\|\varphi \cdot f_n - \mathcal{C}(\varphi \cdot f_n)\|_\infty \longrightarrow 0 \text{ as } n \to \infty.$$

This gives that $A(\mathbf{B}_d)$ is a rich subspace of $C(\mathbf{S}_d)$.

31. The following theorem characterizes weakly compact subsets of X^*, where X is a rich subspace of $C(S, E)$. It is a generalization of Theorem III.C.12.

Theorem. *Suppose that $X \subset C(S, E)$ is a rich subspace and suppose that $K \subset X^*$ is a bounded subset. The following conditions are equivalent.*

(a) *K is not relatively weakly compact.*

(b) *There exists a sequence $(k_n)_{n=1}^{\infty} \subset K$ equivalent to the unit vector basis of ℓ_1.*

(c) *There exists a constant C such that for every N there exists a subset in K, C-equivalent to the unit vector basis in ℓ_1^N.*

(d) *There exists a weakly unconditionally convergent series $\sum_{n=1}^{\infty} \varphi_n$ in X such that*

$$\limsup_{n \to \infty} \sup\{|k(\varphi_n)|: \ k \in K\} > 0.$$

(e) *There exists a sequence $(x_n)_{n=1}^{\infty} \subset X, x_n \xrightarrow{w} 0$ such that*

$$\limsup_{n \to \infty} \sup\{|k(x_n)|: \ k \in K\} > 0.$$

We will prove the following implications (the implications marked $(*)$ are obvious)

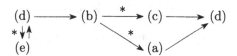

(d)$\Rightarrow$(b). This is still easy. The w.u.c series $(\varphi_n)_{n=1}^{\infty} \subset X$ defines an operator $T\colon X^* \longrightarrow \ell_1$ as $T(x^*) = (x^*(\varphi_n))_{n=1}^{\infty}$. It follows from (d) that $T(K)$ is not norm-compact in ℓ_1 so (b) follows from Theorem III.C.9 (see Exercise III.C.13).

(e)$\Rightarrow$(d). Using (e) let us fix a $\delta > 0$, a sequence $(x_n^*)_{n=1}^{\infty} \subset K$ and a sequence $(x_n)_{n=1}^{\infty} \subset X$ such that $x_n \xrightarrow{w} 0$ and $x_n^*(x_n) > \delta$ and $\|x_n\| \leq 1$ for $n = 1, 2, \ldots$. Let us also fix a sequence of positive numbers $(\varepsilon_n)_{n=1}^{\infty}$ such that $\sum_{n=1}^{\infty} \varepsilon_n < \frac{\delta}{10}$. Let μ_n be the Hahn-Banach extension of x_n^*, $n = 1, 2, \ldots$ and let $\gamma = \sup\{\|\sigma + |\mu_n|\| : n = 1, 2, \ldots\}$. Note also that since $x_n \xrightarrow{w} 0$ we get (see Excercise II.A.3) that $|x_n| \xrightarrow{w} 0$ in $C(S)$. Thus there exists a finite convex combination such that $\|\sum_{n=1}^{N} \lambda_j |x_j| \|_{\infty} < \frac{\varepsilon_1}{\gamma}$ (see II.A.5). From this we infer that there exist n_1, $1 \leq n_1 \leq N$ and an infinite set $\mathbb{N}_1 \subset \mathbb{N}$ with $\min \mathbb{N}_1 > N$ such that

$$\int |x_{n_1}| d(|\mu_n| + \sigma) < \varepsilon_1 \quad \text{for} \quad n \in \mathbb{N}_1. \tag{15}$$

Repeating the same argument we find inductively a sequence $(n_k)_{k=1}^{\infty}$ such that

$$\int |x_{n_k}|d|\mu_{n_s}| < \varepsilon_k \quad \text{for all} \quad s > k, \tag{16}$$

$$\int |x_{n_k}|d\sigma \text{ is so small that } \operatorname{dist}\Big(\prod_{j=1}^{k-1}(1 - |x_{n_j}|)x_{n_k}, X \Big) < \varepsilon_k. \tag{17}$$

Let $\varphi_k \in X$ be such that $\| \prod_{j=1}^{k-1}(1 - |x_{n_j}|)x_{n_k} - \varphi_k \| < \varepsilon_k$. The series $\sum_{k=1}^{\infty} \varphi_k$ is w.u.c. because from (17) we get

$$\sum_{k=1}^{\infty} |\varphi_k| \leq \sum_{k=1}^{\infty} \varepsilon_k + \sum_{k=1}^{\infty} \prod_{j=1}^{k-1}(1 - |x_{n_j}|)|x_{n_k}|$$

$$= \sum_{k=1}^{\infty} \varepsilon_k + \sum_{k=1}^{\infty} \Big(\prod_{j=1}^{k-1}(1 - |x_{n_j}|) - \prod_{j=1}^{k}(1 - |x_{n_j}|) \Big) \leq C.$$

From (17) we also have

$$|x_{n_k}^*(\varphi_k)| = \left| \int \varphi_k d\mu_{n_k} \right| \geq \left| \int \prod_{j=1}^{k-1}(1 - |x_{n_j}|)x_{n_k} d\mu_{n_k} \right| - \varepsilon_k$$

$$\geq \left| \int x_{n_k} d\mu_{n_k} \right| - \left| \int \Big(1 - \prod_{j=1}^{k-1}(1 - |x_{n_j}|)\Big)x_{n_k} d\mu_{n_k} \right| - \varepsilon_k. \tag{18}$$

Since for $0 \leq \alpha_j \leq 1$ we have $\quad 1 - \prod_{j=1}^{s}(1 - \alpha_j) \leq \sum_{j=1}^{s} \alpha_j$, $\quad$ from (18) and (16) we infer that

$$|x_{n_k}^*(\varphi_k)| \geq \delta - \sum_{j=1}^{k} \varepsilon_k \geq \frac{\delta}{2}.$$

(c)$\Rightarrow$(d) and (a)$\Rightarrow$(d). The proof of both implications is practically the same. Both (a) and (c) imply the following.

($*$) There exist elements $(x_1^n, \ldots, x_n^n) \subset X$, and $(k_1^n, \ldots, k_n^n) \subset K$, $n = 1, 2, \ldots$ and numbers C and $\delta > 0$ such that

$$\|k_j^n\| \leq C \qquad \text{for } n = 1, 2, \ldots \text{ and } j = 1, \ldots, n, \tag{19}$$

$$\|x_j^n\| \leq 1 \qquad \text{for } n = 1, 2, \ldots, \text{ and } j = 1, 2, \ldots, n, \tag{20}$$

$$|k_j^n(x_k^n)| \leq \frac{\delta}{3} \qquad \text{if } k > j \text{ for } n = 1, 2, \ldots, \tag{21}$$

$$k_j^n(x_k^n) \geq \delta \qquad \text{if } k \leq j \text{ for } n = 1, 2, \ldots . \tag{22}$$

When we assume (a) then $(*)$ follows from the proof of the Eberlein-Šmulian theorem II.C.3. When we assume (c), then we take $(k_j^n)_{j=1}^n$ to be vectors in K equivalent to the unit vector basis in ℓ_1^n and take x_k^{**} such that $x_k^{**}(k_j^n) = 0$ if $k > j$ and $x_k^{**}(k_j^n) = 1$ if $k \leq j$. Clearly such x_k^{**} with norms uniformly bounded can be found, so the local reflexivity principle II.E.14 (and some renormalization) gives $(*)$.

To show $(*) \Rightarrow (d)$ we will need the following fact.

32 Lemma. *Let H be a Hilbert space and let $f_1, \ldots, f_n \in H$ be vectors with $\|f_j\| \leq 1$ for $j = 1, \ldots, n$. Then there exist non-empty sets $A, B \subset \{1, \ldots, n\}$ with $\max A < \min B$ such that*

$$\left\| \frac{1}{|A|} \sum_{j \in A} f_j - \frac{1}{|B|} \sum_{j \in B} f_j \right\| \leq \frac{4}{\sqrt{\log n}}. \tag{23}$$

Proof: Let us put

$$\alpha_k = \inf \left\{ \left\| \frac{1}{|A|} \sum_{j \in A} f_j \right\| : |A| = 2^k, A \subset \{1, 2, \ldots, n\} \right\}$$

and

$$\beta_k = \inf \left\{ \left\| \frac{1}{|A|} \sum_{j \in A} f_j - \frac{1}{|B|} \sum_{j \in B} f_j \right\| : \right.$$

$$A, B \subset \{1, \ldots, n\}, |A| = |B| = 2^k \text{ and } \max A < \min B \right\}$$

for $k = 0, 1, \ldots, [\log_2 n] - 1$.

Elementary geometry shows that for $x, y, z \in H$ with $x = \frac{(y+z)}{2}$ we have $\max(\|z\|^2, \|y\|^2) \geq \|x\|^2 + \|\frac{(y-z)}{2}\|^2$. This observation implies that $\alpha_k^2 \geq \alpha_{k+1}^2 + \frac{1}{4}\beta_k^2$ so

$$\alpha_0^2 \geq \alpha_{k+1}^2 + \frac{1}{4} \sum_{k=0}^{k} \beta_k^2 \geq \frac{1}{4} \sum_{k=0}^{[\log_2 n]-1} \beta_k^2.$$

Since $\alpha_0 \leq 1$ we get the claim.   ∎

$(*) \Rightarrow (d)$. Let us start with the observation (to be repeatedly used later) that if we divide the set of pairs (n, j), $j = 1, 2, \ldots, n$, $n =$

$1, 2, \ldots$, into a finite number of sets then at least one of those sets will contain a subset of the form

$$\{(n_k, j_s^k): n_k \text{ is an increasing sequence of integers and} \quad (24)$$
$$s = 1, \ldots, k \text{ and } j_1^k < j_2^k < \cdots < j_k^k\}.$$

Let us fix a sequence of positive numbers $(\varepsilon_k)_{k=1}^\infty$ such that $\sum_{k=1}^\infty \varepsilon_k < \frac{\delta}{10}$. Let (as previously) μ_j^n be the Hahn-Banach extension of k_j^n. Using Lemma 32 we find n_1 such that for every (n, j) with $n > n_1$ there are subsets $A, B \subset \{1, \ldots, n\}$ such that

$$\int \left| \frac{1}{|A|} \sum_{j \in A} x_j^{n_1} - \frac{1}{|B|} \sum_{j \in B} x_j^{n_1} \right| d(|\mu_j^n| + \sigma) = \int |z_1| d(|\mu_j^n| + \sigma) < \varepsilon_1. \quad (25)$$

Since there is a only finite number of subsets in $\{1, \ldots, n_1\}$ we find sets A and B such that (25) holds for this pair of subsets A, B on the set of the form (24). Thus (after renumbering) we can assume that there are vectors $(x_1^n, \ldots, x_n^n) \subset X$ and $(k_1^n, \ldots, k_n^n) \subset K$ for $n > n_1$, satisfying (19)-(22). We put $x_1^* = k_s^{n_1}$ where $s = \max A$. In the second step we analogously find $n_2 > n_1$, and $z_2 = |A|^{-1} \sum_{j \in A} x_j^{n_2} - |B|^{-1} \sum_{j \in B} x_j^{n_2}$ such that $\int |z_2| d(|\mu_j^n| + \sigma) < \varepsilon_2$ for all (after passing to appropriate subset as before) n, j with $n > n_2, j = 1, \ldots, n$, and such that $\int |z_2| d\sigma$ is so small that dist$((1 - |z_1|)z_2, X) < \varepsilon_2$. We put $x_2^* = k_s^{n_2}$ where $s = \max A$.

Continuing in this manner we find sequences $(z_j)_{j=1}^\infty \subset X$ and $(x_j^*)_{j=1}^\infty \subset K$ such that

$$|x_j^*(z_j)| > \frac{2}{3}\delta \quad \text{(this follows from (21) and (22)),} \quad (26)$$

$$\text{dist}\left(\prod_{j=1}^{k-1} (1 - |z_j|)z_k, X \right) < \varepsilon_k, \quad (27)$$

$$\int |z_j| |d\mu_k| < \varepsilon_j \quad \text{for} \quad k > j \quad \text{where} \quad (28)$$

μ_k is the Hahn-Banach extension of x_k^*.

Using (26), (27) and (28) we construct the desired weakly unconditionally convergent series exactly like in the proof of (e)$\Rightarrow$(d). ∎

33. Now we wish to cast the above considerations into a more general context. Let us introduce the following concepts. We say that a Banach space X has the Dunford-Pettis property (for short DP) if for every

$x_n \xrightarrow{\omega} 0$ in X and $x_n^* \xrightarrow{\omega} 0$ in X^* we have $x_n^*(x_n) \longrightarrow 0$. Clearly if X^* has DP then also X has DP. We say that a Banach space X has the Pełczyński property (for short P) if for every subset $K \subset X^*$ that is not relatively weakly-compact there exists a weakly unconditionally convergent series $\sum_{n=1}^\infty x_n$ in X such that $\inf_n \sup_{x^* \in K} x^*(x_n) > 0$. Clearly Theorem 31 shows that every rich subspace of $C(S, E)$ has DP and P. Also any $L_1(\mu)$ space has DP.

34.   We have the following, rather routine

Proposition.   *The following conditions on the Banach space X are equivalent:*

(a)   *X has the Dunford-Pettis property;*

(b)   *every weakly compact operator $T\colon X \to Y$ transforms weakly Cauchy sequences into norm Cauchy sequences;*

(c)   *every weakly compact operator $T\colon X \to c_0$ transforms weakly Cauchy sequences into norm Cauchy sequences.*

Proof: (a)$\Rightarrow$(b) Passing to differences it is enough to show that $\|Tx_n\| \to 0$ for every $x_n \xrightarrow{\omega} 0$. But if $x_n \xrightarrow{\omega} 0$ and $\|Tx_n\| \ge \delta > 0$ for $n = 1, 2, \ldots$, then we can take $y_n^* \in Y^*$ with $\|y_n^*\| = 1$ such that $y_n^* T(x_n) \ge \delta$. Passing to a subsequence we can assume that $y_n^* \xrightarrow{\omega^*} y^*$ for some $y^* \in Y^*$. But $y^*(Tx_n) \to 0$ so we can replace y_n^* by $y_n^* - y^*$ and additionally assume that $y_n^* \xrightarrow{\omega^*} 0$. But T^* is weakly compact (see II.C.6(b)) so $T^*(y_n^*) \xrightarrow{\omega} 0$. This contradicts (a) since $T^*(y_n^*)(x_n) = y_n^*(Tx_n) \ge \delta$.
   (b)$\Rightarrow$(c). Obvious.
   (c)$\Rightarrow$(a). Let us take $x_n^* \in X^*$ such that $x_n^* \xrightarrow{\omega} 0$ and define an operator $T\colon X \to c_0$ by $T(x) = \left(x_n^*(x)\right)_{n=1}^\infty$. Clearly $T^{**}(x^{**}) = (x^{**}(x_n^*))_{n=1}^\infty \in c_0$ so T is weakly compact by II.C.6(c). Applying (c) we get that for $x_n \xrightarrow{\omega} 0$ in $X, \|Tx_n\| \to 0$ so in particular $x_n^*(x_n) \to 0$. ∎

35 Proposition.   *Suppose X has the Pełczyński property. Then*

(a)   *X^* is weakly sequentially complete,*

(b)   *for every operator $T\colon X \to Y$ that is not weakly compact there exists a subspace $X_1 \subset X$ such that $X_1 \sim c_0$ and $T|X_1$ is an isomorphic embedding.*

Proof: If K is a subset of X^* that is not relatively weakly compact then there exists a sequence $\{x_n^*\}_{n=1}^\infty \subset K$ which is not relatively weakly compact (see II.C.3). Thus there exists a weakly unconditionally convergent series $\sum_{k=1}^{\infty} x_k$ in X and a subsequence $(x_{n_k}^*)_{k=1}^{\infty}$ such that $|x_{n_k}^*(x_k)| > \delta > 0$ for $k = 1, 2, \ldots$. From II.D.5 we see that we can additionally assume that $(x_k)_{k=1}^{\infty}$ is equivalent to the unit vector basis of c_0.

In order to prove (a) we take $K = \{x_n^*\}_{n=1}^{\infty}$ where x_n^* is weakly Cauchy but not weakly convergent. Let $T \colon c_0 \to X$ be defined by $T(e_k) = x_k$. Then $T^* \colon X^* \to \ell_1$ and one easily sees that $T^*(x_{n_k}^*)$ has no norm Cauchy subsequence, so by III.C.9 $T^*(x_{n_k}^*)$ has no weak Cauchy subsequence. This contradicts the fact that $(x_n^*)_{n=1}^{\infty}$ was weakly Cauchy.

In order to prove (b) let us put $K = T^*(B_{Y^*})$. Then $X_1 = \mathrm{span}(x_n)_{n=1}^{\infty}$ is clearly isomorphic to c_0 and for $x = \sum_{n=1}^{\infty} \alpha_n x_n$ we have

$$\left\| T\left(\sum_{n=1}^{\infty} \alpha_n x_n \right) \right\| = \sup_{x^* \in K} \left| x^*\left(\sum_{n=1}^{\infty} \alpha_n x_n \right) \right| \geq \delta \sup_n |\alpha_n|$$

so $T|X_1$ is an isomorphic embedding.   ∎

Notes and remarks.

The $C(K)$-spaces are among the most widely used Banach spaces. They are also the easiest examples of Banach algebras. As usual in this book we discuss the multiplicative structure only so far as it relates to the linear structure. Thus the well known *Proposition 1* serves as an introduction to the concept of an M-ideal. This concept was introduced by Alfsen-Effros [1972] where *Theorem 4* is also proved. Our proof is a modification of proofs given in Behrends [1979] and Lima [1982]. A similar proof and many applications of the theorem can be found in Gamelin-Marshal-Younis-Zame [1985]. We will present some other applications of the concept of an M-ideal in III.E. The space $H_\infty + C$ was introduced into analysis in the sixties by D. Sarason and A. Devinatz. *Corollary 7* is due to Sarason but the simple proof presented here is from Rudin [1975]; a similar proof is given in Garnett [1980]. The analysis of the particular example $H_\infty + C$ led to the general theory of Douglas algebras, i.e. closed algebras X such that $H_\infty \subset X \subset L_\infty$. $H_\infty + C$ is the smallest such algebra. It is a deep theorem of Marshall and Chang that each such X is the smallest algebra generated by H_∞ and complex conjugates of certain inner functions. For detailed information about all this we refer to Garnett [1980] or Sarason [1979].

Corollary 9 was proved by complicated operator-theoretical arguments by Axler-Berg-Jewell-Shields [1979]. The simple proof given here is due to Luecking [1980]. This started the investigation of M-ideals in Douglas algebras and other spaces connected with function theory. We refer the interested reader to Gamelin-Marshal-Younis-Zame [1985] and to the references quoted there. There are also important applications of the concept of M-ideal to the theory of C^*-algebras; see Choi-Effros [1977] or Alfsen-Effros [1972].

It seems that the first linear extension theorem was proved by Borsuk [1933] where he established a version of our *Corollary 17* and used it to show that $C[0,1] \sim (\Sigma C[0,1])_0$. Borsuk's argument was different and together with later improvements by Dugundji [1951] it gives the following

Theorem A. (Borsuk-Dugundji). *Let S be a closed non-empty subset of a metric space T and let X be a normed vector space. Then there exists a linear extension operator* $u\colon C(S,X) \longrightarrow C(T,X)$ *such that* $\|u\| = 1$ *and for every g in* $C(S,X)$ *the values of the function* $u(g)$ *belong to the convex hull of the set* $g(S)$.

Our *Corollary 15* was proved by Michael-Pełczyński [1967]. Actually it was proved with the additional information that $\|u\| = 1$. We decided not to present this improvement because we are mainly interested in isomorphic theory and our goal is *Theorem 16*. This theorem was proved by Ryll-Nardzewski (unpublished) and the proof we follow here was later given in Pełczyński-Wojtaszczyk [1971]. Davie [1976] used *Proposition 18* in his discussion of classification of operators on Hilbert space. It is his version of the proof that we present. It should be stressed, however, that questions of linear extensions are not limited to spaces with the sup-norm. There is an extensive literature on the existence and non-existence of linear extensions when other norms are involved, in particular Sobolev or Besov norms; see Stein [1970] or Triebel [1978] and the references quoted there.

Proposition 18 and *Theorem 19* are due to A.A. Milutin. He proved them in his Candidate of Sciences dissertation presented to the Moscow State University in 1952. Those results were only published in Milutin [1966]. These are important results. Some reasons for this opinion are as follows.

(a) Results of an isomorphic nature, once established for one 'simple' space K, like $K = \Delta$ or $K = \mathbb{T}$ are valid for $C(K)$ with more complicated compact, uncountable metric spaces K. One modest example of this is the comment made after *Theorem 19*.

(b) This is an important step in the programme of isomorphic classification of $C(K)$-spaces. For separable spaces $C(K)$, i.e. K-compact, metric such classification is known. For countable, metric compact spaces this was done in Bessaga-Pełczyński [1960]. The situation for non-separable spaces is more complicated and only partial results are known.

(c) The Milutin theorem shows that for most important compact sets K the multiplicative structure of $C(K)$ has nothing to do with its linear-topological structure. This contrasts sharply with the isometric situation. Namely we have the following.

Theorem B. *If $d(C(S), C(K)) < 2$ then S is homeomorphic to K, so in fact there exists a linear isometry i: $C(K) \overset{\text{onto}}{\longrightarrow} C(S)$ preserving the multiplication.*

This theorem under the assumption that $C(K)$ and $C(S)$ are actually isometric (with the a description of the isometries) was proved for metric K and S in Banach [1932] and for general K and S in Stone [1937]. This version was given independently by Amir [1965] and Cambern [1967].

The Franklin system was introduced in Franklin [1928], where *Theorem 25* was proved. We follow an approach developed by Ciesielski and Domsta in order to deal with systems of more general spline functions; see Ciesielski-Domsta [1972]. The Franklin system itself was earlier investigated in detail in Ciesielski [1963], [1966] where the fundamental *Proposition 21* was proved. *Proposition 23* was proved by Demko [1977]. The very ingenious proof presented here was given in Demko-Moss-Smith [1984].

Theorem 27 was first proved in Ciesielski [1960] using the Faber-Schauder system and later in Ciesielski [1963] using the Franklin system. The Franklin system is one of the most important orthonormal systems (see Kashin-Saakian [1984]).

The Dunford-Pettis property as defined in **33** was explicitly defined by Grothendieck [1953] who undertook an extensive study of this and related properties. He was directly influenced by the important paper Dunford-Pettis [1940] where among other things it was proved that every weakly compact operator T: $L_1[0,1] \rightarrow X$ maps weakly compact sets into norm-compact sets. Our *Proposition 34* and much more can be found in Grothendieck [1953]. The Pełczyński property (obviously under a different name, property V) appeared first in Pełczyński [1962], where he showed that $C(K)$ has P. The class of rich subspaces of $C(K, E)$ appeared in Bourgain [1984b]. In Bourgain [1983] and [1984b] our *Theorem 31* and *Examples 30* were proved. *Theorem 31* for the particular case

of the disc algebra was shown earlier by Delbaen [1977] and Kislyakov
[1975] (independently and almost simultaneously).

Clearly *Theorem 31* when applied to $C(K)$ gives information about
subsets of $L_1(\mu)$. This information is akin to that given in *Theorem III.C.12*. One can derive Theorem III.C.12 from *Theorem 31* but
even then many of the measure-theoretical arguments from the proof of
III.C.12 have to remain. We have chosen to present a separate proof
of *Theorem III.C.12* because it is an important theorem and the direct
argument is relatively simple. It stresses the important notion of uniform integrability which cannot appear explicitly in the more general
Theorem 31.

The paper Diestel [1980] contains a nice survey and exposition of
the Dunford-Pettis property and related topics. It does not however,
discuss *Theorem 31*.

Exercises

1. Show that the space of compact operators on ℓ_p, $1 < p < \infty$, is an
    M-ideal in $L(\ell_p)$.

2. Show that, if (E, H) has the bounded extension property then $H_0 =
    \{f \in H : f|S = 0\}$ is an M-ideal in H. (The notation agrees with
    10.)

3. Show that every $C(K)$-space, K compact, is a π_1-space.

4. Let $\varphi: \Delta \xrightarrow{\text{onto}} [0,1]$ be the classical Cantor map, i.e. if $\Delta =
    \{-1,1\}^{\mathbb{N}}$ then $\varphi(\xi_j) = \sum_{j=1}^{\infty} (\xi_j + 1)^{-1-j}$. Let $I_\varphi: C[0,1] \to C(\Delta)$
    be given by $I_\varphi(f) = f \circ \varphi$. Show that $I_\varphi(C[0,1])$ is uncomplemented
    in $C(\Delta)$.

5. Find two non-homeomorphic, compact metric spaces K_1 and K_2
    such that $d(C(K_1), C(K_2)) = 2$.

6. A matrix $(\alpha_{jk})_{j,k \geq 0}$ is called a Toeplitz matrix if $\alpha_{j,k} = \varphi(j - k)$.

    (a) Show that a Toeplitz matrix is a matrix of an operator on ℓ_2 if
        and only if $\varphi(s) = \hat{f}(s)$, $s = 0, \pm 1, \pm 2, \ldots$ for some $f \in L_\infty(\mathbb{T})$.

    (b) A matrix $(b_{jk})_{j,k \geq 0}$ is a Schur multiplier if for every matrix
        $(m_{jk})_{j,k \geq 0}$ of a linear operator on ℓ_2 the matrix $(b_{jk} \cdot m_{jk})_{j,k \geq 0}$
        is a matrix of a linear operator on ℓ_2. Show that the Toeplitz
        matrix is a Schur multiplier if and only if $\varphi(s) = \hat{\mu}(s), s =
        0, \pm 1, \pm 2, \ldots$ for some measure μ on $\mathbb{T}$.

(c) Show that the main triangle projection, i.e. the map $(a_{jk})_{j,k\geq0}$ $\to (b_{jk})_{j,k\geq0}$ where

$$b_{jk} = \begin{cases} a_{jk} & \text{if} \quad j \geq k, \\ 0 & \text{otherwise,} \end{cases}$$

is unbounded on $L(\ell_2)$.

7. Show that if $(n_k)_{k=1}^{\infty}$ is a lacunary sequence of integers and the Fourier series $\sum_{k=1}^{\infty} a_k e^{in_k\theta}$ represents a bounded function, then $\sum_{n=1}^{\infty} |a_k| < \infty$.

8. (Korovkin theorem). Suppose that $T_n : C[0,1] \to C[0,1]$ is a sequence of linear operators such that $\|T_n\| \to 1$ as $n \to \infty$ and $T_n(p) \to p$ as $n \to \infty$ for every quadratic polynomial p. Show that $T_n(f) \to f$ in norm for every function $f \in C[0,1]$.

9. For $f \in C[0,1]$ we put $B_n(f)(x) = \sum_{k=0}^{n} f(\frac{k}{n})\binom{k}{n}x^k(1-x)^{n-k}$. The operators B_n are called Bernstein operators.

   (a) Show that each B_n is a linear, norm-1 operator.

   (b) Show that for $f \in C[0,1]$ we have $B_n(f) \to f$ uniformly.

10. For $s > 0$ we define

$$X_s = \{f(z) : f \text{ is analytic in } \mathbb{D} \text{ and } |f(z)|(1-|z|)^s \in L_\infty(\mathbb{D})\}$$

and $X_s^0 = \{f \in X_s : |f(z)| = o(1-|z|)^{-s}\}$. Show that $X_s \sim \ell_\infty$, $X_s^0 \sim c_0$ and $(X_s^0)^{**} = X_s$.

11. Show that there exists a function $f \in A(\mathbb{B}_d), d \geq 1$, such that $\int_{\mathbb{B}_d} |Rf(z)|d\nu(z) = \infty$ where R is the radial derivative (see Exercise III.B.11) and ν is the Lebesgue measure on $\mathbb{B}_d$.

12. A sequence of finite dimensional Banach spaces $(X_n)_{n=1}^{\infty}$ is called a sequence of big subspaces of ℓ_∞^N if there exist constants C and α such that for each n there exists a subspace $\tilde{X}_n \subset \ell_\infty^{N_n}$ such that $d(X_n, \tilde{X}_n) \leq C$ and $N_n \leq \alpha \dim X_n$.

   (a) Let $T_n^\infty \subset C(\mathbb{T})$ be the space of trigonometric polynomials of degree $\leq n, n = 1, 2, \ldots$. Show that $(T_n^\infty)_{n=1}^{\infty}$ is a sequence of big subspaces of ℓ_∞^N.

   (b) Let $W_n^\infty(\mathbb{B}_d) \subset C(\mathbb{B}_d)$ be the space of all polynomials homogenous of degree n. Show that for every $d = 2, 3, \ldots$ the sequence $(W_n^\infty(\mathbb{B}_d))_{n=1}^{\infty}$ is a sequence of big subspaces of ℓ_∞^N.

(c) Show that, if $E \subset \ell_\infty^N, \dim E = n$ and $0 < \delta < 1$, then there exist an integer $k > \frac{\delta n}{2\sqrt{N}}$ and a subspace $G \subset E, \dim G = k$, such that $d(G, \ell_\infty^k) \leq \frac{(1+\delta)}{(1-\delta)}$.

13. Construct the system of quadratic splines analogous to the Franklin system. More precisely, construct an orthonormal system of functions $(g_n)_{n=0}^\infty \subset L_2(\mathbb{T})$ such that each g_n' is a continuous, piecewise linear function with nodes at points t_j, $j = 0, 1, \ldots, n$ for $n = 0, 1, 2, \ldots,$. (The points t_j are defined in **20**.)

14. Let $I \subset \mathbb{T}$ be an interval. The function $a_I(t)$ is defined as

$$
a_I(t) = \begin{cases} 0 \text{ if } t \notin I, \\ |I|^{-1} \text{ if } t \text{ is in the left hand half of } I, \\ -|I|^{-1} \text{ if } t \text{ is in the right hand half of } I. \end{cases}
$$

We define the space B as the space of all functions $f(t)$ which admit a representation $f(t) = \lambda_0 + \sum_{n=1}^\infty \lambda_n a_{I_n}(t)$ for some sequence of intervals $(I_n)_{n=1}^\infty$ and some sequence of scalars $(\lambda_n)_{n=0}^\infty$ with $\sum_{n=0}^\infty |\lambda_n| < \infty$. Then $\inf \sum_{n=0}^\infty |\lambda_n|$ over all representations of f is the norm denoted by $\|f\|_B$.

(a) Show that B is a Banach space.

(b) Show that $f \in B$ if and only if

$$
\sum_{n=0}^\infty (n+1)^{-\frac{1}{2}} \left| \int_{\mathbb{T}} f(t) f_n(t) dt \right| < \infty
$$

where $(f_n)_{n=0}^\infty$ is the Franklin system.

15. The space Λ_* (the Zygmund class) is defined as the space of all functions in $C(\mathbb{T})$ such that

$$
\|f\|_* = \sup \left\{ \left| \frac{f(x-h) + f(x+h) - 2f(x)}{h} \right| : x \in \mathbb{T}, h > 0 \right\} < \infty.
$$

Show that $f \in \Lambda_*$ if and only if $\left| \int_{\mathbb{T}} f(x) f_n(x) dx \right| = 0(n^{-\frac{3}{2}})$, where $(f_n)_{n=0}^\infty$ is the Franklin system.

16. Show that the Franklin system is not an unconditional basic sequence in $Lip_1(\mathbb{T})$.

17. Let $(f_n)_{n=1}^\infty$ be the Franklin system and let $f \in L_1(\mathbb{T})$. Show that the series $\sum_{n=0}^\infty \langle f, f_n \rangle f_n$ converges almost everywhere.

18. Let $A = \mathbb{Z} \cup \frac{1}{2}\mathbb{Z}_-$ and let us consider the subspace $V \subset L_2(\mathbb{R})$ consisting of all continuous, piecewise linear functions on $\mathbb{R}$ with nodes at the points of A. Let τ be a function which is continuous, piecewise linear with nodes at $A_0 = A \cup \{\frac{1}{2}\}$ and orthogonal to V. Assume also that $\|\tau\|_2 = 1$.

   (a) Show that $|\tau(x)| \le Cq^{|x|}$ for some $C > 0$ and $q < 1$.

   (b) Show that the family of functions $\{2^{j/2}\tau(2^j x - k)\}_{(j,k) \in \mathbb{Z} \times \mathbb{Z}}$ is a complete orthonormal system in $L_2(\mathbb{R})$.

19. Suppose that X is a separable Banach space with the Dunford-Pettis property. Assume also that $X \sim Y^*$ for some Banach space Y. Show that X has the Schur property, i.e. weakly convergent sequences converge in norm. It follows that L_1/H_1 and $L_1[0,1]$ are not isomorphic to a dual space.

20. Show that, spaces $L_1(\mu)$ and $C(K)$ do not have complemented, infinite dimensional, reflexive subspaces.

21. Find a Banach space X with the Dunford-Pettis property, such that X^* does not have it.

22. Show that if X is a rich subspace of $C(K)$, then X^* has the Dunford-Pettis property.

23. Show that every operator $T : \ell_\infty \to X$ for X a separable Banach space, is weakly compact.

24. Show that the identity operator $id : C^k(\mathbb{T}^s) \to C^{k-1}(\mathbb{T}^s), k \ge 1$, $s \ge 1$ is compact.

25. Show that $C^1(\mathbb{T}^2)^* = M \oplus F$ where M is isomorphic to $M(\mathbb{T})$ and F is separable.

26. Let us consider $H_\infty(\mathbf{S}) + C(\mathbf{S})$ where $\mathbf{S}$ is the unit sphere in $\mathbb{C}^d, d > 1$. Show that $H_\infty(\mathbf{S}) + C(\mathbf{S})$ is a closed subalgebra.

27. Let us consider the algebraic sum $H_\infty(\mathbb{T}^n) + C(\mathbb{T}^n)$. Show that it is closed in $L_\infty(\mathbb{T}^n)$ but is not a subalgebra if $n > 1$.

III.E. The Disc Algebra

First we study some interpolation problems in the disc algebra A. We describe those subsets $V \subset \overline{\mathbb{D}}$ for which we have $A|V = C(V)$. We also describe the sets $V \subset \mathbb{T}$ such that every $f \in C(V)$ can be extended to a function $F \in A$ with finite Dirichlet integral. We also show that every ℓ_2 sequence is a sequence of lacunary Fourier coefficients of a function in A. Next we show that $A \sim (\Sigma A)_0$. We show that A is not isomorphic to any $C(K)$-space. We present the construction of a Schauder basis in A and give different isomorphic representations of H_∞.

1. Let us recall that the disc algebra A is the space of all functions continuous for $|z| \leq 1$ and analytic for $|z| < 1$, equipped with the norm $\|f\| = \sup_{|z| \leq 1} |f(z)|$. The maximal modulus principle easily implies $\|f\| = \sup_{|z|=1} |f(z)|$ so A can be identified with $\text{span}\{e^{in\theta}\}_{n \geq 0} \subset C(\mathbb{T})$. The disc algebra appears prominently in many parts of analysis; it is the canonical example of a uniform algebra, the von Neumann inequality (see III.F.(25)) gives the functional calculus on A for every contraction on a Hilbert space, etc. It is also a very interesting Banach space.

From the F.-M. Riesz theorem I.B.26 we get immediately that

$$A^* = C(\mathbb{T})^*/A^\perp = L_1/\overline{H}_1^0 \oplus_1 M_s \qquad (1)$$

where M_s denote the space of measures on $\mathbb{T}$ which are singular with respect to the Lebesgue measure.

2. We will discuss some interpolation results for the disc algebra and other related spaces. More precisely, given a space X of analytic functions on $\mathbb{D}$ we look for sets $V \subset \mathbb{D}$ (or even $V \subset \overline{\mathbb{D}}$ if elements of X extend naturally to $\mathbb{T}$) such that the restrictions $X|V$ fill the space $C(V)$ or $\ell_\infty(V)$. We want to impose minimal, sensible conditions. The above description (1) of A^* easily yields the following result.

Proposition. (Rudin-Carleson) Let $\Delta = \overline{\Delta} \subset \mathbb{T}$ be a set of Lebesgue measure 0 and let $\varphi \in C(\mathbb{T})$ be a strictly positive function with $\varphi|\Delta = 1$. Then for every $f \in C(\Delta)$ there exists a $g \in A$ such that $g|\Delta = f$ and $|g(\theta)| \leq \|f\| \cdot \varphi(\theta)$ for all $\theta \in \mathbb{T}$.

Proof: Let us introduce an equivalent norm on A by $\|g\|_\varphi = \sup\{|g(\theta)||\varphi^{-1}(\theta)|: \ \theta \in \mathbb{T}\}$ and let $r: (A, \|\cdot\|_\varphi) \to C(\Delta)$ be the restriction operator $r(g) = g|\Delta$. Since Δ has Lebesgue measure 0 and $\varphi|\Delta = 1$ we infer that r^* is an isometric embedding, so r is a quotient map. Note that r^* assigns to the measure μ on Δ the same measure treated as a measure on $\mathbb{T}$ and considered as a functional on A. Since $(\ker r)^\perp = r^*(C(\Delta)^*)$ we infer that $\ker r$ is an M-ideal. Thus from Theorem III.D.4 we get that there exists $g \in (A, \|\cdot\|_\varphi)$ with $g|\Delta = f$ and $\|g\|_\varphi \leq \|f\|$, i.e. $|g(\theta)| \leq \varphi(\theta)\|f\|$.   ∎

Note that $|\Delta| = 0$ is also a necessary condition for $A|\Delta = C(\Delta)$ for a closed set $\Delta \subset \mathbb{T}$. If $|\Delta| > 0$ then the restriction $r: A \to C(\Delta)$ is 1-1 (It is known that $f \in H_p(\mathbb{T})$ cannot vanish on a set of positive measure. This fact is hidden in the canonical factorisation and the form of an outer function; see I.B.23) and it easily follows from Proposition 2 that it is not an isomorphism. Thus it cannot be onto.

3.   From this proposition and Theorem III.D.16 we get

Corollary.   *If $\Delta \subset \mathbb{T}$ has Lebesgue measure 0 then there exists a linear extension operator $u: C(\Delta) \to A$.*   ∎

4.   Interpolating sequences in the open disc are also of considerable interest. First we consider them for the space H_∞. A sequence $(\lambda_n)_{n=0}^\infty \subset \mathbb{D}$ is called interpolating if the map $f \mapsto (f(\lambda_n))_{n=0}^\infty$ transforms H_∞ onto ℓ_∞. Obviously (see I.B.23) any interpolating sequence has to satisfy the Blaschke condition $\sum_{n=0}^\infty (1 - |\lambda_n|) < \infty$. Thus we can form the Blaschke products $B(z) = \prod_{n=0}^\infty M_n(z)$ and $B_n(z) = B(z)/M_n(z)$ where $M_n(z) = |\lambda_n|\lambda_n^{-1}(\lambda_n - z)(1 - \bar{\lambda}_n z)^{-1}$.
   The following gives some information about interpolating sequences.

Theorem.   *The following conditions on the sequence $(\lambda_n)_{n=0}^\infty \subset \mathbb{D}$ are equivalent:*

(a)   *$(\lambda_n)_{n=0}^\infty$ is an interpolating sequence;*

(b)   *$\inf_{n\geq 0} |B_n(\lambda_n)| = \delta > 0$;*

(c)   *there exists a bounded linear map $T: \ell_\infty \to H_\infty$ such that $T(\xi)(\lambda_k) = \xi_k$ for $k = 0, 1, 2,$ and any $\xi = (\xi_n) \in \ell_\infty$*

Note that condition (c) gives a linear lifting to the map $f \mapsto (f(\lambda_n))_{n=0}^\infty$.

Proof: (a)$\Rightarrow$(b). The open mapping theorem yields a constant C such that for each $n = 0, 1, 2, \ldots$ there exists $f_n \in H_\infty$ with $\|f_n\| \leq C$ and $f_n(\lambda_j) = \delta_{nj}$. From the canonical factorization I.B.23 we get that $f_n = \varphi_n \cdot B_n$ with $\varphi_n \in H_\infty$ and $\|\varphi_n\| = \|f_n\|$. This gives condition (b).

    (b)$\Rightarrow$(c). Let us assume $0 < |\lambda_0| \leq |\lambda_1| \leq \ldots$. We define

$$\phi_n(z) = \left(\frac{1 - |\lambda_n|^2}{1 - \bar{\lambda}_n z}\right)^2 \frac{B_n(z)}{B_n(\lambda_n)} \cdot \exp(\alpha_n(\lambda_n) - \alpha_n(z)) \tag{2}$$

where $\alpha_n(z) = \sum_{k \geq n} (1 + \bar{\lambda}_k z)(1 - \bar{\lambda}_k z)^{-1}(1 - |\lambda_k|^2)$.

    Clearly $\phi_n(\lambda_k) = \delta_{n,k}$ so the operator $T(\xi_n) = \sum_{n \geq 0} \xi_n \phi_n$ satisfies (c) provided it is continuous. This will follow from

$$\sum_{n \geq 0} |\phi_n(z)| \leq C_\delta \quad \text{for} \quad |z| < 1. \tag{3}$$

Since

$$Re\,\alpha_n(z) = \sum_{k \geq n} \frac{(1 - |\lambda_k|^2 |z|^2)(1 - |\lambda_k|^2)}{|1 - \bar{\lambda}_k z|^2} \tag{4}$$

we get

$$Re\,\alpha_n(\lambda_n) = \sum_{k \geq n} \frac{(1 - |\lambda_k|^2 |\lambda_n|^2)(1 - |\lambda_k|^2)}{|1 - \bar{\lambda}_k \lambda_n|^2}$$

$$\leq 2 \sum_{k \geq n} \frac{(1 - |\lambda_n|^2)(1 - |\lambda_k|^2)}{|1 - \bar{\lambda}_k \lambda_n|^2}$$

$$= 2 \sum_{k \geq n} (1 - |M_k(\lambda_n)|^2) \leq 2 - 4 \sum_{k > n} \log |M_k(\lambda_n)| \tag{5}$$

$$\leq 2 - 4 \log \left| \prod_{k \neq n} M_k(\lambda_n) \right| \leq 2 - 4 \log \delta = K_\delta.$$

Using (4), (b) and (5) we get

$$\sum_{n \geq 0} |\phi_n(z)| \leq \sum_{n \geq 0} \left[Re\,\alpha_n(z) - Re\,\alpha_{n+1}(z) \right] \delta^{-1} \exp K_\delta \, \exp(-Re\,\alpha_n(z))$$

$$\leq C_\delta \sum_{n \geq 0} \left[\exp\left(Re\,\alpha_n(z) - Re\,\alpha_{n+1}(z) \right) - 1 \right] \cdot \exp -Re\,\alpha_n(z)$$

$$= C_\delta \sum_{n \geq 0} \left[\exp\left(-Re\,\alpha_{n+1}(z) \right) - \exp\left(-Re\,\alpha_n(z) \right) \right] \leq C_\delta.$$

    (c)$\Rightarrow$(a). is obvious.   ∎

5.   If $(\lambda_n)_{n=0}^{\infty} \subset \mathbb{D}$ is an interpolating sequence then it is quite difficult to describe the set $\{(f(\lambda_n))_{n=0}^{\infty} \colon f \in A\} \subset \ell_{\infty}$. This set clearly depends on the topological structure of the closure of $(\lambda_n)_{n=0}^{\infty}$ in $\overline{\mathbb{D}}$. One can easily construct examples (see Exercise 6) where this set is not closed in ℓ_{∞}. The following theorem characterizes sets of 'free' interpolation for the disc algebra.

Theorem.   *Let $V \subset \overline{\mathbb{D}}$ be a closed subset. Then $A|V = C(V)$ if and only if $|V \cap \mathbb{T}| = 0$ and $V \cap \mathbb{D}$ is an interpolating sequence.*

Proof: If $A|V = C(V)$ then $A|(V \cap \mathbb{T}) = C(V \cap \mathbb{T})$ so by the observation made after Proposition 2 we have $|V \cap \mathbb{T}| = 0$. Clearly $V \cap \mathbb{D}$ is countable and there exists a constant C such that for every finite subset $\{\lambda_1, \ldots, \lambda_n\} \subset V \cap \mathbb{D}$ and any numbers $(\alpha_1, \ldots, \alpha_n)$ there exists $f \in A$ such that $\|f\| \leq C \max_{1 \leq j \leq n} |\alpha_j|$ and $f(\lambda_j) = \alpha_j$. By the standard normal family argument we get that $V \cap \mathbb{D}$ is an interpolating sequence. Conversely let $|V \cap \mathbb{T}| = 0$ and $V \cap \mathbb{D}$ be an interpolating sequence. Let us write $A = A^0 \oplus A^1$ where $A^0 = \{f \in A \colon f|V \cap \mathbb{T} = 0\}$ and $A^1 = u(C(V \cap \mathbb{T}))$ where u is a linear extension operator from $C(V \cap \mathbb{T})$ into A (cf. Corollary 3). Let us also split $C(V) = C^0 \oplus C^1$ where $C^0 = \{f \in C(V) \colon f|V \cap \mathbb{T} = 0\}$ and $C^1 = A^1|V$. In order to prove the theorem it is enough to show that $A^0|V = C^0$. Let $V \cap \mathbb{D} = \{\lambda_n\}_{n \geq 0}$ with $|\lambda_0| \leq |\lambda_1| \leq |\lambda_2| \leq \ldots$. Let $(\phi_n)_{n \geq 0}$ be given by (2). It is known (and easily follows from the standard proof that the Blaschke product converges) that each $B_n(z), n = 0, 1, 2, \ldots$ is continuous on $\overline{\mathbb{D}} \backslash (V \cap \mathbb{T})$. Also the $\alpha_n(z), n = 0, 1, 2,$ are continuous on $\overline{\mathbb{D}} \backslash (V \cap \mathbb{T})$ so all the $(\phi_n)_{n \geq 0}$ are continuous on $\overline{\mathbb{D}} \backslash (V \cap \mathbb{T})$. Let $\varphi_n \in A$ be such that $\|\varphi_n\| \leq 2$, $\varphi_n|V \cap \mathbb{T} = 0$ and $\varphi_n(\lambda_n) = 1$ for $n = 0, 1, 2, \ldots$. The existence of such φ_n's easily follows from Proposition 2. It follows from (3) that for $\psi_n = \varphi_n \cdot \phi_n$, $n = 0, 1, 2, \ldots$ we have

$$\sum_{n \geq 0} |\psi_n(z)| \leq C \quad \text{for} \quad |z| < 1$$

and $\psi_n(\lambda_k) = \delta_{n,k}$. For $f \in C^0$ we define $F = \sum_{n \geq 0} f(\lambda_n) \psi_n$. Since $F \in A$ and $F|V = f$ the theorem is proved.   ■

6.   Let us consider now a more specialized interpolation result. Before we proceed we have to recall the notion of a Dirichlet integral (Dirichlet norm) which is instrumental in proving the Dirichlet principle by variational methods. We restrict our attention to analytic functions only.

We define the Dirichlet space

$$D = \left\{ f(z) \colon f(z) \text{ is analytic for } |z| < 1 \text{ and } \left(\int_D |f'(z)|^2 d\nu(z) \right)^{\frac{1}{2}} < \infty \right\}.$$

An easy computation shows that $f(z) = \sum_{n \geq 0} \hat{f}(n) z^n \in D$ if and only if $\sum_{n \geq 0} n |\hat{f}(n)|^2 < \infty$. We are interested in $D \cap A$ or, more precisely the sets $E \subset \mathbb{T}$ for which we have $D \cap A | E = C(E)$. Clearly E has to have measure zero but this is not enough. The proper condition involves the notion of capacity which we will now recall. For a closed set $E \subset \mathbb{T}$ and a Borel measure μ on E we define the energy of μ as

$$\mathcal{E}(\mu) = \int_E \int_E \log \frac{2}{|x - y|} d\mu(x) d\mu(y), \tag{6}$$

where this integral is understood as the Lebesgue integral on ExE with respect to the measure $\mu x \mu$.

Since for $x = e^{i\theta}$ and $y = e^{i\varphi}$, $|x - y| = 2|\sin \frac{1}{2}(\theta - \varphi)|$ we can write the integral (6) as

$$\int_{\mathbb{T}} \int_{\mathbb{T}} \log \frac{1}{|\sin \frac{\theta - \varphi}{2}|} d\mu(\theta) d\mu(\varphi).$$

Writing $\log |\sin \frac{1}{2}\theta|^{-1} \sim \sum_{-\infty}^{+\infty} \gamma_n e^{in\theta}$ we get $\mathcal{E}(\mu) = \sum_{-\infty}^{+\infty} \gamma_n |\hat{\mu}(n)|^2$. Integration by parts yields $\gamma_n = \frac{c}{|n|} + o(\frac{1}{|n|})$. Moreover one can show that $\gamma_n \geq 0$. This gives $\mathcal{E}(\mu) \geq 0$ for every Borel measure μ on $\mathbb{T}$ and

$$\mathcal{E}(\mu) < \infty \text{ if and only if } \sum_{n \neq 0} |n|^{-1} |\hat{\mu}(n)|^2 < \infty. \tag{7}$$

Note also that if $\mathcal{E}(\mu) < \infty$ and g is a bounded function on $\mathbb{T}$ then $\mathcal{E}(g \cdot \mu) < \infty$.

The capacity of E, denoted by $\gamma(E)$ is defined as

$$\gamma(E) = \exp - \inf \{ \mathcal{E}(\mu) \colon \mu \text{ is a probability measure on } E \}.$$

If $|E| > 0$ then $\gamma(E) > 0$ because the Lebesgue measure restricted to E has finite energy.

Now we are read to state:

Theorem. *Let $E \subset \mathbb{T}$ be a closed set. Then $(D \cap A)|E = C(E)$ if and only if $\gamma(E) = 0$.*

For the proof we will need two lemmas.

7 Lemma. Let $\mu \in M(\mathbb{T})$ and $\mathcal{E}(\mu) < \infty$. Then for every closed set $E \subset \mathbb{T}$ with $\gamma(E) = 0$ we have $\mu(E) = 0$.

Proof: It is enough to check for positive μ only. Now if $\mu(E) > 0$, then, since $\gamma(E) = 0$ we have

$$\infty = \mathcal{E}(\mu|E) = \int_E \int_E \log \frac{2}{|x-y|} d\mu(x) d\mu(y)$$
$$\leq \int_{\mathbb{T}} \int_{\mathbb{T}} \log \frac{2}{|x-y|} d\mu(x) d\mu(y) = \mathcal{E}(\mu) < \infty$$

which is absurd.   ∎

8 Lemma. Let μ be a measure in $M(\mathbb{T})$. If $\sum_{n=1}^{\infty} \left(\frac{|\hat{\mu}(n)|^2}{n} \right) < \infty$ then $\sum_{n=-\infty, n\neq 0}^{+\infty} \left(\frac{|\hat{\mu}(n)|^2}{|n|} \right) < \infty$.

Proof: We will actually show the inequality

$$\left| \sum_{n=1}^{\infty} \frac{|\hat{\mu}(n)|^2}{n} - \sum_{-\infty}^{-1} \frac{|\hat{\mu}(n)|^2}{|n|} \right| \leq \frac{\|\mu\|^2}{4\pi}. \tag{8}$$

It is enough to establish (8), which obviously implies the lemma, for absolutely continuous μ. If $\mu = f d\lambda$ we put $f^*(\theta) = \overline{f(-\theta)}$ and $g = f * f^*$. Clearly $\hat{g}(n) = 2\pi|\hat{f}(n)|^2 = 2\pi|\hat{\mu}(n)|^2$. It is known and easily checked that the function $h(\theta) = \frac{1}{2}(\pi - |\theta|) \, \mathrm{sgn}\theta$ defined for $-\pi \leq \theta \leq \pi$ has the Fourier series $\sum_1^{\infty} n^{-1} \sin n\theta$. We have

$$\frac{\pi}{2}\|\mu\|^2 \geq \frac{\pi}{2}\|g\|_1 = \|h\|_\infty \|g\|_1$$
$$\geq \left| \int_{-\pi}^{\pi} h(\theta)g(\theta)d\theta \right| = \left| 2\pi^2 \sum_{n=1}^{\infty} \frac{1}{n}(|\hat{\mu}(n)|^2 - |\hat{\mu}(-n)|^2) \right|$$

so (8) follows.   ∎

Proof of the Theorem. We identify $D \cap A$ with the subspace X of $D \oplus A$ consisting of pairs (f, f). This allows us to describe $(D \cap A)^*$ as $D^* \oplus A^*/X^\perp$. Let r denote the restriction operator from $D \cap A$ into $C(E)$. It is enough to consider E with $|E| = 0$. For $\mu \in M(E) = C(E)^*$ we get $r^*(\mu) = (0, \mu) + X^\perp$. Thus, by duality, r is onto if and only if there exists a $c > 0$ such that

$$\inf\{\|(0, \mu) - (S, \nu)\|\colon (S, \nu) \in X^\perp\} \geq C\|\mu\| \quad \text{for all} \quad \mu \in M(E). \tag{9}$$

Assume now that $\gamma(E) = 0$. Since $(e^{in\theta}, e^{in\theta}) \in X$ for $n = 0, 1, 2, \ldots$ we get that for $(S, \nu) \in X^{\perp}$ we have $S(e^{in\theta}) + \hat{\nu}(-n) = 0$ for $n = 0, 1, 2, \ldots$. For $S \in D^*$ we get easily that $\sum_{n=1}^{\infty} n^{-1}|S(e^{in\theta})|^2 < \infty$ so for $(S, \nu) \in X^{\perp}$ we have $\sum_{n=1}^{\infty} n^{-1}|\hat{\nu}(-n)|^2 < \infty$. From Lemma 8 and (7) we get $\mathcal{E}(\nu) < \infty$. From Lemma 7 we get $\nu(E) = 0$. Thus for $(S, \nu) \in X^{\perp}$ and $\mu \in M(E)$ we have $\|(0, \mu) - (S, \nu)\| = \|S\|_{D^*} + \|\mu - \nu\| \geq \|\mu\|$ so by (9) r is onto.

Assume now $\gamma(E) > 0$. From (7) we get that for every $\mu \in M(\mathbb{T})$ with $\mathcal{E}(\mu) < \infty$ the functional S_{μ} defined on D as $S_{\mu}(f) = \sum_{n \geq 0} \hat{f}(n)\hat{\mu}(n)$ is continuous and $\|S_{\mu}\|_{D^*} \leq C\mathcal{E}(\mu)^{\frac{1}{2}}$. In particular if $\mathcal{E}(\mu) < \infty$, then $(-S_{\mu}, \mu) \in X^{\perp}$. Since $\gamma(E) > 0$ there exists a probability measure μ on E with $\mathcal{E}(\mu) < \infty$. Let $\varphi_n \in L_{\infty}(\mu)$ be a sequence of functions such that $|\varphi_n| = 1$ μ-a.e. and $\varphi_n \to 0$ as $n \to \infty$ in the $\sigma(L_{\infty}(\mu), L_1(\mu))$-topology. From (6) we infer that $\mathcal{E}(\varphi_n\mu) \to 0$ as $n \to \infty$. So

$$\inf\{\|(0, \varphi_n\mu)) - (S, \nu)\|\colon (S, \nu) \in X^{\perp}\}$$
$$\leq \|(0, \varphi_n\mu) - (-S_{\varphi_n\mu}, \varphi_n\mu)\| = \|S_{\varphi_n\mu}\|_{D^*}$$
$$\leq C\mathcal{E}(\varphi_n\mu)^{\frac{1}{2}} \to 0 \text{ as } n \to \infty.$$

This shows that (9) is violated, so r is not onto. ∎

9. In the previous sections we have been mostly interested in interpolation taking into account values of the function. The other very natural and important problem is to impose some conditions on Fourier coefficients. The following result is interesting in itself and will be used in Chapter III.F.

Theorem. Let $(n_k)_{k=0}^{\infty}$ be a sequence of positive integers such that $n_{k+1} \geq 2n_k$ for all k in $\mathbb{N}$ and let $(v_k)_{k=0}^{\infty} \in \ell_2$. Then there exists $g \in A$ with $g(z) = \sum_{n=0}^{\infty} \hat{g}(n)z^n$ and $\hat{g}(n_k) = v_k$ and $\|g\|_{\infty} \leq C\|(v_k)\|_2$.

Since $(\hat{g}(n))_{n \geq 0} \in \ell_2$ for every $g \in A$ one cannot relax the condition on the sequence $(v_k)_{k=0}^{\infty}$. Thus the theorem can be rephrased as $\{\hat{f}(n_k)\colon f \in A\} = \ell_2$.

Proof: Let us consider only numbers z with $|z| = 1$. Assume also $\sum_{k=1}^{\infty} |v_k|^2 = 1$. We define inductively two sequences of polynomials $g_k(z)$ and $h_k(z)$, $k \geq 0$ as follows:

$$g_0(z) = v_0 z^{n_0} \quad \text{and} \quad h_0(z) = 1 \tag{10}$$

and for $k > 0$ we put

$$g_k(z) = g_{k-1}(z) + v_k z^{n_k} h_{k-1}(z),$$

$$h_k(z) = h_{k-1}(z) - \bar{v}_k \bar{z}^{n_k} g_{k-1}(z). \tag{11}$$

Using the elementary identity

$$|a + vb|^2 + |b - \bar{v}a|^2 = (1 + |v|^2)(|a|^2 + |b|^2)$$

valid for all complex numbers a, b, v we inductively obtain that

$$|g_k(z)|^2 + |h_k(z)|^2 = \prod_{j=0}^{k}(1 + |v_j|^2) \le C. \tag{12}$$

We also obtain inductively that

$$g_k = \sum_{j=0}^{n_k} \hat{g}_k(j) z^j \quad \text{and} \quad h_k = \sum_{j=-n_k}^{0} \hat{h}_k(j) z^j.$$

Since $n_{k+1} \ge 2n_k$ we infer that there is no cancellation of Fourier coefficients in (11). In particular we get $\hat{h}_k(0) = 1$ for all k and thus $\hat{g}_k(n_s) = v_s$ for $s \le k$. Thus (12) and the open mapping theorem I.A.5 give the claim.   ∎

10.   We have seen projections in A whose image is a $C(K)$-space. Now we will investigate projections whose image is isomorphic to A. This will lead to the proof of Theorem 12. Given a positive number ε and an interval I in $\mathbb{T}$ we say that a function $f \in A$ is ε-supported on I if $|f(t)| \le \varepsilon\|f\|$ for $t \in \mathbb{T}\backslash I$. We say that a subspace $X \subset A$ is ε-supported on I if every $f \in X$ is ε-supported on I.
    We have the following.

Proposition.   *For every $\varepsilon > 0$ and an interval $I \subset \mathbb{T}$ there exists a subspace $X \subset A$ and a projection $P \colon A \xrightarrow{\text{onto}} X$ such that*

(a)   *X is ε-supported on I,*

(b)   *$d(A, X) \le 1 + \varepsilon$,*

(c)   *$P1 = 0$ and $\|P\| \le 1 + \varepsilon$,*

(d)   *for every $g \in A$, δ-supported on $\mathbb{T}\backslash I$ we have $\|Pg\| \le (\varepsilon + \delta)\|g\|$.*

Proof: This is an interplay between averaging projections and con-
formal mappings. Every conformal map $\varphi \colon \mathbb{D} \to \mathbb{D}$ induces an isometry
$I_\varphi \colon A \to A$ defined by $I_\varphi(f) = f \circ \varphi$. Suppose the proposition holds
for some $\varepsilon > 0$ and some interval I. Then it holds for the same $\varepsilon > 0$
and any other interval I_1. To see this let us take a conformal map φ
such that $\varphi(I_1) = I$. Then $I_\varphi(X)$ is ε-supported on I and $I_\varphi P I_{\varphi^{-1}}$ is a
projection onto $I_\varphi(X)$ and (a)-(d) hold. Let

$$Q_n(f)(z) = \frac{1}{n} \sum_{k=1}^{n} f(e^{2\pi i \frac{k}{n}} z) e^{2\pi i \frac{k}{n}}. \tag{13}$$

This is a norm-1 projection and $ImQ_n \cong A$.

For a positive number $\xi < \frac{1}{2}$ we find a function $F \in A$, with $\|F\| \leq 1$
such that

$$|F(e^{i\theta}) - 1| < \xi \quad \text{for} \quad \frac{2\pi}{n} < \theta < 2\pi - \frac{2\pi}{n} \text{ and } F(1) = 0.$$

Such a function is easy to construct using conformal mappings as before.
Observe that for $f \in ImQ_n$ we have

$$\|f\| \geq \|F \cdot f\| \geq (1 - \xi)\|f\| \tag{14}$$

and

$$\|Q_n(F \cdot f) - f\| = \|Q_n(f \cdot (F - 1))\| \leq 2\xi\|f\|. \tag{15}$$

Let I be such that $|F(t)| < \xi$ for $\mathbb{T}\backslash I$. From (14) we infer that
$X = F \cdot ImQ_n$ is a closed subspace of A, $(\frac{\xi}{(1-\xi)})$-supported on I, with
$d(X, A) \leq (1 - \xi)^{-1}$. The condition (15) shows that $Q_n|X$ is an isomor-
phism between X and ImQ_n with $\|(Q_n|X)^{-1}\| \leq (1 - 2\xi)^{-1}$. We define
$P = (Q_n|X)^{-1}Q_n$. This is a projection onto X with $\|P\| \leq (1 - 2\xi)^{-1}$.
Since $Q_n 1 = 0$ $P(1) = 0$ as well. If g is a function in A which is
δ-supported on $\mathbb{T}\backslash I$ then

$$\|Pg\| \leq (1 - 2\xi)^{-1}\|Q_n g\| \leq (1 - 2\xi)^{-1}\left(\delta + \frac{1}{n}\right)\|g\|.$$

If ξ was chosen small enough and n big enough we see that (a)-(d)
hold.   ∎

11 Proposition.   *The space A contains a complemented subspace isomorphic to $(\Sigma A)_0$.*

Proof: Let us take a sequence of disjoint closed intervals $(I_n)_{n=1}^\infty$ in $\mathbb{T}$ and a sequence of positive numbers $(\varepsilon_n)_{n=1}^\infty$ with $\sum_{n=1}^\infty \varepsilon_n = \varepsilon < 0.1$. Using Proposition 10 we construct subspaces $X_n \subset A$, ε_n-supported on I_n for $n = 1, 2, \ldots$, and projections P_n from A onto X_n satisfying (a)-(d). It is routine to show that for $x_n \in X_n, n = 1, 2, \ldots$

$$(1 - 2\varepsilon) \sup \|x_n\| \leq \left\| \sum_{n=1}^\infty x_n \right\| \leq (1 + \varepsilon) \sup \|x_n\| \qquad (16)$$

so $X = \text{span}\{X_n\}_{n=1}^\infty \sim (\Sigma A)_0$. Let $R(f) = \sum_{n=1}^\infty P_n f$. In order to show that R is continuous it is enough to check that for every $f \in A$ we have $\|P_n f\| \to 0$. Fix $t_n \in I_n$. Since f is uniformly continuous in $\overline{\mathbb{D}}$ we get that $f - f(t_n)$ is δ_n-supported on $\mathbb{T} \backslash I_n$ for some $\delta_n \to 0$. From Proposition 10 (c) and (d) we get $\|P_n f\| = \|P_n(f - f(t_n))\| \leq (\delta_n + \varepsilon_n)\|f\|$, and this yields the continuity of R. Clearly $R\colon A \to X$. For $x = \sum_{n=1}^\infty x_n \in X$ with $x_n \in X_n$ and $\|x\| = 1$ we define $h_n = \sum_{k=1, n \neq k}^\infty x_k$. Since $Rx - x = \sum_{n=1}^\infty P_n h_n$ we get from (16) that $\|Rx - x\| \leq (1 + \varepsilon) \sup_n \|P_n h_n\|$. Once more using (16) we get $\|x_n\| \leq (1 - 2\varepsilon)^{-1}$ for $n = 1, 2, \ldots$ so $\|h_n\| \leq (1 + \varepsilon)(1 - 2\varepsilon)^{-1}$. Since each x_n is ε_n-supported on I_n we get that for $t \in I_n$ we have $|h_n(t)| \leq \varepsilon(1 - 2\varepsilon)^{-1}$. Proposition 10 (d) gives $\|P_n(h_n)\| \leq [\varepsilon_n + \varepsilon(1 - 2\varepsilon)^{-1}]\|h_n\|$ so we conclude that

$$\|Rx - x\| < 0.8.$$

This shows that $R|X$ is an isomorphism of X (see II.B.14) and one checks that $(R|X)^{-1}$ is a continuous projection from A onto X. ∎

12.   The decomposition method (see II.B.21) and Proposition 11 yield immediately

Theorem.   *The disc algebra A is isomorphic to its infinite c_0-sum.*∎

13 Remark.   Since the projections constructed in Proposition 10 have the property that $P^*(L_1/H_1) \subset L_1/H_1$ we easily see that $(\Sigma L_1/H_1)_1 \sim L_1/H_1$ and passing to the duals we get $(\Sigma H_\infty)_\infty \sim H_\infty$.

14.   Most of the results in this chapter show the analogy between A and $C(K)$-spaces. We would like to point out however that A is not a $C(K)$-space.

To see this we need to observe that every $C(K)$-space has the following extension property:

There exists a constant C such that for every Banach space X, every subspace Y of it and every finite rank operator $T\colon Y \to C(K)$ there exists an operator $\widetilde{T}\colon X \to C(K)$ such that $\widetilde{T}|Y = T$ and $\|\widetilde{T}\| \le C\|T\|$. $\quad(*)$

This follows (for any $C > 1$) directly from II.E.5(c) and III.B.2 (or the Hahn-Banach theorem).

Let $f = \sum_{-N}^{N} a_n e^{in\theta}$ be a trigonometric polynomial. The operator $T_f\colon A \to A$ defined by $T_f(g) = f * g$ is a finite dimensional operator and $\|T_f\| \le \|f\|_1$. The operator T_f has a unique rotation invariant extension $\widetilde{T}_f\colon C(\mathbb{T}) \to A$ which is given by $\widetilde{T}_f(g) = \mathcal{R}f * g$ where $\mathcal{R}$ is the Riesz projection (see I.B.20). The standard averaging argument shows that for any extension T we have $\|T\| \ge \|\widetilde{T}_f\|$. But $\|\widetilde{T}_f\| = \|\mathcal{R}f\|_1$. Since the Riesz projection is unbounded on $L_1(\mathbb{T})$ (see I.B.20 and 25) we see that A does not have the above property $(*)$. Since if Z has $(*)$ its complemented subspaces also have $(*)$ we get

Theorem. *The disc algebra A is not isomorphic to any complemented subspace of any $C(K)$-space.*

Despite this theorem and some other striking differences between A and $C(K)$-spaces, which we will exploit e.g. in III.F.7, the general impression is that A is quite similar to a $C(K)$-space. Actually the idea of comparing A to $C(K)$ underlies most of the results about the disc algebra A presented in this book.

15. Another proof of the above theorem, which is different, although similar in spirit, follows from the Lozinski-Kharshiladze theorem III.B.22 and the following

Proposition. *Let T_n^∞ denote the space of all trigonometric polynomials of degree $\le n$ with the sup-norm. There exists a sequence of projections $(P_n)_{n=1}^{\infty}$ in the disc algebra such that*

(a) $\|P_n\| \le C$,

(b) $d(Im P_n, T_n^\infty) < C$
for some constant C and all $n=1,2,3,\ldots$.

Proof: The proof depends on the properties of the Fejér kernels; see I.B.16. Let $A_n = \mathrm{span}(1, z, \ldots, z^{2n}) \subset A$. We define $i_n\colon A_n \to A \oplus_\infty A$

by $i_n \left(\sum_{k=0}^{2n} a_k z^k \right) = \left(\sum_{k=0}^{2n} a_k z^k, \sum_{k=0}^{2n} a_k z^{2n-k} \right)$. Clearly $\|i_n\| \leq 1$.
We also define $\phi_n \colon A \oplus A \to A_n$ by

$$\phi_n \left(\sum_{k=0}^{\infty} a_k z^k, \sum_{k=0}^{\infty} b_k z^k \right) = \sum_{k=0}^{2n} a_k \left(1 - \frac{k}{2n} \right) z^k + \sum_{k=0}^{2n} b_k \left(1 - \frac{k}{2n} \right) z^{2n-k}$$

$$= \sum_{k=0}^{2n} a_k \left(1 - \frac{k}{2n} \right) z^k + \sum_{k=0}^{2n} b_{2n-k} \frac{k}{2n} z^k.$$

Using the properties of Fejér operators once more we get $\|\phi_n\| \leq 2$. One checks that $\phi_n i_n = id_{A_n}$ so $i_n \phi_n$ is a projection of norm ≤ 2 onto $i_n(A_n)$, and $d(A_n, i_n(A_n)) \leq 2$. Since T_n^∞ is isometric to A_n and A is isomorphic to $A \oplus_\infty A$ the proposition follows.   ■

Let us note that the same argument works if we replace A by H_∞. This implies that H_∞ also is not isomorphic to any complemented subspace of any $C(K)$-space.

16.   Let A_r denote the space of complex valued functions $f \in C(\mathbb{T})$ such that $\tilde{f}$ is also in $C(\mathbb{T})$. Let us recall (see I.B.22) that $\tilde{f}$ is the trigonometric conjugate of the function f. The norm in A_r is naturally defined by $\|f\| = \|f\|_\infty + \|\tilde{f}\|_\infty$.

Proposition.   *The space A_r is isomorphic to the disc algebra A.*

Proof: The desired isomorphism from A onto A_r is defined by

$$i \left(\sum_{n=0}^{\infty} a_n e^{in\theta} \right) = \sum_{k=0}^{\infty} a_{2k+1} \sin k\theta + \sum_{k=0}^{\infty} a_{2k} \cos k\theta.   \blacksquare$$

17.   We intend to show that the disc algebra has a Schauder basis. Clearly the monomials $(z^n)_{n=0}^{\infty}$ are not a Schauder basis in A. The previous proposition gives us a different representation of the disc algebra. In this representation we can forget for a while about analyticity. It also makes the proof of our next theorem more transparent.

Theorem.   *The Franklin system is a Schauder basis in A_r.*

The definition and properties of the Franklin system have been given in III.D.20-27. In the proof of Theorem 17 we will freely use notation from there.

Proof: If $\mathcal{F}_n$ is the n-th Fejér kernel (see I.B.16) then for $f \in A_r$ we have $\|f - f * \mathcal{F}_n\| \to 0$ as $n \to \infty$ so the trigonometric polynomials are dense in A_r. For every trigonometric polynomial $\varphi(\theta)$ there exists a sequence φ_n of finite sums of Franklin functions (φ_n can be taken to interpolate φ in $t_0, t_1, \ldots, t_n$) such that $\varphi_n \to \varphi$ in Lip_1. Since, as is well known, $\|\tilde{\varphi}_n - \tilde{\varphi}\|_\infty \le C\|\varphi_n - \varphi\|_{Lip_1}$ we see that the Franklin system in linearly dense in A_r. Thus our theorem follows from Theorem III.D.25 and the following inequality which is the main point of the Proof:

There exists a constant C such that for every N and every $f \in C(\mathbb{T})$

$$\left\| \sum_{n=0}^{N} \langle f, f_n \rangle \tilde{f}_n \right\|_\infty \le C(\|f\|_\infty + \|\tilde{f}\|_\infty). \tag{17}$$

For a fixed $x \in \mathbb{T}$ and $f \in C(\mathbb{T})$ we have

$$S_N(f)(x) \overset{df}{=} \sum_{n=0}^{N} \langle f, f_n \rangle \tilde{f}_n(x)$$

$$= \int_{\mathbb{T}} \int_0^\pi \cot \frac{t}{2} \cdot (K_N(x-t, y) - K_N(x+t, y)) dt f(y) dy \tag{18}$$

where $K_N(x, y) = \sum\limits_{n=0}^{N} f_n(x) f_n(y)$.

If we define

$$\Lambda_{N,x}(t) = \begin{cases} 0 & \text{if } |t - x| \le \frac{1}{N}, \\ \cot \frac{t-x}{2} & \text{if } |t - x| > \frac{1}{N} \end{cases}$$

then we can write

$$S_N(f)(x) = \int_{\mathbb{T}} f(y) \int_0^{\frac{1}{N}} \cot \frac{t}{2} \cdot \left(K_N(x-t, y) - K_N(x+t, y) \right) dt dy$$

$$+ \int_{\mathbb{T}} f(y) \left[\int_{\mathbb{T}} \Lambda_{N,x}(t) K_N(t, y) dt - \Lambda_{N,x}(y) \right] dy \tag{19}$$

$$+ \int_{\mathbb{T}} f(y) \Lambda_{N,x}(y) dy = I_1 + I_2 + I_3.$$

For every y, $K_N(x, y)$ is a piecewise linear function with nodes at least $\frac{1}{2N}$ apart so Proposition II.D.21(b) yields the estimate for the slope which gives

$$I_1 \leq \|f\|_\infty \cdot \int_{\mathbb{T}} \int_0^{\frac{1}{N}} \cot \frac{t}{2} \cdot |K_N(x-t,y) - K_N(x+t,y)| dt dy$$

$$\leq \|f\|_\infty \int_{\mathbb{T}} \int_0^{\frac{1}{N}} \cot \frac{t}{2} \cdot C(N+1)^2 q^{N dist(x,y)} t dt dy \qquad (20)$$

$$\leq C\|f\|_\infty (N+1) \int_{\mathbb{T}} q^{N dist(x,y)} dy \leq C\|f\|_\infty.$$

The estimate

$$I_3 \leq C\|\tilde{f}\|_\infty \qquad (21)$$

follows immediately from the known properties of the trigonometric conjugation operator; see Zygmund [1968] p. 92 or Katznelson [1968] Corollary III.2.6.

In order to estimate I_2 we write

$$I_2 \leq \int_{\mathbb{T}} \left| \int_{\mathbb{T}} \Lambda_{N,x}(t) K_N(t,y) dt - \Lambda_{N,x}(y) \right| dy = \|P_N \Lambda_{N,x} - \Lambda_{N,x}\|_1 \quad (22)$$

where P_N is the N-th partial sum projection with respect to the Franklin system. Let $\varphi_{N,x}$ be a piecewise linear function in $C(\mathbb{T})$ with nodes at $t_0, t_1, \ldots, t_N$ such that $\varphi_{N,x}(t_j) = \Lambda_{N,x}(t_j)$ for $j = 0, 1, \ldots, N$. One checks (draw the picture) that

$$\|\Lambda_{N,x} - \varphi_{N,x}\|_1 \leq C. \qquad (23)$$

Since the Franklin system is a basis in $L_1(\mathbb{T})$ (see III.D.26) from (23) we get

$$\|P_N \Lambda_{N,x} - \Lambda_{N,x}\|_1 \leq \|P_N \Lambda_{N,x} - P_N \varphi_{N,x}\|_1 + \|\varphi_{N,x} - \Lambda_{N,x}\|_1 \leq C.$$

From (20), (21), (22) and (23), (17) follows immediately. ∎

18.   Now we want to give different isomorphic representations of H_∞. Having different isomorphic representations of an important space is generally useful because each representation carries with it different intuitions, and even the possibility of using different analytical tools.

Let $(f_j)_{j=0}^\infty$ be the basis in A which exists by Theorem 17 and let $H_\infty^n = \text{span}\{f_j\}_{j \leq n}$. Let us recall also that $A_n = \text{span}(1, \ldots, z^{2n}) \subset A$ (see **15**).

Theorem. *The spaces* $\left(\sum_{n=1}^{\infty} H_{\infty}^n\right)_{\infty}$, $\left(\sum_{n=1}^{\infty} A_n\right)_{\infty}$ *and* H_{∞} *are isomorphic.*

Proof: First note that $\left(\sum_{n=1}^{\infty} H_{\infty}^n\right)_{\infty}$ is isomorphic to its infinite ℓ_{∞}-sum. This follows from II.B.24. A standard perturbation argument (see II.E.12) shows that for certain λ each H_{∞}^n is λ-isomorphic to a λ-complemented subspace of $A_{k(n)}$ and analogously it follows from Proposition 15 that each A_n is λ-isomorphic to a λ-complemented subspace of $H_{\infty}^{k(n)}$. This yields that $\left(\sum_{n=1}^{\infty} H_{\infty}^n\right)_{\infty}$ is isomorphic to a complemented subspace of $\left(\sum_{n=1}^{\infty} A_n\right)_{\infty}$ and also $\left(\sum_{n=1}^{\infty} A_n\right)_{\infty}$ is isomorphic to a complemented subspace of $\left(\sum_{n=1}^{\infty} H_{\infty}^n\right)_{\infty}$ so our first observation and II.B.24 give $\left(\sum_{n=1}^{\infty} H_{\infty}^n\right)_{\infty} \sim \left(\sum_{n=1}^{\infty} A_n\right)_{\infty}$. Remark 13 yields that $\left(\sum_{n=1}^{\infty} H_{\infty}^n\right)_{\infty}$ is isomorphic to a complemented subspace of H_{∞}. To complete the proof we need to show that $\left(\sum_{n=1}^{\infty} A_n\right)_{\infty}$ contains a complemented copy of H_{∞}. Let $\mathcal{F}_n$ be the Fejér kernel and define $i\colon H_{\infty} \to \left(\sum_{n=1}^{\infty} A_n\right)_{\infty}$ by $i(f) = (f * \mathcal{F}_n)_{n=1}^{\infty}$. The properties of the Fejér kernel (see I.B.16) give that i is an isometry. To define the projection onto $i(H_{\infty})$ we use a compactness argument. Let B denote the unit ball of H_{∞} with $\sigma(H_{\infty}, L_1/H_{\infty})$-topology, and let B_n be the unit ball in A_n. We define maps $\pi_n\colon \prod_{n=1}^{\infty} B_n \to B$ by $\pi_n(h_1, h_2, \ldots) = h_n$. Since the space of all maps from $\prod_{n=1}^{\infty} B_n$ into B is compact we take π to be a cluster point of $\{\pi_n\}_{n=1}^{\infty}$. One checks that π is homogenous and additive so it extends to a continuous linear map $\pi\colon \left(\sum_{n=1}^{\infty} A_n\right)_{\infty} \to H_{\infty}$. Moreover

$$\pi i(f) = \pi\left((f * \mathcal{F}_n)_{n=1}^{\infty}\right) = \sigma(H_{\infty}, L_1/H_1)\text{-}\lim_{k \to \infty} \pi_k\left((f * \mathcal{F}_n)_{n=1}^{\infty}\right)$$

$$= \sigma(H_{\infty}, L_1/H_1)\text{-}\lim_{k} f * \mathcal{F}_k = f.$$

This shows that $i\pi$ is a norm-one projection onto $i(H_{\infty})$. ∎

Notes and remarks.
As noted in **1** the disc algebra is an important space. It is a prototypic uniform algebra, so much information about it can be found in Gamelin [1969], Hoffman [1962] or Garnett [1981]. The closely related space H_{∞} is even more fascinating; the whole book Garnett [1981] deals with it.

A more Banach space oriented exposition is in Pełczyński [1977]. The connection with operator theory hinted in **1** is presented in detail in the beautiful lectures of Nikolskiĭ [1980]. The theory of peak sets and peak-interpolation sets is a well developed topic in uniform algebra theory; see Gamelin [1969] II.12. Our *Proposition 2* is a prototype of this theory. It was proved by Rudin [1956] and Carleson [1957]. The appeal to Theorem III.D.4 can be avoided but it saves some calculations.

The problem of characterizing interpolating sequences for H_∞ was an object of very intense study in the late 50's; Hoffman's book [1962] contains a nice presentation of these early results. By now it has grown into a vast area (see Garnett [1981]). The beautiful and simple proof of (b)$\Rightarrow$(c) in Theorem 4 is due to Peter Jones (see Gorin-Hruščov-Vinogradov [1981]).

Our Theorem 5 is a particular case of results in Casazza-Pengra-Sundberg [1980] where complemented ideals in A are fully described. The description of ideals in A is contained in Hoffman [1962]. This says that every closed ideal I in A is of the form $A_K \cdot F$ where $K \subset \mathbb{T}$ is a closed subset of Lebesgue measure 0 and $A_K = \{f \in A: f|K = 0\}$ and F is an inner function such that $\overline{F^{-1}(0)} \cap \mathbb{T} \subset K$ and the measure determining the singular part of F is supported on K.

The result of Casazza-Pengra-Sundberg [1980] asserts that I is complemented in A if and only if F is a Blaschke product whose zeros form an interpolating sequence. In paragraph 6 we gave a crash course in elementary potential theory for subsets of $\mathbb{T}$. Chapter III of Kahane-Salem [1963] contains everything we state and use. Theorem 6 is one of the results contained in Hruščov-Peller [1982]. Our presentation follows Koosis [1981]. The direct proof of Theorem 9 is taken from Fournier [1974]. Much more general theorems are proved in Vinogradov [1970]. In particular he has shown that given $(v_k)_{k=1}^\infty \in \ell_2$ there exists $f(z) = \sum_{k=0}^\infty a_n z^n$ such that $a_{2^k} = v_k$ and $f(z)$ is holomorphic in G and continuous in $\overline{G}$ where G is any region in $\mathbb{C}$ with smooth boundary and $\partial G \cap \mathbb{T}$ contains an interval.

Theorem 12 and its proof are taken from Wojtaszczyk [1979]. There are many proofs that A is not isomorphic to any $C(K)$-space. We will see some more in III.I. The fact was first observed by Pełczyński [1964a] with basically the same proof as the one given in 14. The argument was extended to the context of ordered groups by Rosenthal [1966].

Proposition 15 is an unpublished observation of J. Bourgain and A. Pełczyński. The question if A has a Schauder basis was asked by Banach [1932]. The answer was given by Bočkariov [1974]. The use of the Franklin system in the construction was rather unexpected.

Theorem 18 is a rather routine consequence of previous results. It was observed in Wojtaszczyk [1979a]. It shows in particular that H_∞ is isomorphic to the second dual of a Banach space. Note that with the natural duality we have $H_\infty \cong (L_1/H_1)^*$. This is the unique isometric predual of H_∞ (see Ando [1978] and Wojtaszczyk [1979a]).

The space L_1/H_1 in its turn is not isomorphic to a dual Banach space; this was noted in Pełczyński [1977] and Wojtaszczyk [1979a]; see Exercise III.D.19.

Exercises

1. Let $\Delta \subset \mathbb{T}$ be a compact set of measure zero and let $(n_k)_{k=1}^{\infty}$ be a lacunary sequence of natural numbers. Given $f \in C(\Delta)$ and a sequence $(a_k) \in \ell_2$ show that there exists $h \in A(\mathbb{D})$ such that $h|\Delta = f$ and $\hat{h}(n_k) = a_k$ for $k = 1, 2, \dots$.

2. Suppose that $x^* \in A^*$. Show that there exists only one measure on $\mathbb{T}$ such that $\|\mu\| = \|x^*\|$ and $\mu|A = x^*$.

3. Let $V \subset A^*$ be a relatively weakly compact subset. For each $v \in V$ let $\check{v} \in M(\mathbb{T})$ be its norm-preserving extension (see Exercise 2). Show that $\check{V} = \{\check{v} : v \in V\}$ is relatively weakly compact in $M(\mathbb{T})$.

4. (a) Let $f \in D$ (the Dirichlet space). Show that f induces a functional on $H_1(\mathbb{T})$, i.e. we have an inequality $\left| \int_{\mathbb{T}} g(e^{i\theta}) \overline{f(e^{i\theta})} d\theta \right| \leq C_f \cdot \|g\|_1$.

    (b) Show that there are unbounded functions in D.

5. The matrix $(\alpha_{ij})_{i,j \geq 0}$ is called a Hankel matrix if $\alpha_{ij} = \varphi(i+j)$ for some φ. An operator on ℓ_2 is called a Hankel operator if its matrix in the natural unit vector basis $(e_j)_{j=0}^{\infty}$ is a Hankel matrix.

    (a) Show that $L_\infty(\mathbb{T})/H_\infty^0$ is isometric to the space of all Hankel operators.

    (b) Show that $C(\mathbb{T})/A_0$ is isometric to the space of all compact Hankel operators.

    (c) Show that for every Hankel operator T there exists a best approximation by a compact Hankel operator.

6. (a) Suppose that $(z_k)_{k=1}^{\infty} \subset \mathbb{D}$ is such that $\frac{(1-|z_k|)}{(1-|z_{k-1}|)} < c < 1$ for $k = 1, 2, 3, \dots$. Show that $(z_k)_{k=1}^{\infty}$ is an interpolating sequence. Show that if $(z_k)_{k=1}^{\infty}$ are positive real numbers then the above condition is also necessary for $(z_k)_{k=1}^{\infty}$ to be interpolating.

    (b) Find an interpolating sequence $(\lambda_n)_{n=1}^{\infty} \subset \mathbb{D}$ such that $\{(f(\lambda_n))_{n=1}^{\infty} : f \in A\}$ is not closed in ℓ_∞.

7. Let $\mathbb{P}_r = \{z \in \mathbb{C} : r \leq |z| \leq r^{-1}\}$ for $0 < r < 1$ and let $A(\mathbb{P}_r)$ denote the space of all functions continuous in $\mathbb{P}_r$ and analytic in the interior.

(a) Show that $A(\mathbb{P}_r)$ contains a 1-complemented isometric copy of $A(\mathbb{D})$.

(b) Show that $A(\mathbb{D})$ contains a complemented copy of $A(\mathbb{P}_r)$ with the constants independent of r.

(c) Every $f \in A(\mathbb{P}_r)$ can be written as $\sum_{-\infty}^{+\infty} a_n z^n$. Show that the map $P_r\left(\sum_{-\infty}^{+\infty} a_n z^n\right) = \sum_{n=0}^{\infty} a_n z^n$ is a projection from $A(\mathbb{P}_r)$ onto $A(r^{-1}\mathbb{D})$. Show that $\sup_{1 < r < 1} \|P_r\| = \infty$.

8. (a) Show that $f \in H_\infty(\mathbb{D})$ is a Blaschke product if and only if $|f(z)| \le 1$ and $\lim_{r \to 1} \int_{-\pi}^{\pi} \log |f(re^{i\theta})| d\theta = 0$.

(b) Suppose $f \in H_\infty(\mathbb{D})$ is an inner function. Show that for every ρ with $0 < \rho < 1$ the functions

$$\omega_\varphi(z) = \frac{f(z) - \rho e^{i\varphi}}{1 - \rho e^{-i\varphi} f(z)}$$

are Blaschke products for almost all $\varphi, 0 < \varphi \le 2\pi$.

(c) Show that every $f \in H_\infty(\mathbb{D})$ is a limit in the topology of uniform convergence on compact subsets of $\mathbb{D}$, of a sequence of finite Blaschke products.

(d) Show that the closed unit ball in A is the closed convex hull of finite Blaschke products.

(e) (von Neumann inequality). Let $T: H \to H$ (H a Hilbert space) be a contraction, i.e. $\|T\| \le 1$. Show that for every polynomial $p(z)$ we have $\|p(T)\| \le \sup_{z \in \mathbb{D}} |p(z)|$.

9. (a) Show that if $P: A \to A$ is a norm-one, finite dimensional projection with $\dim \operatorname{Im} P > 1$ then $\operatorname{Im} P^* \subset \{\mu: \mu \perp m\}$ where m is the Lebesgue measure on $\mathbb{T}$.

(b) Show that the disc algebra is not a π_1-space.

10. (a) Suppose $f \in L_\infty(\mathbb{T})$. Show that there exists $g \in H_\infty(\mathbb{T})$ such that $\|f - g\| = \inf\{\|f - h\|: h \in H_\infty(\mathbb{T})\}$.

(b) Suppose $f \in C(\mathbb{T})$. Show that $\operatorname{dist}(f, H_\infty) = \operatorname{dist}(f, A)$.

(c) Suppose that $f \in C(\mathbb{T})$. Show that there exists only one $g \in H_\infty(\mathbb{T})$ such that $\|f - g\| = \operatorname{dist}(f, H_\infty)$, and that $|g - f| = \text{const}$.

(d) Show that there exists $f \in C(\mathbb{T})$ such that its best approximation in $H_\infty(\mathbb{T})$, i.e. $g \in H_\infty(\mathbb{T})$ such that $\|f - g\| = \operatorname{dist}(f, H_\infty)$, is not continuous.

III.F. Absolutely Summing And Related Operators.

We discuss in detail p-absolutely summing operators. The Pietsch factorization theorem, which is basic to this theory, is proved. The fundamental Grothendieck theorem is proved in its three most useful forms. Later we improve it and show the Grothendieck-Maurey theorem, that every operator from any L_1-space into a Hilbert space is p-absolutely summing for all $p > 0$. We present the trace duality and show that the p'-nuclear norm is dual to the p-absolutely summing norm. We also introduce and discuss p-integral operators. We show the connection between cotype 2 and the coincidence of classes of p-absolutely summing operators for various p's. The extrapolation result for p-absolutely summing operators is proved. We apply Grothendieck's theorem to exhibit examples of power bounded but not polynomially bounded operators on a Hilbert space and to give some estimates for the norm of a polynomial of a power-bounded operator. We also present many applications to harmonic analysis: we construct good local units in $L_1(G)$, we prove the classical Orlicz-Paley-Sidon theorem and give some characterizations of Sidon sets.

1. In this chapter we will discuss several important classes of operators, namely p-absolutely summing, p-integral and p-nuclear operators. All these classes have some ideal properties so we will introduce the general concept of an operator ideal. We are given an operator ideal if for each pair of Banach spaces X, Y we have a class of operators $I(X, Y)$ such that

(1) $I(X, Y)$ is a linear subspace (not necessarily closed) of $L(X, Y)$ containing all finite rank operators,

(2) if $T \in I(X, Y)$, $A \in L(Z, X)$ and $B \in L(Y, V)$ that $BTA \in I(Z, V)$ for all Banach spaces X, Y, Z, V and all operators A, B.

An operator ideal is a Banach (quasi-Banach) operator ideal if on each $I(X, Y)$ we have a norm (quasi-norm) i such that

(3) $(I(X, Y), i)$ is complete for each X, Y

(4) $i(BTA) \le \|B\| i(T) \|A\|$ whenever the composition makes sense

(5) for every rank-one operator $T\colon X \to Y$ we have $i(T) = \|x^*\| \cdot \|y\|$,
where $T(x) = x^*(x) \cdot y$.

Actually the reader has already encountered some examples of Banach operator ideals. Compact operators (see I.A.15) and weakly compact operators (see II.C.6) form Banach operator ideals with the operator norm.

2.   An operator $T\colon X \to Y$ is p-absolutely summing, $0 < p \le \infty$ (we write $T \in \Pi_p(X,Y)$), if there exists a constant $C < \infty$ such that for all finite sequences $(x_j)_{j=1}^n \subset X$ we have

$$\left(\sum_{j=1}^n \|Tx_j\|^p\right)^{\frac{1}{p}} \le C \sup\left\{\left(\sum_{j=1}^n |x^*(x_j)|^p\right)^{\frac{1}{p}} : x^* \in X^*, \|x^*\| \le 1\right\}.$$
(6)

We define the p-summing norm of an operator T by

$$\pi_p(T) = \inf\{C\colon (6)\text{ holds for all } (x_j)_{j=1}^n \subset X, n = 1, 2, \ldots\}. \quad (7)$$

Let us observe that for $p = 1$ the condition (6) is equivalent to the fact that T transforms weakly unconditionally convergent series into absolutely convergent series.

3.   We have the following

Theorem.   *For $0 < p \le \infty$ the p-absolutely summing operators form a quasi-Banach (Banach if $1 \le p \le \infty$) operator ideal when considered with the p-absolutely summing norm $\pi_p(\cdot)$.*

The proof of this theorem consists of routine verification of conditions (1)-(5) and is omitted.   ∎

4.   One easily checks that for all X, Y we have $\Pi_\infty(X,Y) = L(X,Y)$ and $\pi_\infty(T) = \|T\|$. For $p < \infty$ the situation is less trivial. The identity $id\colon \ell_2 \to \ell_2$ is not p-absolutely summing for any $p < \infty$. To see that condition (6) fails it is enough to consider finite orthonormal sets. The canonical example of a p-absolutely summing map is given as follows: Let μ be a probability Borel measure on a compact space K. Let

id_p: $C(K) \to L_p(K,\mu)$ be the formal identity. Then for $1 \le p < \infty$ we have $\pi_p(id_p) = 1$. Simply we have

$$\left(\sum_{j=1}^{n} \|id_p(f_j)\|^p \right)^{\frac{1}{p}} = \left(\sum_{j=1}^{n} \int |f_j|^p d\mu \right)^{\frac{1}{p}}$$

$$\le \left(\sup_{k \in K} \sum_{j=1}^{n} |f_j(k)|^p \right)^{\frac{1}{p}} \qquad (8)$$

$$= \sup \left\{ \left(\sum_{j=1}^{n} \left| \int f_j d\nu \right|^p \right)^{\frac{1}{p}} : \nu \in M(K), \|\nu\| = 1 \right\}.$$

so $\pi_p(id_p) \le 1$, but taking the one element family consisting of a constant function we see that $\pi_p(id_p) = 1$. A slight but useful variation of this example is the map $f \mapsto f \cdot g$ defined as a map from $L_\infty(\mu)$ into $L_p(\mu)$ (clearly $g \in L_p(\mu)$). Here we do not assume that μ is a probability measure; it can be arbitrary. The same calculation as in (8) gives $\pi_p(f \mapsto f \cdot g) = \|g\|_p$.

5. The following proposition describes some formal but useful properties of p-absolutely summing operators.

Proposition. (a) If $T \in \Pi_p(X,Y)$, $0 < p < \infty$ and $X_1 \subset X$ is a closed subspace then $T|X_1 \in \Pi_p(X_1,Y)$ and $\pi_p(T|X_1) \le \pi_p(T)$.

(b) If $T \in \Pi_p(X,Y)$, $0 < p < \infty$ and $Y_1 \subset Y$ is a closed subspace and $T(X) \subset Y_1$ then $T \in \Pi_p(X,Y_1)$ and the norms of T in both spaces are the same.

(c) If $(X_\gamma)_{\gamma \in \Gamma}$ is a net of subspaces of X directed by inclusion such that $\bigcup_{\gamma \in \Gamma} X_\gamma$ is dense in X and $T: X \to Y$ then $\pi_p(T) = \sup_\gamma \pi_p(T|X_\gamma)$ for $0 < p \le \infty$.

Proof: Parts (a) and (b) are obvious from the definition. Part (c) requires a simple approximation argument (see II.E.12) and is omitted. ∎

6. Now we will give some very important examples of p-summing maps. Actually these are one operator acting between different spaces. Later on we will call this operator the Paley operator.

Proposition. For $f \in L_1(\mathbb{T})$ let $P(f) = (\hat{f}(2^n))_{n=1}^{\infty}$.

(a) $P: A \to \ell_2$ is 1-absolutely summing and onto.

(b) $P: C(\mathbb{T}) \to \ell_2$ is p-absolutely summing for $1 < p < \infty$ and onto.

(c) $P\colon C(\mathbb{T}) \to c_0$ *is 1-absolutely summing.*

Proof: It is clear that P is continuous in all the cases (a), (b) and
(c). That P is onto in cases (a) and (b) follows directly from Theorem
III.E.9. In order to see (a) let us factor

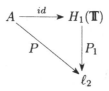

where P_1 is the Paley projection, i.e. the operator P acting on $H_1(\mathbb{T})$.
It follows from Paley's inequality I.B.24 that P_1 is continuous. We see
from Proposition 5 (a), (b) and from **4** that id is 1-absolutely summing
so P is also (see (4)). To see (b) we consider the factorization

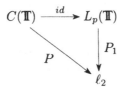

where $P_1\left(\sum_{-\infty}^{+\infty} \hat{f}(n)e^{in\theta}\right) = (\hat{f}(2^n))_{n=1}^{\infty}$. The operator P_1 is bounded
by the remark after Proposition III.A.7 so P is p-absolutely summing
by (4) and **4**. For (c) we use the factorization

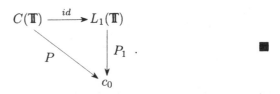

7.   The above proposition easily yields the fundamental theorem of
Grothendieck.

Theorem. (Grothendieck).   *Every operator* $T\colon L_1(\mu) \to H$, *where*
H *is a Hilbert space, is 1-absolutely summing.*

Remark.   It follows from the closed graph theorem or from the proof
given below that there exists a constant K such that $\pi_1(T) \le K\|T\|$ for
all $T\colon L_1(\mu) \to H$. The smallest such constant is called the Grothendieck
constant and denoted by K_G.

Proof: Let us start with an operator T: $\ell_1 \to \ell_2$. Using Proposition 6 (a) we have a commutative diagram

where φ is defined by $\varphi(e_j) = f_j$ where $(e_j)_{j=1}^{\infty}$ are unit vectors in ℓ_1 and $f_j \in A$ are such that $\|f_j\| \le C\|T\|$ and $P(f_j) = T(e_j), j = 1, 2, \ldots$. Since P is onto, such f_j's can be chosen. Clearly φ is continuous with $\|\varphi\| \le C\|T\|$. Thus we get $\pi_1(T) \le \pi_1(P)\|\varphi\| \le C_1\|T\|$. This implies in particular that for every N and every operator T: $\ell_1^N \to \ell_2$ we have $\pi_1(T) \le C_1\|T\|$. The theorem for arbitrary $L_1(\mu)$ follows from this observation, Proposition 5(c) and II.E.5(a)   ∎

Surprisingly this is a very powerful theorem and is probably the most important result of this chapter. We will discuss this theorem in greater detail later on. For now, we will continue our study of general properties of p-absolutely summing maps.

8.   The following result, known as the Pietsch factorization theorem, shows that the operators described in **4** are models for general p-absolutely summing operators.

Theorem.   Let $0 < p < \infty$. The following are equivalent for an operator T: $X \to Y$.

(a) T is p-absolutely summing with $\pi_p(T) \le C$.

(b)   There exist a Borel probability measure μ on $(B_{X^*}, \sigma(X^*, X))$ and a constant C such that

$$\|Tx\| \le C\left(\int_{B_{X^*}} |x(x^*)|^p d\mu(x^*) \right)^{\frac{1}{p}} \quad \text{for} \quad x \in X. \qquad (9)$$

Moreover if $1 \le p < \infty$ conditions (a) and (b) are equivalent to

(c)   for every (equivalently for some) isometric embedding i: $X \to C(K)$ there exists a Borel probability measure μ on K and a constant C such that

$$\|Tx\| \le C\left(\int_K |i(x)|^p d\mu \right)^{\frac{1}{p}} \quad \text{for} \quad x \in X. \qquad (10)$$

Let us rephrase the condition (10) (or (9)). Denote by $X_p \subset L_p(K, \mu)$ the closure of $i(X)$ in $L_p(K, \mu)$. Then (10) means that T induces a continuous linear operator $\tilde{T}: X_p \to Y$. Thus we have the commutative diagram

$$
\begin{array}{ccc}
X & \xrightarrow{T} & Y \\
i \downarrow & & \uparrow \tilde{T} \\
i(X) & \xrightarrow{id} & X_p \\
\downarrow & & \downarrow \\
C(K) & \xrightarrow{id} & L_p(K, \mu)
\end{array} \qquad (10')
$$

Proof: The fact that (b) and (c) imply (a) follows directly from (10'). One simply has to check that $\pi_p(id) < \infty$. For $p \geq 1$ this was done in (8). For $K = B_{X^*}$ the same argument works also for $p < 1$. We will prove that (a) implies (b) and (c). Let us take any isometric embedding $i: X \to C(K)$. Let $V \subset C(K)$ be the set of all functions $\phi(k)$ of the form

$$
\phi(k) = C^p \sum_j |i(x_j)(k)|^p - \sum_j \|Tx_j\|^p \quad \text{for some finite } (x_j) \subset X.
$$

The set V is a convex cone in $C(K)$. Let C^- be the open, convex cone of all negative functions in $C(K)$. If $K = B_{X^*}$ then (6) yields $V \cap C^- = \emptyset$ for $0 < p < \infty$. For other K's and $p \geq 1$ we get

$$
\sup \left\{ \sum |x^*(x_j)|^p: \ x^* \in X^*, \|x^*\| \leq 1 \right\}
$$
$$
= \sup \left\{ \sum \left| \int i(x_j) d\mu \right|^p : \ \mu \in M(K), \|\mu\| \leq 1 \right\}
$$
$$
= \sup \left\{ \sum |i(x_j)(k)|^p: \ k \in K \right\}
$$

since $\varphi(\mu) = \sum_j |\int i(x_j) d\mu|^p$ is a $\sigma(M(K), C(K))$-continuous, convex function on $B_{M(K)}$, so it attains its maximum on an extreme point. Thus $V \cap C^- = \emptyset$ in this case also. By the Hahn-Banach (see I.A.11) and the Riesz (see I.B.11) theorems there exists a measure μ on K with $\|\mu\| = 1$ such that

$$
\int_K \phi(k) d\mu \geq 0 \quad \text{for all} \quad \phi \in V \qquad (11)
$$

and

$$
\int_K f(k) d\mu \leq 0 \quad \text{for all} \quad f \in C^-. \qquad (12)
$$

Since C^- consists of all negative functions (12) implies that μ is a probability measure. Taking ϕ corresponding to the one-element family $\{x\}$ we get from (11)

$$\|Tx\|^p \le C^p \int |i(x)|^p d\mu \quad \text{for all} \quad x \in X.$$

This shows that (a) implies (b) and (c).   ∎

In general we have almost no control on the measure μ. Let us however consider the following special situation. Let G be a compact, abelian group with the Haar measure m. Assume that G acts as a group of homeomorphisms of a compact space K, i.e. for each $g \in G$ we have a homeomorphism i_g of K such that $i_g \cdot i_h = i_{g \cdot h}$. For a function $f \in C(K)$ and $g \in G$ we define $I_g f(k) = f(i_g(k))$. Assume that $X \subset C(K)$ is a closed invariant subspace, i.e. $I_g(X) \subset X$ for all $g \in G$. Assume that $T \colon X \to Y$ is a p-absolutely summing operator, $1 \le p \le \infty$ such that $\|TI_g x\| = \|Tx\|$ for all $g \in G$ and $x \in X$. Then there exists a measure μ on K such that

$$\|Tx\| \le \pi_p(T)\left(\int |x|^p d\mu \right)^{\frac{1}{p}} \text{ and } i_g(\mu) = \mu \text{ for all } g \in G. \qquad (13)$$

Let $\tilde{\mu}$ be a probability measure on K given by Theorem 8. We have

$$\|Tx\|^p = \int_G \|TI_g x\|^p dm(g)$$

$$\le \pi_p(T)^p \int_G \int_K |x(i_g(k))|^p d\tilde{\mu}(k) dm(g)$$

$$= \pi_p(T)^p \int_K |x(k)|^p d\mu(k)$$

where for $A \subset K$ we have $\mu(A) = \int_G \tilde{\mu}(i_g(A)) dm(g)$, so (13) follows.

Let us also note that (a) does not imply (c) if $p < 1$. To see this let us take the operator (really a functional) $T \colon C[0,1] \to \mathbb{R}$ defined by $Tf = \int_0^1 f(s) ds$. Clearly it is p-summing for every $p > 0$ with $\pi_p(T) = 1$. Suppose that there is a probability measure μ on $[0,1]$ satisfying (10), so we have

$$\left| \int_0^1 f(s) ds \right|^p \le C \int |f(s)|^p d\mu(s) \quad \text{for all } f \in C[0,1]. \qquad (14)$$

One checks that if (14) holds for some measure μ it also holds for the part of it which is absolutely continuous with respect to the Lebesgue measure, μ_c. But μ_c is a non-zero, non-atomic measure and (14) says that T induces a non-zero linear functional on $L_p([0,1], \mu_c)$. Since $p < 1$ this is impossible; see I.B.4.

9.   From Pietsch's theorem we can derive some properties of p-absolutely summing operators.

Corollary.   (a) If $0 < p < q < \infty$ then $\Pi_p(X,Y) \subset \Pi_q(X,Y)$ for all Banach spaces X, Y. Also $\pi_p(\cdot) \geq \pi_q(\cdot)$.
   (b) Every p-absolutely summing operator $0 < p < \infty$ is weakly compact and maps relatively weakly compact sets onto norm-pre-compact sets.
   (c) If X is a subspace of X_1 and $T \in \prod_p(X,Y)$ with $p \geq 2$, then T extends to an operator $T_1 \in \prod_2(X_1,Y)$.

Proof: (a) and (b) follow directly from (10') and the fact that $id: i(x) \rightarrow X_p$ is q-absolutely summing and satisfies (b). For (c) observe that by (a) we can assume $p = 2$. Then use (10') and the fact that every subspace in a Hilbert space is complemented.

10.   All this may seem rather trivial and quite abstract. In order to convince the reader that these are important and powerful concepts, before investigating them any further, let us give some applications of the results already obtained.

Theorem.   Let X be a complemented subspace of $L_1(\mu)$. Assume also that $(x_n, x_n^*)_{n=1}^\infty$ is a normalized unconditional basis in X. Then there exists a constant C such that $\sum_{n=1}^\infty |x_n^*(x)| \leq C\|x\|$ for all $x \in X$, so $(x_n)_{n=1}^\infty$ is equivalent to the unit vector basis in ℓ_1.

Proof: Since X has cotype 2 (see III.A.21-23) or by the Orlicz theorem (see II.D.6) we get that the map $T: X \rightarrow \ell_2$ defined by $T(x) = (x_n^*(x))_{n=1}^\infty$ is continuous. Let $P: L_1(\mu) \overset{\text{onto}}{\longrightarrow} X$ be a projection. By Theorem 7 the operator TP is 1-absolutely summing so $T = TP|X$ is also 1-absolutely summing. Thus for every finite sequence of scalars $(a_n)_{n=1}^N$ we have

$$\sum_{n=1}^N |a_n| = \sum_{n=1}^N \|T(a_n x_n)\| \leq \pi_1(T) \sup_{\|x^*\| \leq 1} \sum_{n=1}^N |x^*(a_n x_n)|$$

$$= \pi_1(T) \sup_{\|x^*\| \leq 1} \sup_{|\varepsilon_n|=1} \left\| x^* \left(\sum_{n=1}^N \varepsilon_n a_n x_n \right) \right\|$$

$$\leq \pi_1(T) ubc(x_n) \left\| \sum_{n=1}^{N} a_n x_n \right\|. \qquad \blacksquare$$

Let us note some special instances of this theorem

(a) Since $L_1[0,1]$ is not isomorphic to ℓ_1 (see III.A.5 and III.A.7) we get a new proof that $L_1[0,1]$ does not have an unconditional basis (see II.D.10).

(b) The spaces $L_p[0,1]$ and $\ell_p, 1 < p \leq 2$, are not isomorphic to complemented subspaces of $L_1[0,1]$, although they are isomorphic to subspaces of $L_1[0,1]$ (see Notes and remarks to III.A). The same conclusion can be easily derived from III.D.34(b) (see Exercise III.D.20). Note however the important difference between these two arguments. Using III.D.34(b) we get no information about the complementation of ℓ_p^n in $L_1[0,1]$. A careful reading of the above proof gives that if P is a projection from $L_1(\mu)$ onto a subspace isometric to ℓ_p^n then $\|P\| \geq Cn^{1-\frac{1}{p}}, 1 \leq p \leq 2$. This reflects the difference between the global, infinite dimensional approach and local, more quantitative one.

(c) ℓ_1 has only one unconditional basis in the sense that every normalized unconditional basis in ℓ_1 is equivalent to the unit vector basis.

11. Let $W_1^l(\mathbb{T}^2)$, $l \geq 1$, denote the Sobolev space (see I.B.30). By the Sobolev embedding Theorem I.B.30 the identity operator $id: W_1^l(\mathbb{T}^2) \to W_2^{l-1}(\mathbb{T}^2)$ is continuous. Note that $W_2^{l-1}(\mathbb{T}^2)$ is a Hilbert space. We want to show that it is not 2-absolutely summing. Since $W_2^l(\mathbb{T}^2) \subset W_1^l(\mathbb{T}^2)$ it suffices to show that $id: W_2^l(\mathbb{T}^2) \to W_2^{l-1}(\mathbb{T}^2)$ is not 2-absolutely summing. Let us put $f_{n,m}(\theta, \varphi) = (m+n)^{-l}e^{in\theta}e^{im\varphi}$, $n, m > 0$. We have

$$\sum_{0<m+n\leq N} \|id(f_{n,m})\|^2$$

$$\geq \sum_{0<n+m\leq N} \sum_{j=0}^{l-1} \left[\frac{n^j m^{l-1-j}}{(n+m)^l} \right]^2$$

$$= \sum_{r=0}^{N} r^{-2l} \sum_{n+m=r} \sum_{j=0}^{l-1} \left[n^j m^{l-1-j} \right]^2$$

$$\geq C_l \sum_{r=0}^{N} r^{-2l} \cdot r \cdot r^{2(l-1)} \qquad (15)$$

$$= C_l \sum_{r=1}^{N} \frac{1}{r}$$

$$= C_l \log N .$$

On the other hand $\|f_{n,m}\|_{W_2^l(\mathbf{T}^2)} \leq C$ and $(f_{n,m})$ is orthogonal in $W_2^l(\mathbf{T}^2)$ so we get

$$\sup_{\|x^*\| \leq 1} \sum_{0 < n+m \leq N} |x^*(f_{n,m})|^2 \leq C^2. \tag{16}$$

Clearly (15) and (16) show that id is not 2-absolutely summing.

Comparing the above argument with Theorem 7 we infer that $W_1^l(\mathbf{T}^2)$ is not isomorphic to any $L_1(\mu)$-space. Averaging over $k - 2$ variables we see that $W_1^l(\mathbf{T}^2)$ is complemented in $W_1^l(\mathbf{T}^k)$ for $l \geq 1$ and $k \geq 2$. Thus we obtain

Theorem.   *For $l \geq 1$ and $k \geq 2$ the Sobolev space $W_1^l(\mathbf{T}^k)$ is not isomorphic to a complemented subspace of any $L_1(\mu)$-space.*   ∎

12.   We also want to show how these general concepts apply to some concrete problems in harmonic analysis. We will see more of this later. Let G be a compact abelian group with Haar measure m.

Proposition.   *An invariant operator $T: C(G) \to C(G)$ is 1-absolutely summing if and only if $Tf = f * h$ for some $h \in L_\infty(G)$. We have $\pi_1(T) = \|h\|_\infty$.*

Proof: Suppose $h \in L_\infty(G)$. As is well known and easy to check (see I.B.13) $f * h \in C(G)$ for all $f \in L_1(G)$. Thus every $T: C(G) \to C(G)$ which is of the form $Tf = f * h$ for some $h \in L_\infty(G)$ is 1-absolutely summing with $\pi_1(T) \leq \|h\|_\infty$. Conversely if $T: C(G) \to C(G)$ is translation invariant and $\pi_1(T) = 1$, from Theorem 8 and (13) we get

$$\|Tf\|_\infty \leq \int_G |f(g)| dm(g) \quad \text{for every} \quad f \in C(G). \tag{17}$$

Since T is translation invariant, there exists a measure μ such that $Tf = f * \mu$. From (17) we infer that $\|f * \mu\|_\infty \leq \|f\|_1$ for all $f \in L_1(G)$. An easy limiting argument shows that $\|\nu * \mu\|_\infty \leq \|\nu\|_1$ for all $\nu \in M(G)$. Taking as ν the Dirac measure at the neutral element of G we get that $\mu = hdm$ for some $h \in L_\infty(G)$, with $\|h\|_\infty \leq 1$.   ∎

This proposition can yield some very concrete results.

13 Theorem.   *Suppose $\Lambda \subset \mathbb{N}$ is a set of cardinality k. Given $\varepsilon > 0$ there exists a trigonometric polynomial f such that*

$$\hat{f}(\ell) = 1 \quad \text{for all} \quad \ell \in \Lambda, \tag{18}$$

$$\|f\|_1 \leq 1 + \varepsilon, \tag{19}$$

$$|\{\ell \in \mathbb{N}\colon \hat{f}(\ell) \neq 0\}| \leq \left(\frac{c}{\varepsilon}\right)^{2k} \text{ for some constant } c. \tag{20}$$

Proof: In operator-theoretical terms we are looking for a rotation invariant operator $T\colon C(\mathbb{T}) \to C(\mathbb{T})$ such that

$$T|E = id \quad \text{where } E = \operatorname{span}\{e^{in\theta}\}_{n\in\Lambda}, \tag{18'}$$

$$\|T\| \leq 1 + \varepsilon, \tag{19'}$$

$$\operatorname{rank} T \leq \left(\frac{c}{\varepsilon}\right)^{2k}. \tag{20'}$$

Let us fix a number δ such that $0 < \delta < 1$. From II.E.13 there exist a natural number $N < (\frac{3}{\delta})^{2k}$ and an embedding $u\colon E \to \ell_\infty^N$ with $\|u\| \cdot \|u^{-1}\| \leq 1 + \delta$. Let $\tilde{u}\colon C(\mathbb{T}) \to \ell_\infty^N$ be an extension of u with $\|u\| = \|\tilde{u}\|$ (see III.B.1) and let $v\colon \ell_\infty^N \to C(\mathbb{T})$ be an extension of u^{-1} with $\|v\| \leq (1+\delta)\|u^{-1}\|$. Observe that by II.E.5(c) there exists a subspace F with $E \subset F \subset C(\mathbb{T})$ such that $d(F, \ell_\infty^{\dim F}) \leq 1 + \delta$, so such an extension exists (see also III.E.14). Let us put $T_1 = v\tilde{u}\colon C(\mathbb{T}) \to C(\mathbb{T})$. Clearly $T_1|E = id$ and

$$\pi_1(T_1) \leq \|v\| \cdot \|\tilde{u}\| \pi_1(id\colon \ell_\infty^N \to \ell_\infty^N)$$
$$\leq \|v\| \cdot \|\tilde{u}\| \cdot N \leq (1+\delta)^2 N.$$

We now define

$$T_2 = \frac{1}{2\pi} \int_0^{2\pi} I_\theta T_1 I_{-\theta}\, d\theta$$

where I_θ is the rotation by the angle θ. The operator T_2 is rotation invariant and satisfies

$$T_2|E = id, \tag{18''}$$

$$\|T_2\| \leq \|T_1\| \leq (1+\delta)^2, \tag{19''}$$

$$\pi_1(T_2) \leq \pi_1(T_1) \leq (1+\delta)^2 N. \tag{20''}$$

From Proposition 12 we know that T_2 is convolution with a certain function h which satisfies

$$\hat{h}(\ell) = 1 \quad \text{for} \quad \ell \in \Lambda, \tag{18'''}$$

$$\|h\|_1 \leq (1+\delta)^2, \tag{19'''}$$

$$\|h\|_\infty \leq (1+\delta)^2 N. \tag{20'''}$$

Observe that (19''') and (20''') give $\|h\|_2 \leq (1+\delta)^2\sqrt{N}$. We define $g = h * h * h$ and put

$$f(\theta) = \sum_{s:|\hat{g}(s)|>N^{-4}} \hat{g}(s)e^{is\theta}.$$

We have

$$\|f\|_1 \leq \|g\|_1 + \|g - f\|_1 \leq (1+\delta)^6 + \sum_{s:|\hat{g}(s)|<N^{-4}} |\hat{g}(s)|$$

$$\leq (1+\delta)^6 + N^{-\frac{4}{3}}\sum |\hat{h}(s)|^2 = (1+\delta)^6 + N^{-\frac{4}{3}}\|h\|_2^2$$

$$\leq (1+\delta)^6 + (1+\delta)^4 N^{-\frac{1}{3}}$$

and also

$$|\{k \in \mathbb{N}: \hat{f}(k) \neq 0\}| = |\{k:|\hat{h}(k)| > N^{-\frac{4}{3}}\}|$$

$$\leq \|h\|_2^2 N^{\frac{8}{3}} \leq (1+\delta)^4 N^4.$$

Since $\hat{f}(k) = 1$ for $k \in \Lambda$ we see that for appropriately chosen δ the polynomial f satisfies (18)-(20). ∎

Clearly the same argument works for any compact abelian group.

14. Now we will present a different form of Grothendieck's theorem, the so-called Grothendieck inequality. We will derive it from Theorem 7. Actually it is also easy to derive Theorem 7 from the Grothendieck inequality.

Theorem. (Grothendieck's inequality). *Let $(a_{n,m})_{n,m=1}^N$ be a finite or infinite matrix such that for every two sequences of scalars $(\alpha_n)_{n=1}^N$ and $(\beta_m)_{m=1}^N$ we have*

$$\left|\sum_{n,m=1}^N a_{n,m}\alpha_n\beta_n\right| \leq \sup_n |\alpha_n| \cdot \sup_n |\beta_n|. \tag{21}$$

Then for any two sequences $(h_n)_{n=1}^N$ and $(k_m)_{n=1}^N$ in an arbitrary Hilbert space H we have

$$\left|\sum_{n,m=1}^N a_{n,m}\langle h_n, k_m\rangle\right| = K_G \sup_n \|h_n\| \cdot \sup_n \|k_m\| \tag{22}$$

where K_G is the Grothendieck constant.

Proof: Let us define an operator $T \colon \ell_1^N \to H$ by $T(e_n) = h_n$ for $n = 1, 2, \ldots, N$. Clearly $\|T\| = \sup_n \|h_n\|$. For every $n = 1, 2, \ldots, N$ let us put $z_n = (a_{n,m})_{m=1}^N \in \ell_1^N$. From Theorem 7 and (21) we get

$$\sum_{n=1}^N \|Tz_n\|$$

$$\leq K_G\|T\| \sup \left\{ \sum_{n=1}^N |\langle v, z_n\rangle|\colon v \in \ell_\infty^N, \|v\|_\infty \leq 1 \right\}$$

$$= K_G\|T\| \sup \left\{ \sum_{n=1}^N \left| \sum_{m=1}^N \beta_m a_{n,m} \right| \colon \right.$$

$$\left. |\beta_m| \leq 1 \text{ for } m = 1, 2, \ldots, N \right\} \qquad (23)$$

$$\leq K_G\|T\| \sup \left\{ \sum_{n,m=1}^N \alpha_n \beta_m a_{n,m}\colon |\alpha_n| \leq 1, |\beta_m| \leq 1 \right\}$$

$$= K_G \sup_n \|h_n\|.$$

On the other hand

$$\sum_{n=1}^N \|Tz_n\| = \sum_{n=1}^N \left\| \sum_{m=1}^N a_{nm} h_m \right\|$$

$$\geq \frac{1}{\sup \|k_n\|} \left| \sum_{n=1}^N \langle k_n, \sum_{m=1}^N a_{nm} h_m \rangle \right| \qquad (24)$$

$$= \frac{1}{\sup \|k_n\|} \left| \sum_{n,m=1}^N a_{n,m} \langle k_n, h_m \rangle \right|.$$

Comparing (23) and (24) we get (22).   ■

Note that (21) means simply that the matrix $(a_{n,m})$ defines a norm-one operator from ℓ_∞^N into ℓ_1^N.

15.   The Hilbert space figuring so prominently in Grothendieck's theorem, it is not surprising that it has important applications to Hilbert space operators. We will discuss some of them presently.

An operator $T \colon H \to H$ is said to be power bounded if $\sup_{n \geq 0} \|T^n\| < \infty$. Clearly every operator of the form $T = VT_1V^{-1}$

where $\|T_1\| \le 1$ and V is an isomorphism of H, is power bounded. Actually the classical von Neumann inequality proved in von Neumann [1951] (see Exercise III.E.8 or Exercise III.H.19 or Nagy-Foias [1967]) gives that for every such T and every polynomial φ

$$\|\varphi(T)\| \le C \sup_{|z| \le 1} |\varphi(z)|. \tag{25}$$

An operator satisfying (25) is said to be polynomially bounded. We will discuss those notions a little. Let us start with some examples.

Proposition. *For every sequence* $c = (c_n)_{n=0}^\infty$ *with* $c_n = \pm 1$ *if* $n = 2^k$ *and* $c_n = 0$ *otherwise, there exists an operator* $T_c\colon H \to H$ *such that*

$$\sup_c \sup_n \|T_c^n\| < \infty, \tag{26}$$

there are $\xi, \eta \in H$ *such that* $\langle T_c^n \xi, \eta \rangle = c_n$ *for* $n = 0, 1, 2, \dots$. (27)

Let us observe that not every operator T_c is polynomially bounded. For a polynomial φ we have

$$\|\varphi(T_c)\| \, \|\xi\| \, \|\eta\| \ge |\langle \varphi(T_c)\xi, \eta \rangle| = \left| \sum_n \hat\varphi(n) \langle T_c^n \xi, \eta \rangle \right| \tag{28}$$

$$= \left| \sum_n \hat\varphi(n) c_n \right| = \left| \sum_k c_{2^k} \hat\varphi(2^k) \right|.$$

Observe that Theorem III.E.9 gives polynomials φ with $\|\varphi\|_\infty \le 1$ but $\left| \sum_k c_{2^k} \hat\varphi(2^k) \right|$ as big as we please.

Proof of the proposition.   Let us put

$$a(n, m) = \begin{cases} c_{n+m} & \text{if} \quad 0 \le n \le m, \\ 0 & \text{if} \quad n > m \ge 0, \end{cases}$$

and

$$b(n, m) = \begin{cases} c_{n+m} & \text{if} \quad n > m \ge 0, \\ 0 & \text{if} \quad m \ge n \ge 0. \end{cases}$$

One easily checks that $a(n, m)$ is a matrix of a bounded operator A on ℓ_1 while $b(n, m)$ is a matrix of a bounded operator B on ℓ_∞ with norms bounded by a constant independent of c. On sequences $(\xi_n)_{n=0}^\infty$ we formally define operators

$$S_k((\xi_n)_{n=0}^\infty) = (\xi_k, \xi_{k+1}, \dots) \text{ and } S_k^*((\xi_n)_{n=0}^\infty) = (\underbrace{0, \dots, 0}_{k \text{ times}}, \xi_0, \xi_1, \dots).$$

One checks that for finite sequences $(\xi_n)_{n=0}^{\infty}$ one has $AS_k^* - S_kA = S_kB - BS_k^* \stackrel{\text{def}}{=} V_k$. From these relations we infer that V_k acts on both ℓ_1 and ℓ_{∞} with norms uniformly bounded in k. Thus by interpolation (see I.B.6) V_k acts also on ℓ_2 and $\|V_k\| \leq C$. We now define H as a completion of $\ell_1 \oplus \ell_1$ with respect to the scalar product

$$\langle\langle (f_1, g_1), (f_2, g_2) \rangle\rangle = \langle f_1, f_2 \rangle + \langle A(f_1) + g_1, A(f_2) + g_2 \rangle$$

where $\langle \cdot, \cdot \rangle$ is the usual scalar product. We define $T_c \colon H \to H$ by $T_c(f, g) = (S_1^* f, S_1 g)$. Clearly $T_c^k(f, g) = (S_k^* f, S_k g)$. Since

$$
\begin{aligned}
\|T_c^k(f, g)\|_H^2 &= \|S_k^* f\|_2^2 + \|AS_k^* f + S_k g\|_2^2 \\
&\leq \|f\|_2^2 + (\|V_k f\|_2 + \|S_k Af + S_k g\|_2)^2 \\
&\leq \|f\|_2^2 + (C\|f\|_2 + \|Af + g\|_2)^2 \leq C\|(f, g)\|_H^2
\end{aligned}
$$

we get (26). With $\xi = (e_0, 0)$ and $\eta = (0, e_0)$ we get for $n = 0, 1, 2, \ldots$

$$\langle\langle T_C^n \xi, \eta \rangle\rangle = \langle\langle (e_n, 0), (0, e_0) \rangle\rangle = \langle A(e_n), e_0 \rangle = a(n, 0) = c_n.$$

So (27) also holds. ∎

16. Since the estimate (25) does not hold for every power-bounded operator one has to seek other estimates for $\|\varphi(T)\|$. For a polynomial $\varphi(z) = \sum_{n \geq 0} \hat{\varphi}(n) z^n$ we put

$$\alpha(\varphi) = \inf\{\||(a_{n,m})_{n,m \geq 0}\||\colon \sum_{n+m=k} a_{nm} = \hat{\varphi}(k), \quad k = 0, 1, 2, \ldots\}$$

where

$$\||(a_{n,m})_{n,m \geq 0}\|| = \sup\left\{\left|\sum_{n,m} \varepsilon_n \eta_m a_{nm}\right|\colon |\varepsilon_n| = 1, |\eta_m| = 1\right\}.$$

Theorem. Let $T \colon H \to H$ be such that $\sup_{n \geq 0} \|T^n\| = C$. Then for every polynomial $\varphi(z)$ we have

$$\|\varphi(T)\| \leq C^2 K_G \alpha(\varphi). \tag{29}$$

Proof: For arbitrary $x, y \in H$ with $\|x\| = \|y\| = 1$ and arbitrtary $(a_{n,m})_{n,m \geq 0}$ with $\sum_{n+m=k} a_{nm} = \hat{\varphi}(k)$ we have

$$\langle \varphi(T)x, y \rangle = \langle \sum_{k \geq 0} \hat{\varphi}(k)T^k x, y \rangle$$

$$= \sum_k \hat{\varphi}(k) \langle T^k x, y \rangle = \sum_{n,m} a_{nm} \langle T^n T^m x, y \rangle \qquad (30)$$

$$= \sum_{n,m} a_{nm} \langle T^m x, (T^n)^* y \rangle.$$

Since $\|T^m x\| \leq C$ and $\|(T^n)^* y\| \leq C$ for all m and n and all x, y as above the Grothendieck inequality (Theorem 14) applied to (30) yields

$$|\langle \varphi(T)x, y \rangle| \leq C^2 K_G \| |(a_{n,m})| \|. \qquad (31)$$

Since x, y are arbitrary vectors with $\|x\| = \|y\| = 1$ and $(a_{n,m})$ is also arbitrary, (31) implies (29). ∎

17.   In order to apply Theorem 16 we need some method of computing or estimating $\alpha(\varphi)$. This is generally a non-trivial matter. The following lemma will nevertheless give some interesting information.

Lemma.   Let $w_n(t)$ for $n = 1, 2, \ldots$ be the piecewise linear function determined by the conditions $w_n(t) = 0$ for $t \leq 2^{n-1}$ and $t \geq 2^{n+1}$ and $w_n(2^n) = 1$. Let us define a sequence of polynomials $(W_n(z))_{n \geq 0}$ as follows. We put $W_0(z) = 1 + z$ and for $n \geq 1$ we put $W_n(z) = \sum_{k \geq 0} w_n(k)z^k$. Then for every polynomial $\varphi(z)$ we have

$$\alpha(\varphi) \leq \sum_{n \geq 0} \|\varphi * W_n\|_\infty. \qquad (32)$$

Proof: For every polynomial φ we have $\varphi = \sum_{n \geq 0} \varphi * W_n$. Since $\alpha(\cdot)$ is a rotation invariant norm on polynomials one gets $\alpha(f * \varphi) \leq \|f\|_1 \alpha(\varphi)$. As is well known, $\|W_n\|_1 \leq 2$ for $n = 0, 1, 2, \ldots$ (this can be seen by writing W_n as a sum of shifted Fejér kernels; see I.B.16), so it is enough to show that for φ such that supp $\hat{\varphi} \subset [2^{n-1}, 2^{n+1}]$ one has $\alpha(\varphi) \leq C\|\varphi\|_\infty$. Given such a φ let us define $\psi = \sum k^{-1} \hat{\varphi}(k)z^k$. We can write $\psi = z^{2^{n-1}} \cdot [(\varphi \cdot z^{-2^{n-1}}) * g_n]$ where $g_n = \sum_{k=-\infty}^{+\infty} (2^{n-1} + |k|)^{-1} e^{ik\theta}$. Comparing Fourier coefficients we see that

$$g_n = \sum_{k=1}^\infty k\left(\frac{1}{2^{n-1} + k - 1} + \frac{1}{2^{n-1} + k + 1} - \frac{2}{2^{n-1} + k} \right) \mathcal{F}_{k-1}$$

where $\mathcal{F}_k$ is a Féjer kernel. This shows that $g_n \geq 0$ and

$$\|g_n\|_1 \leq \sum_{k=1}^{\infty} k \left(\frac{1}{2^{n-1} + k - 1} + \frac{1}{2^{n-1} + k + 1} - \frac{2}{2^{n-1} + k} \right) = \frac{1}{2^{n-1}}$$

so $\|\psi\|_\infty \leq \|\varphi\|_\infty \cdot \|g_n\|_1 \leq 2^{-n+1} \|\varphi\|_\infty$. Let $(a_{n,m})_{n,m \geq 0}$ be the matrix given by $a_{n,m} = \hat{\psi}(n + m)$. It is known (see Exercise III.E.5 or Duren [1970]) that this matrix acts on ℓ_2 with norm $\leq \|\psi\|_\infty \leq 2^{-n+1} \|\varphi\|_\infty$. Observe also that $\alpha(\varphi) \leq \|(a_{n,m}): \ell_\infty \to \ell_1\|$. Since actually $(a_{n,m}): \ell_\infty^{2^{n+1}} \to \ell_1^{2^{n+1}}$ we have the factorization

$$\ell_\infty^{2^{n+1}} \xrightarrow{id} \ell_2^{2^{n+1}} \xrightarrow{(a_{n,m})} \ell_2^{2^{n+1}} \longrightarrow \ell_1^{2^{n+1}}$$

so

$$\begin{aligned}
\alpha(\varphi) &\leq \|(a_{n,m}): \ell_\infty \to \ell_1\| \\
&\leq \|id: \ell_\infty^{2^{n+1}} \to \ell_2^{2^{n+1}}\| \cdot \|(a_{n,m}): \ell_2 \to \ell_2\| \cdot \|id: \ell_2^{2^{n+1}} \to \ell_1^{2^{n+1}}\| \\
&\leq 2\|\varphi\|_\infty. \quad \blacksquare
\end{aligned}$$

Remark. The estimate for $\|g_n\|_1$ is a repetition of a well known theorem in the theory of Fourier series (see Katznelson [1968] Th. I.4.1).

18 Corollary. *If φ is a polynomial of degree n and T is a power-bounded operator on a Hilbert space then $\|\varphi(T)\| \leq C \cdot \log(n+2) \cdot \|\varphi\|_\infty$.*

Proof: This follows directly from (29) and (32). $\quad \blacksquare$

It seems to be unknown if this is the best estimate. The operators constructed in Proposition 15 can be used to show that at least $\sqrt{\log(n + 2)}$ is needed.

There is also a direct proof of Corollary 18. Using the canonical factorization I.B.23 we write the Dirichlet kernel $\sum_{k=0}^n e^{ik\theta}$ as a product of two functions $f = \sum_{k=0}^\infty \alpha_k e^{ik\theta}$ and $g = \sum_{k=0}^\infty \beta_k e^{ik\theta}$ such that

$$\left(\sum_{k=0}^\infty |\alpha_k|^2 \right)^{\frac{1}{2}} \cdot \left(\sum_{k=0}^\infty |\beta_k|^2 \right)^{\frac{1}{2}} = \frac{1}{2\pi} \int_0^{2\pi} |\sum_{k=0}^n e^{ik\theta}| d\theta \leq C \log(n + 2).$$

Then we can write

$$p(T) = \frac{1}{2\pi} \int_0^{2\pi} p(t) \left(\sum_{k=0}^\infty \alpha_k e^{-ikt} T^k \right) \left(\sum_{k=0}^\infty \beta_k e^{-ikt} T^k \right) dt$$

so for every $x, y \in H$ with $\|x\| = \|y\| = 1$ we have

$$|\langle p(T)x, y\rangle| \leq \|p\|_\infty \cdot \frac{1}{2\pi} \int_0^{2\pi} \left\|\sum_{k=0}^\infty \alpha_k e^{-ikt} T^k x\right\| \cdot \left\|\sum_{k=0}^\infty \beta_k e^{ikt} (T^k)^* y\right\| dt$$

$$\leq \|p\|_\infty \cdot \left(\frac{1}{2\pi} \int_0^{2\pi} \left\|\sum_{k=0}^\infty \alpha_k e^{-ikt} T^k x\right\|^2 dt\right)^{\frac{1}{2}}$$

$$\cdot \left(\frac{1}{2\pi} \int_0^{2\pi} \left\|\sum_{k=0}^\infty \beta_k e^{ikt} (T^k)^* y\right\|^2 dt\right)^{\frac{1}{2}}.$$

Representing the Hilbert space H as $L_2(\Omega, \mu)$ and using the orthonormality of the sequence $(e^{ikt})_{k=-\infty}^{+\infty}$ in $L_2(\mathbb{T})$ we get

$$\left(\frac{1}{2\pi} \int_0^{2\pi} \left\|\sum_{k=0}^\infty \alpha_k e^{-ikt} T^k x\right\|^2 dt\right)^{\frac{1}{2}} = \left(\sum_{k=0}^\infty |\alpha_k|^2 \|T^k x\|^2\right)^{\frac{1}{2}}$$

and analogously

$$\left(\frac{1}{2\pi} \int_0^{2\pi} \left\|\sum_{k=0}^\infty \beta_k e^{ikt} (T^k)^* y\right\|^2 dt\right)^{\frac{1}{2}} = \left(\sum_{k=0}^\infty |\beta_k|^2 \|(T^k)^* y\|^2\right)^{\frac{1}{2}}.$$

Putting this together we get

$$|\langle p(T)x, y\rangle|$$

$$\leq \|p\|_\infty \left(\sum_{k=0}^\infty |\alpha_k|^2\right)^{\frac{1}{2}} \cdot \left(\sum_{k=0}^\infty |\beta_k|^2\right)^{\frac{1}{2}} \cdot \max_k \|T^k x\| \max_k \|(T^k)^* y\|$$

$$\leq C\|p\|_\infty \log(n+2).$$

19.   We hope that the previous discussion justifies the opinion that a more detailed study of p-absolutely summing operators is interesting and important. Before we resume such a study we would like to introduce two new families of Banach operator ideals.

An operator $T: X \to Y$ (X, Y arbitrary Banach spaces) is p-nuclear, $1 \leq p \leq \infty$, if it can be written in the form

$$T(x) = \sum_{j=1}^\infty x_j^*(x) y_j \qquad (33)$$

with

$$\left(\sum_{j=1}^{\infty}\|x_j^*\|^p\right)^{\frac{1}{p}} \sup_{y^*\in Y^*,\|y^*\|\leq 1}\left(\sum_{j=1}^{\infty}|y^*(y_j)|^{p'}\right)^{\frac{1}{p'}} < \infty$$

$$\text{where}\quad \frac{1}{p}+\frac{1}{p'}=1. \tag{34}$$

Obvious modifications are made in (34) for $p=1$ and $p=\infty$. We define the p-nuclear norm $n_p(T)$ as the infimum of the quantities (34) over all representations (33). The class of all p-nuclear operators from X into Y is denoted by $N_p(X,Y)$.

An equivalent definition of p-nuclear operators runs as follows: an operator $T\colon X\to Y$ is p-nuclear if it admits a factorization

$$
\begin{array}{ccc}
X & \xrightarrow{\;T\;} & Y \\
u\downarrow & & \uparrow v \\
\ell_\infty & \xrightarrow{\;\Delta\;} & \ell_p
\end{array}
\tag{35}
$$

where u,v are continuous operators and Δ is a diagonal operator, i.e. $\Delta(\xi_j)=(\delta_j\xi_j)$ for some $(\delta_j)\in\ell_p$.

Given a representation (33) satisfying (34) we define $u(x) = \left(\frac{x_j^*(x)}{\|x_j^*\|}\right)_{j=1}^{\infty}$ and $(\delta_j)_{j=1}^{\infty}=(\|x_j^*\|)_{j=1}^{\infty}$ and $v(\xi_j)=\sum_j\xi_j y_j$. The diagram (35) then commutes. Conversely given (35) the above formulas easily yield (33) and (34). Note also that

$$n_p(T) = \inf\{\|u\|\cdot\|v\|\cdot\|\Delta\|\colon\ u,v,\Delta\ \text{satisfy (35)}\}.$$

We can interpret $\Delta\colon\ell_\infty\to\ell_p$ as id: $\ell_\infty\to L_p(\mathbf{N},\mu)$ with the measure μ defined by $\mu(S)=\sum_{j\in S}|\delta_j|^p$. This shows that $N_p(X,Y)\subset\Pi_p(X,Y)$ for $1\leq p\leq\infty$ with $\pi_p(\cdot)\leq n_p(\cdot)$. It is also pretty clear from (35) that $N_p(X,Y)$ is an operator ideal. Actually the proof of the following is routine.

20 Theorem. *The p-nuclear operators with p-nuclear norm $n_p(\cdot)$, $1\leq p\leq\infty$ form a Banach operator ideal.* ∎

The most important case is that of the 1-nuclear operators, simply called nuclear or trace class operators. Note also that $N_\infty(X,Y)$ is simply the class of operators admitting a factorization through ℓ_∞. Thus if E is a (finite dimensional) Banach space and id_E denotes the identity operator on E, then $n_\infty(id_E) = \gamma_\infty(E)$, where $\gamma_\infty(E)$ is as defined and studied in III.B.3-5.

21.   We say that an operator $T\colon X \to Y$ is p-integral, $1 \le p \le \infty$, and write $T \in I_p(X,Y)$ if it admits a factorization

$$
\begin{array}{ccc}
X & \xrightarrow{\ \ T\ \ } & Y \\[4pt]
\alpha \big\downarrow & & \big\uparrow \beta \\[4pt]
C(K) & \xrightarrow{\ \ id\ \ } & L_p(K,\mu)
\end{array}
\qquad (36)
$$

where μ is a probability measure on a compact space K and α,β are continuous linear operators. We define the p-integral norm

$$i_p(T) = \inf\{\|\alpha\| \cdot \|\beta\|\colon \text{there exists factorization (36)}\}.$$

We have the following routine theorem whose proof is omitted.

Theorem.   *The p-integral operators $I_p(X,Y)$ with norm $i_p(T)$, $1 \le p \le \infty$, form a Banach operator ideal.*   ∎

22.   A comparison between (36), (19) and (10′) gives

Proposition.   *For all Banach spaces X,Y and numbers p with $1 \le p \le \infty$ we have*
$$N_p(X,Y) \subset I_p(X,Y) \subset \Pi_p(X,Y)$$
with the corresponding inequality for norms, i.e.
$$\pi_p(\cdot) \le i_p(\cdot) \le n_p(\cdot). \qquad\qquad ∎$$

Let us also note that the Pietsch theorem (Theorem 8) implies that for every Y and every p with $1 \le p \le \infty$ we have $I_p(C(K),Y) = \Pi_p(C(K),Y)$ with equality of norms. Note also that it follows from the Pietsch theorem (see 10′) and the fact that every subspace of a Hilbert space is 1-complemented, that for all Banach spaces X,Y we have $I_2(X,Y) = \Pi_2(X,Y)$. With equality of norms.

23 Theorem.   *Let $T\colon X \to Y$ be a p-integral operator, $1 \le p \le \infty$. Assume also that $i\colon X \to C(S)$ is an isomorphic embedding. Then T*

admits a factorization

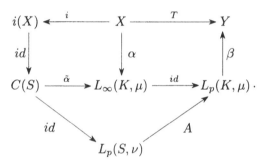

where ν is a probability measure on S and B is a continuous linear operator with $\|B\| \leq i_p(T) \cdot \|i\| \cdot \|i^{-1}\| + \varepsilon$.

Proof: The proof is best illustrated by the following diagram where K, μ, α, β have the same meaning as in (36).

On $i(X) \subset C(S)$ we consider the operator $\alpha i^{-1} \colon i(X) \to L_\infty(K, \mu)$. Using III.B.2 we find an extension $\tilde{\alpha} \colon C(S) \to L_\infty(K, \mu)$ with $\|\tilde{\alpha}\| = \|\alpha \cdot i^{-1}\| \leq \|\alpha\| \|i^{-1}\|$. Since $id\tilde{\alpha}$ is p-absolutely summing with $\pi_p(id\tilde{\alpha}) \leq \|\tilde{\alpha}\|$ the Pietsch factorization theorem (Theorem 8) gives a probability measure ν on S and an operator A, with $\|A\| \leq \|\tilde{\alpha}\|$. The desired factorization is given by $B = \beta A$. ∎

24 Corollary. *Let X be a finite dimensional Banach space and let $T \colon X \to Y$. Then $i_p(T) = n_p(T)$ for $1 \leq p \leq \infty$.*

Proof: It is enough to show that $n_p(T) \leq i_p(T)$ (see Proposition 22). Given $\varepsilon > 0$ we get from II.E.13 an isomorphic embedding $i \colon X \to \ell_\infty^N$ with $\|i\| \cdot \|i^{-1}\| \leq 1 + \varepsilon$. Since $\ell_\infty^N = C(S)$ where $S = \{1, 2, \ldots, n\}$ we apply Theorem 23 to get a measure ν. Since $id \colon C(S) \to L_p(S, \nu)$ is the same as $\Delta \colon \ell_\infty^N \to \ell_p^N$ with $\Delta = (\delta_n)_{n=1}^N$ where $\delta_n = \nu(\{n\})^{\frac{1}{p}}$ we get the p-nuclear factorization. ∎

25 Corollary. *If $T \colon X \to Y$ with X and Y finite dimensional, then*

$$\pi_2(T) = i_2(T) = n_2(T).$$

Proof: From remarks made after Proposition 22 we know that $\pi_2(\cdot) = i_2(\cdot)$ (always) so the claim follows from Corollary 24.   ∎

The reader should consult the exercises to find examples showing that the above classes are different in general. One of the reasons that p-nuclear and p-integral operators are important is that they are connected with p-absolutely summing operators via duality. Before we proceed we have to discuss duality as applied to Banach operator ideals.

26.   Given a Banach operator ideal $I(X, Y)$ we want to describe the dual space $I(X, Y)^*$. In general this subject is quite involved; the theory of general tensor products and the approximation property play important roles. We will discuss it only for finite dimensional Banach spaces X and Y. Despite this restriction, the results can be applied in the study of p-absolutely summing operators on infinitive dimensional spaces. This follows from Proposition 5.

If $T: X \to X$ (X finite dimensional) is a linear operator then T has a representation (non-unique of course) in the form $T(x) = \sum_j x_j^*(x)x_j$. The trace of T is defined as $tr(T) = \sum_j x_j^*(x_j)$. As is well known and easily checked this definition is correct, i.e. it does not depend on the particular representation of T. Obviously the trace is a linear functional on $L(X, X)$. Given $T: X \to Y$ and $S: Y \to X$ we see that $tr(ST) = tr(TS)$. For every operator $S: Y \to X$ the formula $\varphi_S(T) = tr(TS)$, for $T \in L(X, Y)$, defines a linear functional on $L(X, Y)$. Counting dimensions we realize that we can identify $L(X, Y)^*$ with $L(Y, X)$ if the duality is given by the trace, i.e. $\langle S, T \rangle = tr(ST)$. All this is elementary linear algebra and can be found in most textbooks on the subject.

If we have a norm i on $L(X, Y)$ then the norm i^* on $L(Y, X)$ is dual to i (with respect to trace duality) if for every $T \in L(X, Y)$ we have

$$i(T) = \sup\{tr(ST): \ i^*(S) \leq 1\}. \tag{37}$$

27.   The following important theorem identifies π_p^*.

Theorem.   *Let X and Y be finite dimensional Banach spaces. Then $(\Pi_p(X, Y), \pi_p)^* = (N_{p'}(Y, X), n_{p'}), 1 \leq p \leq \infty$, with the trace duality.*

Proof: Let us fix $T \in L(X, Y)$. For $S: Y \to X$ with the representation $S(y) = \sum_j y_j^*(y) \cdot x_j$ we have

$$tr(TS) = \sum_j y_j^*(Tx_j) \leq \sum_j \|y_j^*\| \cdot \|Tx_j\|$$

$$\leq \left(\sum_j \|y_j^*\|^{p'} \right)^{\frac{1}{p'}} \left(\sum_j \|Tx_j\|^p \right)^{\frac{1}{p}} \tag{38}$$

$$\leq \left(\sum_j \|y_j^*\|^{p'} \right)^{\frac{1}{p'}}$$

$$\pi_p(T) \sup \left\{ \left(\sum_j |x^*(x_j)|^p \right)^{\frac{1}{p}} : x^* \in X^*, \|x^*\| \leq 1 \right\}.$$

Since (38) holds for every representation of S we get $|tr(TS)| \leq n_{p'}(S) \cdot \pi_p(T)$.

On the other hand let us fix $x_1, \ldots, x_n$ in X such that $\sum_j |x^*(x_j)|^p \leq 1$ for every $x^* \in X^*$ with $\|x^*\| \leq 1$ and such that $\left(\sum_j \|Tx_j\|^p \right)^{\frac{1}{p}} \geq (1 - \varepsilon)\pi_p(T)$. Let us choose $y_j^* \in Y^*$ such that $\|y_j^*\| = 1$ and $y_j^*(Tx_j) = \|Tx_j\|$ for $j = 1, 2, \ldots, n$. Let us also fix numbers $a_j \geq 0$, $j = 1, 2, \ldots, n$ such that $\sum_j a_j^{p'} = 1$ and $\sum_j a_j \|Tx_j\| = \left(\sum_j \|Tx_j\|^p \right)^{\frac{1}{p}}$. We define $S \colon Y \to X$ by the formula $S(y) = \sum_{j=1}^n a_j y_j^*(y)x_j$. Since

$$n_{p'}(S) \leq \left(\sum_j a_j^{p'} \right)^{\frac{1}{p'}} \cdot \sup \left\{ \left(\sum_j |x^*(x_j)|^p \right)^{\frac{1}{p}} : x^* \in X^*, \|x^*\| \leq 1 \right\} \leq 1$$

and

$$tr(TS) = \sum_j a_j y_j^*(Tx_j) \geq (1 - \varepsilon)\pi_p(T)$$

we get the claim.   ∎

28.   Our aim now is to establish the dual version of Grothendieck's theorem. Because of later applications in III.I we give a more abstract presentation than is really necessary.

Proposition. *Let X be a finite dimensional Banach space. The following conditions are equivalent:*

(a)   *for every $T \colon X \to \ell_1$ we have $\pi_2(T) \leq C\|T\|$;*

(b)   *for every $T \colon X^* \to \ell_2$ we have $\pi_1(T) \leq C_1\|T\|$.*
More precisely if (a) holds then $C_1 \leq K_G \cdot C$. If (b) holds then $C \leq C_1$.

Proof: (a)⇒(b). Clearly we can restrict our attention to $T \colon X^* \to \ell_2^N$. From Theorem 27 we see that we have to estimate $tr(TS)$ for every

$S: \ell_2^N \to X^*$ with $n_\infty(S) = 1$. Using the standard approximation (see Proposition II.E.12) we get the diagram

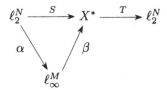

with $\|\alpha\| \cdot \|\beta\| \le 1 + \varepsilon$.

Dualizing this diagram we get

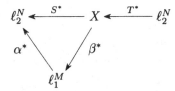

From Corollary 25 and Theorem 7 we get using (a)

$$|tr(TS)| = |tr(S^*T^*)| \le \pi_2(\alpha^*)\pi_2(\beta^*T^*)$$
$$\le \pi_2(\alpha^*)\pi_2(\beta^*)\|T^*\| \le K_G\|\alpha^*\|C\|\beta^*\| \cdot \|T^*\|$$
$$\le K_G C(1 + \varepsilon)\|T\|.$$

Since ε was arbitrary we get (b).

(b)$\Rightarrow$(a). As before we use Theorem 27 so we have to look for $tr(TS)$ where we have the following diagram:

$$\begin{array}{ccccc}
\ell_1^N & \xrightarrow{\ S\ } & X & \xrightarrow{\ T\ } & \ell_1^N \\
{\scriptstyle\alpha}\downarrow & & {\scriptstyle\beta}\uparrow & & \\
\ell_\infty^M & \xrightarrow{\ \Delta\ } & \ell_2^M & &
\end{array}$$

with $\|\alpha\| \cdot \|\beta\| \cdot \|\Delta\| \le 1 + \varepsilon$.

Dualizing this diagram we get

$$\begin{array}{ccccc}
\ell_\infty^N & \xleftarrow{\ S^*\ } & X^* & \xleftarrow{\ T^*\ } & \ell_\infty^N \\
{\scriptstyle\alpha^*}\uparrow & & {\scriptstyle\beta^*}\downarrow & & \\
\ell_1^M & \xleftarrow{\ \Delta^*\ } & \ell_2^M & &
\end{array}$$

Using (b) we get $\pi_1(S^*T^*) \leq \|\alpha^*\|\|\Delta^*\|\|T^*\|\pi_1(\beta^*) \leq C_1(1+\varepsilon)\|T\|$. But S^*T^* is an operator on ℓ_∞^N so by remarks made after Proposition 22 we have $\pi_1(S^*T^*) = i_1(S^*T^*)$. Corollary 24 gives $i_1(S^*T^*) = n_1(S^*T^*)$ so we have

$$|tr(TS)| = |tr(S^*T^*)| \leq n_1(S^*T^*) \leq C_1(1+\varepsilon)\|T\|.$$

This gives (a). ∎

29 Theorem. *For any $C(K)$-space (in particular for $L_\infty(\mu)$) and $1 \leq p \leq 2$ we have*

$$L(C(K), L_p) = \Pi_2(C(K), L_p) \text{ with } \pi_2(\cdot) \leq K_G\|\cdot\|.$$

Proof: From Proposition 5 and II.E.5 we infer that it is enough to show $\pi_2(T) \leq K_G\|T\|$ for every $T\colon \ell_\infty^N \to \ell_p^M, 1 \leq N, M < \infty$. The Grothendieck theorem (Theorem 7) says that $X = \ell_1^N$ satisfies (b) of Proposition 28 so we get

$$\pi_2(T) \leq K_G\|T\| \text{ for every } T\colon \ell_\infty^N \to \ell_1^M. \tag{39}$$

If $1 < p \leq 2$ then ℓ_p^M is isometric to a subspace of $L_1[0,1]$ (see III.A.16). From II.E.12 and an easy approximation we see that we can assume $\ell_p^M \subset \ell_1^{M'}$ so (39) and Proposition 5 give the theorem. ∎

30. This dual version of the Grothendieck Theorem also has some nice applications. We will discuss some of these, connected with harmonic analysis. Let G be a compact, abelian group with dual group Γ and Haar measure m. The following is a classical result of Orlicz, Paley and Sidon.

Theorem. *Let $\Lambda = (\lambda_\gamma)_{\gamma \in \Gamma}$ be a function on Γ. The map $\Lambda(f) = (\lambda_\gamma \hat{f}(\gamma))_{\gamma \in \Gamma}$ maps $C(G)$ into $\ell_1(\Gamma)$ if and only if $\Lambda \in \ell_2(\Gamma)$.*

Proof: If $\Lambda \in \ell_2(\Gamma)$ then Λ acts not only from $C(G)$ into $\ell_1(\Gamma)$ but also from $L_2(G)$ into $\ell_1(\Gamma)$, simply because characters form a complete orthonormal system in $L_2(G)$. Conversely if $\Lambda\colon C(G) \to \ell_1(\Gamma)$, then by Theorem 29 $\pi_2(\Lambda) \leq K_G\|\Lambda\|$. The Pietsch factorization theorem (Theorem 8) and (13) give

$$\|\Lambda f\| \leq K_G\|\Lambda\| \cdot \left(\int_G |f(g)|^2 dm(g)\right)^{\frac{1}{2}}.$$

But this means that

$$\sum_{\gamma\in\Gamma}|\lambda_\gamma|\,|\hat{f}(\gamma)| \le K_G\|\Lambda\|\left(\sum_{\gamma\in\Gamma}|\hat{f}(\gamma)|^2\right)^{\frac{1}{2}}$$

for every $(\hat{f}(\gamma))_{\gamma\in\Gamma}\in\ell_2(\Gamma)$ so $\left(\sum_{\gamma\in\Gamma}|\lambda_\gamma|^2\right)^{\frac{1}{2}}\le K_G\|\Lambda\|.$ ■

31. Let us recall that a subset $S\subset\Gamma$ is called a Sidon set if $\{\gamma\}_{\gamma\in S}\subset C(G)$ is equivalent to the unit vector basis in $\ell_1(S)$. For $S\subset\Gamma$ the symbol $C_S(G)$ will denote span$\{\gamma:\ \gamma\in S\}\subset C(G)$. The following theorem shows that the Banach space structure of $C_S(G)$ determines if S is Sidon or not.

Theorem. *If $C_S(G)^*$ is isomorphic to some $C(K)$ space than S is a Sidon set.*

Proof: Let us fix $f\in C_S(G)$ and let us consider the operator $T_f\colon M(G)\to C(G)$ given by the formula $T_f(\mu)=f*\mu$. We can factorize this operator as follows:

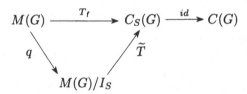

where q is the quotient map and $I_S=\{\mu\in M(G)\colon\hat{\mu}(\gamma)=0$ for $\gamma\in S\}$. Clearly $M(G)/I_S\cong C_S(G)^*\sim C(K)$ and also the space $C_S(G)$ can be considered as a subspace of $C_S(G)^{**}\sim M(K)$ so by Theorem 29 we get $\pi_2(\widetilde{T})<\infty$. This implies $\pi_2(T_f)<\infty$. But Grothendieck's theorem implies that also $\pi_1(T_f)<\infty$. The proof is completed by an application of the following.

32 Lemma. *If an operator $T\colon M(G)\to C(G)$ given by $T(\mu)=\sum_{\gamma\in\Gamma}a_\gamma\hat{\mu}(\gamma)\gamma$ is 1-absolutely summing then $\sum_{\gamma\in\Gamma}|a_\gamma|<\infty$, and conversely.*

Proof: Let $L\subset\Gamma$ be a finite set and let $E\subset C(G)$ be such that $d(E,\ell_\infty^N)\le 1+\varepsilon$ and $\gamma\in E$ for every $\gamma\in L$. Let P be a projection from $C(G)$ onto E such that $\|P\|\le 1+\varepsilon$. Let $L_1^L=$ span$\{\gamma:\ \gamma\in L\}\subset L_1(G)\subset M(G)$. For the operator $S\colon L_1^L\to E$ defined by $S(\mu)=PT(\mu)$

we have $\pi_1(S) \leq (1+\varepsilon)\pi_1(T)$. For every sequence $\eta = (\eta_\gamma)_{\gamma \in L}$ with $|\eta_\gamma| = 1$ we define a map $u_\eta: E \to L_1^L$ by $u_\eta(f) = \left(\sum_{\gamma \in L} \eta_\gamma \hat{f}(\gamma) \cdot \gamma\right)$. Since $\| \sum_{\gamma \in L} \eta_\gamma \hat{f}(\gamma) \cdot \gamma \|_1 \leq \| \sum_{\gamma \in L} \eta_\gamma \hat{f}(\gamma) \cdot \gamma \|_2 \leq \|f\|_2 \leq \|f\|_\infty$ we get $\|u_\eta\| \leq 1$ and thus also $n_\infty(u_\eta) \leq 1 + \varepsilon$.

From Theorem 27 we get

$$\left| \sum_{\gamma \in L} \eta_\gamma a_\gamma \right| = |tr(u_\eta S)| \leq \pi_1(S) \cdot n_\infty(u_\eta) \leq (1+\varepsilon)^2 \pi_1(T).$$

Since this holds for every $\varepsilon > 0$, every finite set $L \subset \Gamma$ and every η we get $\sum_{\gamma \in \Gamma} |a_\gamma| \leq \pi_1(T)$. The converse is obvious.   ∎

33.   Now we return to the study of p-absolutely summing operators. One lesson to be learnt from our previous discussions is that it is very useful to have equalities of the form $\Pi_p(X, Y) = L(X, Y)$. This is only rarely true so we try to look for equalities of the form $\Pi_p(X, Y) = \Pi_q(X, Y)$. These are also useful (see III.H.12). Before we proceed let us state some easy but useful facts.

Proposition.   (a) *Let $1 \leq p \leq \infty$ and let $T: X \to Y$. Then*

$$\pi_p(T) = \sup\{\pi_p(TS): \quad S: \ell_{p'}^m \to X, \ \|S\| \leq 1, \ m = 1, 2, \ldots\}. \quad (40)$$

(b)   *If $T: X \to Y$ is an operator and $0 < p \leq \infty$ and (Ω, μ) is any measure space and $f(\omega)$ is an X-valued Bochner integrable function (see III.B.28) then*

$$\left(\int \|Tf(\omega)\|^p d\mu(\omega) \right)^{\frac{1}{p}} \leq \pi_p(T) \sup_{x^* \in X^*, \|x^*\| \leq 1} \left(\int |x^*(f(\omega))|^p d\mu(\omega) \right)^{\frac{1}{p}}. \quad (41)$$

Proof: (a) Clearly the quantity on the right hand side of (40) does not exceed $\pi_p(T)$ (see (4) and Theorem 3). On the other hand let us take $x_1, \ldots, x_m \subset X$ such that $\sup \left\{ \left(\sum_{j=1}^m |x^*(x_j)|^p \right)^{\frac{1}{p}}: \|x^*\| \leq 1 \right\} = 1$ and $\left(\sum_{j=1}^m \|T(x_j)\|^p \right)^{\frac{1}{p}} \geq \pi_p(T) - \varepsilon$. Let us define $S: \ell_{p'}^m \to X$ by $S(e_j) = x_j$

for $j = 1, 2, \ldots, m$. We have

$$\|S\| = \sup\left\{\left\|\sum_{j=1}^{m}\alpha_j x_j\right\| : \sum_{j=1}^{m}|\alpha_j|^{p'} \le 1\right\}$$

$$= \sup\left\{\sum_{j=1}^{m}\alpha_j x^*(x_j) : \|x^*\| \le 1 \text{ and } \sum_{j=1}^{m}|\alpha_j|^{p'} \le 1\right\}$$

$$= \sup\left\{\left(\sum_{j=1}^{m}|x^*(x_j)|^p\right)^{\frac{1}{p}} : \|x^*\| \le 1\right\} = 1.$$

Also

$$\pi_p(T) - \varepsilon \le \left(\sum_{j=1}^{m}\|T(x_j)\|^p\right)^{\frac{1}{p}} = \left(\sum_{j=1}^{n}\|TS(e_j)\|^p\right)^{\frac{1}{p}} \le \pi_p(TS).$$

(b) By a standard approximation argument it is enough to check (41) for step functions of the form $\sum_{j=1}^{m} x_j \xi_{A_j}$ for disjoint sets A_j, $j = 1, 2, \ldots, m$. If for some j we have $\mu(A_j) = \infty$ then both sides of (34) are infinite. If for all j we have $\mu(A_j) < \infty$ then we put $y_j = \mu(A_j)^{\frac{1}{p}} x_j$ for $j = 1, 2, \ldots, m$ and we see that (41) takes the form

$$\left(\sum_{j=1}^{n}\|Ty_j\|^p\right)^{\frac{1}{p}} \le \pi_p(T) \sup_{\|x^*\|\le 1, x^* \in X^*}\left(\sum_{j=1}^{n}|x^*(x_j)|^p\right)^{\frac{1}{p}}$$

which is clearly true.   ■

34.   We now return to the investigation of the identity $\Pi_p(X, Y) = \Pi_q(X, Y)$. We will prove an important extrapolation type result. It is analogous to Exercise III.A.2. We will use it in III.H and it also permits some improvements of our previous results. It is also the first result, besides the Pietsch factorization theorem, which deals with p-absolutely summing operators for $p < 1$.

Theorem.   Let X be a Banach space and let $1 \le q \le 2$. Suppose that for some number p, with $0 < p < q$ we have

$$\Pi_q(X, Y) = \Pi_p(X, Y) \text{ for all Banach spaces } Y. \tag{42}$$

Then for all Banach spaces Y and for all numbers p such that $0 < p < q$ we have $\Pi_p(X, Y) = \Pi_q(X, Y)$.

Proof: First note that (42) implies that there exists a constant C such that for every Y and for every operator $T\colon X \to Y$ we have

$$\pi_p(T) \le C\pi_q(T). \tag{43}$$

If (43) does not hold then there are $T_n\colon X \to Y_n$ with $\pi_q(T_n) = 1$ but $\pi_p(T_n) > 4^n, n = 1, 2, 3, \ldots$. Then the operator $T\colon X \to \left(\sum_{n=1}^{\infty} Y_n\right)_2$ defined by $T(x) = (2^{-n}T_n(x))_{n=1}^{\infty}$ is q-absolutely summing but not p-absolutely summing.

Let $\mathcal{K}$ denote $(B_{X^*}, \sigma(X^*, X))$ and let P denote the set of all probability measures on $\mathcal{K}$. We will identify X with its canonical image in $C(\mathcal{K})$. Using Theorem 8 and (43) we see that (42) is equivalent to

for every $\lambda \in \mathsf{P}$ there exists $\lambda_1 \in \mathsf{P}$ such that

$$\left(\int_{\mathcal{K}} |x(x^*)|^q d\lambda(x^*)\right)^{\frac{1}{q}} \le C\left(\int_{\mathcal{K}} |x(x^*)|^p d\lambda_1(x^*)\right)^{\frac{1}{p}} \text{ for } x \in X. \tag{44}$$

Analogously we see that in order to show the theorem it is enough to show that for every $r < 1, r < p$ and for every $\lambda \in \mathsf{P}$ there exists $\mu \in \mathsf{P}$ and $C_r > 0$ such that

$$\left(\int_{\mathcal{K}} |x(x^*)|^q d\lambda(x^*)\right)^{\frac{1}{q}} \le C_r\left(\int_{\mathcal{K}} |x(x^*)|^r d\mu(x^*)\right)^{\frac{1}{r}} \text{ for all } x \in X. \tag{45}$$

Note that if $\lambda = \lambda_1$ in (44) then the Hölder inequality (see Exercise III.A.2) gives (45) for all $r, 0 < r < q$ with $\mu = \lambda$. In general however $\lambda \ne \lambda_1$. We put $\lambda = \lambda_0$ and inductively applying (44) we get a sequence $\lambda_0, \lambda_1, \lambda_2, \ldots$ in P such that for all $x \in X$,

$$\left(\int |x|^q d\lambda_n\right)^{\frac{1}{q}} \le C\left(\int |x|^p d\lambda_{n+1}\right)^{\frac{1}{p}}.$$

We fix $\theta, 0 < \theta < 1$ such that $\frac{1}{p} = \frac{\theta}{r} + \frac{(1-\theta)}{q}$ and put $a_n = 2^{-n-1}$. For

$x \in X$ we have

$$\sum_{n \geq 0} a_n \|x\|_{L_q(\lambda_n)}$$

$$\leq C \sum_{n \geq 0} a_n \|x\|_{L_p(\lambda_{n+1})}$$

$$\leq C \sum_{n \geq 0} a_n \|x\|_{L_r(\lambda_{n+1})}^{\theta} \cdot \|x\|_{L_q(\lambda_{n+1})}^{1-\theta}$$

$$\leq C \left(\sum_{n \geq 0} a_n \|x\|_{L_r(\lambda_{n+1})} \right)^{\theta} \left(\sum_{n \geq 0} a_n \|x\|_{L_q(\lambda_{n+1})} \right)^{1-\theta}$$

$$= C \left(2 \sum_{n \geq 0} a_n \|x\|_{L_r(\lambda_n)} \right)^{\theta} \left(2 \sum_{n \geq 0} a_n \|x\|_{L_q(\lambda_n)} \right)^{1-\theta}.$$

Thus

$$\sum_{n \geq 0} a_n \|x\|_{L_q(\lambda_n)} \leq (2C)^{\frac{1}{\theta}} \sum_{n \geq 0} a_n \|x\|_{L_r(\lambda_n)}. \tag{46}$$

Since $r \leq 1$ we get

$$\sum_{n \geq 0} 2^{-n-1} \left(\int |x|^r d\lambda_n \right)^{\frac{1}{r}} \leq \left(\sum_{n \geq 0} 2^{-r(n+1)} \int |x|^r d\lambda_n \right)^{\frac{1}{r}}. \tag{47}$$

From (46) and (47) we obtain

$$\left(\int |x|^q d\lambda \right)^{\frac{1}{q}} \leq 2 \sum_{n \geq 0} a_n \|x\|_{L_q(\lambda_n)} \leq C_r \left(\sum_{n \geq 0} 2^{-r(n+1)} \int |x|^r d\lambda_n \right)^{\frac{1}{r}}$$

$$= C_r \left(\int |x|^r d\mu \right)$$

where $\mu = \sum_{n \geq 0} 2^{-r(n+1)} \lambda_n$. Normalizing μ appropriately we get (45). ∎

35 Corollary. (Grothendieck-Maurey).   *Every operator from $L_1(\mu)$ into a Hilbert space H is p-absolutely summing for every $p > 0$.*

Proof: From Theorem 7 and Corollary 9 we get $\Pi_2(L_1(\mu), H) = \Pi_1(L_1(\mu), H) = L(L_1(\mu), H)$. Theorem 34 gives the claim. ∎

Note that this is a strengthening of Theorem 7 which does not follow from our proof of it, simply because the operator $P: A \to \ell_2$ (see

Proposition 6(a)) is *not p*-absolutely summing for $p < 1$, (see Exercise III.I.2). Note also that for the above argument to work we do not need the full power of Theorem 7. It is enough to know that $L(L_1(\mu), H) = \Pi_p(L_1(\mu), H)$ for some $p < 2$. This fact can be derived from Proposition 6(b) exactly like Theorem 7 was derived from Proposition 6(a). If we avoid the use of Proposition 6 we can obtain some equalities of the form $\Pi_p(X, Y) = \Pi_q(X, Y)$ without knowing that $\Pi_p(X, Y) = L(X, Y)$. The following theorem is a useful example of such situation.

36 Theorem. *If X is a Banach space of cotype 2 then for any Banach space Y*

$$\Pi_p(X, Y) = \Pi_2(X, Y) \text{ for all } p \leq 2.$$

Let us start with the following.

37 Lemma. *If Y is a Banach space of cotype 2, then for any Banach space X and any $p \geq 2$ we have $\Pi_p(X, Y) = \Pi_2(X, Y)$, and $\pi_2(T) \leq C_2(Y) \cdot C_p \cdot \pi_p(T)$.*

Proof: Let us take $T \in \Pi_p(X, Y)$ and $x_1, \ldots, x_n \in X$. From the definition of cotype (see III.A.17) and Kahane's inequality (see III.A.20) we get

$$\left(\sum_{j=1}^{n} \|T x_j\|^2 \right)^{\frac{1}{2}} \leq C \left(\int \left\| \sum_{j=1}^{n} r_j(t) T(x_j) \right\|^p dt \right)^{\frac{1}{p}} \tag{48}$$

$$\leq C \left(\int \left\| T \left(\sum_{j=1}^{n} r_j(t) x_j \right) \right\|^p dt \right)^{\frac{1}{p}},$$

with $C = C_2(Y) \cdot C_p$.

Applying Proposition 33(b) and Khintchine's inequality to (48) we get

$$\left(\sum_{j=1}^{n} \|T x_j\|^2 \right)^{\frac{1}{2}} \leq C \pi_p(T) \sup_{\|x^*\| \leq 1} \left(\int \left| x^* \left(\sum_{j=1}^{n} r_j(t) x_j \right) \right|^p dt \right)^{\frac{1}{p}}$$

$$\leq C \pi_p(T) \sup_{\|x^*\| \leq 1} \left(\sum |x^*(x_j)|^2 \right)^{\frac{1}{2}},$$

so $\pi_2(T) \leq C \pi_p(T)$, and the constant is of the right form. The remaining inclusion is always true. ∎

Proof of Theorem 36.   We know from Theorem 34 that it is enough to show the theorem for a fixed $p, 1 < p < 2$. Also it is enough to consider X, Y finite dimensional and to keep track of the constants (see Proposition 5). Under these assumptions, for $T: X \to Y$ Theorem 27 and Lemma 37 give

$$\pi_p(T) = \sup\{tr(ST) : S: Y \to X, \; n_{p'}(S) \le 1\}$$
$$\le \sup\{tr(ST) : S: Y \to X, \; \pi_{p'}(S) \le 1\}$$
$$\le \sup\{tr(ST) : S: Y \to X, \; \pi_2(S) \le C_2(Y) \cdot C_p\}$$
$$= C_2(Y) \cdot C_p \cdot \sup\{tr(ST) : S: Y \to X, \; \pi_2(S) \le 1\}.$$

Applying Corollary 25 we get

$$\pi_p(T) \le C_2(Y)C_p\pi_2(T). \qquad \blacksquare$$

Notes and Remarks.
Much of this chapter, as well as much of modern Banach space theory, is the outgrowth of Grothendieck [1956]. The work of Grothendieck was phrased, however, in the language of tensor products and bilinear forms. This language, although still used by some and known by many, seems to have been generally replaced by the language of operators. In our book we adhere to this usage and avoid tensor products almost entirely.

The notion of Banach operator ideal emerged in the late 60's, mainly as a result of many attempts to understand Grothendieck [1956]. From this time on, A. Pietsch and his students and collaborators have studied many aspects of the abstract concept, contributing greatly to the creation of the theory of operator ideals as presented in Pietsch [1978]. The p-absolutely summing operators (probably the most important operator ideal) were introduced by Pietsch [1967] as a generalization of Grothendieck's 'application semiintégrale à droite' which are now called 1-absolutely summing operators. In this paper A. Pietsch proved the basic properties of p-absolutely summing operators, in particular the fundamental *Theorem 8* (the idea of using the separation argument in the proof is due to S. Kwapień) and *Corollary 9*. The Grothendieck theorem (*Theorem 7*) was proved by Grothendieck [1956] who called it the fundamental theorem of the metric theory of tensor products. The proof was understood and presented in the language of 1-summing operators by Lindenstrauss-Pełczyński [1968]. These authors proved the Grothendieck inequality (our *Theorem 14*) directly and derived *Theorem 7* from it. The proof presented here was found by A. Pełczyński with some help from the present author and was published in Pełczyński

[1977]. Numerous other proofs have been published for various versions of the Grothendieck theorem. We refer to Pisier [1986], Haagerup [1987] and Jameson [1987] for references. Considerable effort has gone into evaluating the Grothendieck constant K_G. It is known that this constant is different for real and complex scalars. The most precise estimates for the complex case are $1.338 \leq K_G^C \leq 1.40491$ (see Haagerup [1987] for the proof and a discussion of the known results).

There exist also C^*-algebra versions of the Grothendieck theorem. They are quite involved, but useful in the theory of C^*-algebras. As an example let us quote the following

Theorem A. *If A is a C^*-algebra and Y is a Banach space of cotype 2, then every linear operator T: $A \to Y$ factors through a Hilbert space.*

This was proved in Pisier [1986a]. This theorem is rather in the spirit of Chapter III.H but it is also the most Banach space theoretical statement. A more detailed presentation of even the most important results in this area requires, quite naturally, some familiarity with the theory of C^*-algebras. We refer the interested reader to Pisier [1986a] for the proof of this result and for a detailed description of and references to the earlier works.

Theorem 10 was proved in Lindenstrauss-Pełczyński [1968]. *Theorem 11* is a special case of a result proved by Kislyakov [1976]. This theorem should be compared with Proposition III.A.3. It is a Banach space manifestation of a phenomenon common in harmonic analysis that certain continuity results for important operators which hold for $1 < p < \infty$ fail for $p = 1$ or $p = \infty$. Incidentally one gets a rather crazy proof that the multipliers considered in Lemma III.A.4 are not continuous in $L_1(\mathbb{T}^2)$. The fact (Exercise III.G.13) that $C^1(\mathbb{T}^2)$ is not isomorphic to any $C(K)$-space was stated by Grothendieck [1956a] and the first proof was published by Henkin [1967]. Actually Henkin proved the much stronger result that $C^k(\mathbb{T}^\ell)$ for $k \geq 1$ and $\ell > 1$ is not homeomorphic to any $C(K)$-space with the homeomorphism and its inverse being uniformly continuous. The basic idea of the proof of *Theorem 11* has been applied in a similar but much more general context in Pełczyński-Senator [1986]. *Theorem 13* is taken from Bourgain [1987]. Its main point is the estimate (20). Apart from this estimate facts of this type are well known and much used in harmonic analysis (see Rudin [1962a] 2.6).

Problems centred around the von Neumann inequality (25) are among the most interesting in operator theory on Hilbert space. Even our small sample shows the remarkable variety of methods used. The first example of a power-bounded but not polynomially bounded operator was given by Lebow [1968]. He showed that an example constructed by Foguel [1964] of a power bounded operator which is not similar to any contraction has this property. We present here some results taken from Peller [1982] and Bożejko [1987]. More precisely the proof of *Proposition 15* is taken from Bożejko [1987] while *Theorem 16* and *Lemma 17* and *Corollary 18* are due to Peller. The direct proof of *Corollary 18* was communicated to the author by G. Pisier. Observe that estimates of the type $\|\phi(T)\| \leq \beta(\phi)$ where β is some norm on polynomials lead to a functional calculus for T on some class of functions. This is a very important subject in operator theory. It has many connections with other branches of analysis. The reader may consult Nikolskiĭ [1980] for a more complete picture. This subject has also a branch in the theory of Banach algebras. A nice result (once more relying on the Grothendieck theorem) is a theorem of Varopoulos [1975] that any Banach algebra X isomorphic as a Banach space to a $C(K)$-space is algebraically and topologically isomorphic to some subalgebra of the algebra of all operators on a Hilbert space.

It is an open problem asked by P. Halmos if every polynomially bounded operator T is similar to a contraction, i.e. is of the form $T = VT_1V^{-1}$ with $\|T_1\| \leq 1$ and V an isomorphism of an underlying Hilbert space. The best partial result in this direction seems to be contained in Bourgain [1986a]. He proved

Theorem B.   *If $T: H \rightarrow H$ satisfies $\|p(T)\| \leq M\|p\|_\infty$ for every polynomial p and if $\dim H = N < \infty$ then there exists $S: H \rightarrow H$ such that $\|STS^{-1}\| \leq 1$ and $\|S\|\|S^{-1}\| \leq M^4 \log N$.*

The proof is quite complicated and uses, among other things, Grothendieck's inequality and Theorem III.I.10. For applications of the Grothendieck theorem in the theory of stochastic processes the reader may consult Rao [1982]. For applications to interpolation theory the papers of V.I. Ovchinnikov should be consulted (e.g. Ovchinnikov [1976] and [1985]).

The notion of 1-nuclear operator goes back to Grothendieck. It is a generalization of σ_1 operators on a Hilbert space (see Remarks after III.G.18). The theory of p-nuclear operators has been developed in Chevet [1969] and Persson-Pietsch [1969]. The important and useful *Theorem 27* can be found in Persson-Pietsch [1969]. It is only a small sample of various duality results for other operator ideals. There is also a (more complicated) duality theory for operators on infinite dimensional spaces. *Proposition 28* is folklore. It is a formalization of the connection between *Theorem 7* and its dual form, *Theorem 29*. This connection was already known to Grothendieck [1956] and was quite explicit in Lindenstrauss-Pełczyński [1968]. The notion of p-integral operator and all our results about them can be traced to Persson-Pietsch [1969]. Actually our p-integral operators are quite often called in the literature *strictly p-integral*, with the name 'p-integral operator' reserved for operators $T\colon X \to Y$ such that iT is strictly p-integral (i.e. integral according to the definition given in **21**) where i is the canonical embedding of Y into Y^{**}. Such operators appear naturally in the duality theory for operator ideals when the spaces are infinite dimensional.

   Theorem 30 is a classical result of the theory of Fourier series. It is very similar in spirit to Theorem III.A.25. The theory of Sidon sets is an interesting part of commutative harmonic analysis. The standard, but a bit outdated, reference is Lopez-Ross [1975]. More recent advances in this area are connected with the use of Banach space methods. Our *Theorem 31* is one such example. It is a variant of a result of Varopoulos [1976]. A stronger result proved in Bourgain-Milman [1985] is

Theorem C. *If G is a compact abelian group with dual group Γ and if for $S \subset \Gamma$ the space $C_S(G)$ has cotype p for some finite p, then S is a Sidon set.*

   Theorem 34 and *36* are due to Maurey [1974]. Our proof of *Theorem 34* follows Pisier [1986] and is due to Maurey and Pisier. These theorems will be useful later on in Chapters III.H and III.I.

   There are many places where the theory of p-absolutely summing, p-integral and p-nuclear operators is presented. We conclude these remarks by listing some of them: Pietsch [1978], Pisier [1986], Tomczak-Jaegermann [1989], Kislyakov [1977], Jameson [1987].

Exercises

1. Let $p, q, r \geq 1$ be such that $\frac{1}{p} + \frac{1}{q} = \frac{1}{r}$ and let $T \in \Pi_p(X, Y)$ and $S \in \Pi_q(Y, Z)$. Show that $ST \in \Pi_r(X, Z)$ and $\pi_r(ST) \leq \pi_p(T)\pi_q(S)$.

2. Show that $id\colon \ell_1 \to \ell_2$ is not 1-integral. Is it p-integral for some $p > 1$? Show that $id\colon \ell_1 \to \ell_\infty$ is 1-integral.

3. (a) Let $T\colon L_1[0,1] \to C[0,1]$ be given by $Tf(x) = \int_0^x f(t)dt$. Show that T is not 1-absolutely summing.

(b) Let $\sigma\colon \ell_1^N \to \ell_\infty^N$ be defined as $\sigma((\xi_j)_{j=1}^N) = \left(\sum_{j=1}^n \xi_j\right)_{n=1}^N$. Show that

$$\pi_1(\sigma) \sim C \log(N+1).$$

4. For $f = \sum_{n=0}^\infty a_n z^n \in H_1(\mathbb{D})$ we define $T(f) = \left(\frac{a_n}{\sqrt{n+1}}\right)_{n=0}^\infty$. Show that $T\colon H_1(\mathbb{D}) \to \ell_2$ is bounded but not 1-absolutely summing.

5. Let $K(x,y)$ be a measurable function on $[0,1] \times [0,1]$. Let $Tf(x) = \int_0^1 K(x,y)f(y)dy$. Find necessary and sufficient conditions for $K(x,y)$ so that the operator T maps $C[0,1]$ into itself and is 1-absolutely summing.

6. Show that $id\colon C[0,1] \to L_p[0,1]$ is not q-absolutely summing for any $q < p$.

7. (a) Show that every p-nuclear operator, $1 \le p < \infty$, is compact.

   (b) Show that if $1 \le p < \infty$ and X is reflexive then $I_p(X,Y) = N_p(X,Y)$.

8. Show that if $id\colon X \to X$ is p-absolutely summing for some $p < \infty$, then X is finite dimensional. In particular in every infinite dimensional Banach space there exists an unconditionally convergent series which is not absolutely convergent.

9. (a) Let G be an infinite, compact, metrisable abelian group and let $\mu \in M(G)$ be such that $\hat{\mu}(\gamma) \to 0$ as $\gamma \to \infty, \mu \ge 0$ and μ is singular with respect to the Haar measure m. (Note that III.C.6 shows that such a μ exists.) Let $T_\mu\colon C(G) \to L_1(G,m)$ be defined by $T_\mu(f) = f * \mu$. Show that T is a compact, 1-integral but not 1-nuclear operator.

   (b) Let $\varphi \in L_\infty(T)\backslash C(T)$. Show that $Tf = f * \varphi$ considered as an operator on $C(T)$ is compact and 1-integral but not 1-nuclear.

10. Show Theorem 7 assuming Theorem 14.

11. Suppose that $f \in L_1(\mathbb{T})$ is such that $\hat{f}(2^k) = 1$ for $k = 1, 2, \ldots, N$ and $\int_{\mathbb{T}} |f| \le 1 + \varepsilon$. Show that for every $\alpha, 0 < \alpha < 1$, there exists a constant $C = C(\alpha, \varepsilon) > 0$ such that

$$|\{\ell\colon \hat{f}(\ell) \ne 0\}| \ge CN^{\alpha \ln N}.$$

12. (a) Suppose that X is a Banach space with unconditional basis $(x_n)_{n=1}^\infty$. Show that every 1-absolutely summing operator $T\colon X \to Y$ factors through ℓ_1.

   (b) Show that $C[0,1]$ does not have an unconditional basis. This is a special case of II.D.12 but try to prove it using (a).

(c) Suppose that $F \subset \mathbb{N}$ is a $\Lambda(2)$ set (see I.B.14) and suppose that $L_p^F = \operatorname{span}\{e^{in\theta}\}_{n \in F} \subset L_p(\mathbb{T})$, $p > 2$, has an unconditional basis. Show that the characters are unconditional in L_p^F. The same holds for $C_F = \operatorname{span}\{e^{in\theta}\}_{n \in F} \subset C(\mathbb{T})$.

13. Show that for every $p \neq 2$, there exists a subset $F \subset \mathbb{N}$ such that $id_F \colon C_F \to L_p^F$ (for notation compare Exercise 12 (c)) is p-absolutely summing but not p-integral.

III.G. Schatten-Von Neumann Classes

In this chapter we consider Schatten-von Neumann classes of operators on a Hilbert space and their applications in the theory of Banach spaces. We start with the notion of an approximation number of an operator between Banach spaces. We prove that the approximation numbers of an operator and its adjoint are the same. Then we study operators on Hilbert space. We prove the Weyl inequality and basic facts connecting eigenvalues, s-numbers and approximation numbers. Various characterizations of Hilbert-Schmidt operators are presented. We also show the classical Fredholm-Bernstein-Szasz theorem about Fourier coefficients of Hölder continuous functions. Next we give results about summability of eigenvalues of p-absolutely summing operators on a general Banach space and apply them to eigenvalues of Hille-Tamarkin integral operators.

1. Given an operator $T\colon X \to Y$ we define its approximation numbers

$$\alpha_n(T) = \inf\{\|T - T_n\|\colon\ T_n\colon X \to Y,\ \mathrm{rank}\ T_n < n\}, \quad n = 1, 2, \ldots .$$

Clearly $\alpha_1(T) = \|T\|$ and the sequence $(\alpha_n(T))_{n=1}^\infty$ is decreasing. If $\alpha_n(T) \to 0$ then T is a norm limit of finite dimensional operators, thus compact. One proves routinely that if Y has b.a.p. (see II.E.2) and $T\colon X \to Y$ is compact then $\alpha_n(T) \to 0$.

2 Proposition. *The following inequalities hold for every $n, m \geq 1$ and all operators T, S:*

$$\alpha_{n+m-1}(T + S) \leq \alpha_n(T) + \alpha_m(S); \tag{1}$$

$$\alpha_{n+m-1}(T \circ S) \leq \alpha_n(T) \cdot \alpha_m(S). \tag{2}$$

Proof: The argument for (1) is obvious. To prove (2) let us take any T_n with rank $T_n < n$ and any S_m with rank $S_m < m$. Then

$$\|TS - (T_nS + TS_m - T_nS_m)\| = \|(T - T_n)(S - S_n)\| \leq \|T - T_n\|\|S - S_m\|.$$

Since

$$\mathrm{rank}(T_nS + TS_m - T_nS_m) \leq\ \mathrm{rank}(T_n(S - S_m)) +\ \mathrm{rank}(TS_m)$$
$$< n + m - 1$$

we get

$$\alpha_{n+m-1}(TS) \leq \inf_{T_n,S_m} \|T - T_n\| \|S - S_m\| = \alpha_n(T)\alpha_n(S). \qquad \blacksquare$$

3. For $0 < p < \infty$ we define $A^p(X,Y)$ to be the set of all operators $T: X \to Y$ such that $\sum_{n=1}^{\infty} \alpha_n(T)^p < \infty$. We denote $(\sum_{n=1}^{\infty} \alpha_n(T)^p)^{\frac{1}{p}}$ as $a_p(T)$.

Proposition. *The quantity $a_p(T)$ is a quasi-norm on $A^p(X,Y)$ for $0 < p < \infty$. $A^p(X,Y)$ with this quasi-norm is a quasi-Banach operator ideal.*

Proof: Since $\alpha_n(T) = 0$ if rank $T < n$ we get that finite rank operators are in $A^p(X,Y)$. From (1) we get

$$
\begin{aligned}
a_p(T + S) = \left(\sum_{n=1}^{\infty} \alpha_n(T + S)^p \right)^{\frac{1}{p}} &\leq \left(2 \sum_{n=1}^{\infty} \alpha_{2n-1}(T + S)^p \right)^{\frac{1}{p}} \\
&\leq 2^{\frac{1}{p}} \left(\sum_{n=1}^{\infty} [\alpha_n(T) + \alpha_n(S)]^p \right)^{\frac{1}{p}} \leq C_p[a_p(T) + a_p(S)].
\end{aligned}
\tag{3}
$$

Since $a_p(\lambda T) = |\lambda| a_p(T)$ we get that $a_p(T)$ is a quasi-norm and III.F (1) holds. The conditions III.F (2), (4) and (5) are obvious. The proof of completeness follows the usual lines and is left to the reader. $\qquad \blacksquare$

We would like to point out that $a_p(T)$ is not a norm even for $p \geq 1$. In general on $A^p(X,Y)$ there is no equivalent norm (see Exercise 6), so $A^p(X,Y)$ cannot be made into a Banach operator ideal. The detailed analysis of the convexity of $A^p(X,Y)$ can be found in Pietsch [1987] 2.3.7.

4 Proposition. *If $T \in A^p(X,Y)$ and $S \in A^q(Y,Z)$ then $ST \in A^s(X,Z)$ with $\frac{1}{s} = \frac{1}{p} + \frac{1}{q}$.*

Proof: Using (2) and Hölder's inequality we get

$$
\begin{aligned}
a_s(ST) = \left(\sum_{n=1}^{\infty} \alpha_n(ST)^s \right)^{\frac{1}{s}} &\leq 2^{\frac{1}{s}} \left(\sum_{n=1}^{\infty} \alpha_{2n-1}(ST)^s \right)^{\frac{1}{s}} \\
&\leq 2^{\frac{1}{s}} \left(\sum_{n=1}^{\infty} \alpha_n(S)^s \cdot \alpha_n(T)^s \right)^{\frac{1}{s}} \leq 2^{\frac{1}{s}} a_p(T) a_q(S). \qquad \blacksquare
\end{aligned}
$$

5 Proposition. *If* $T: X \to Y$ *is a compact operator then* $\alpha_n(T) = \alpha_n(T^*)$, $n = 1, 2, \ldots$.

Proof: Since $\alpha_n(T^*) \leq \alpha_n(T)$ it is enough to check that $\alpha_n(T^{**}) \geq \alpha_n(T)$, $n = 1, 2, \ldots$. For a fixed n and arbitrary $\varepsilon > 0$ let us take V, a finite ε-net in $T(B_X)$. Since T is compact $T(B_X)$ is norm-dense in $T^{**}(B_{X^{**}})$ so V is also an ε-net in $T^{**}(B_{X^{**}})$. Fix also an operator $T_n: X^{**} \to Y^{**}$ with rank $T_n < n$ and $\|T^{**} - T_n\| \leq \alpha_n(T^{**}) + \varepsilon$. Put $F = \mathrm{span}\{V \cup T_n(X^{**})\}$. From the principle of local reflexivity II.E.14 we get an operator $\varphi: F \to X$ with $\|\varphi\| \leq 1 + \varepsilon$ and $\varphi \mid F \cap X = id$. In particular $\varphi|V = id$. For $x \in X$ with $\|x\| \leq 1$ let us fix $v \in V$ such that $\|Tx - v\| \leq \varepsilon$. Then we have

$$\|Tx - \varphi T_n x\| \leq \varepsilon + \|v - \varphi T_n x\| \leq \varepsilon + (1 + \varepsilon)\|v - T_n x\|$$
$$\leq \varepsilon + (1 + \varepsilon)(\|v - Tx\| + (\|T^{**}x - T_n x\|)$$
$$\leq \varepsilon + (1 + \varepsilon)(2\varepsilon + \alpha_n(T^{**})).$$

Since φT_n has rank less than n and ε was arbitrary we obtain $\alpha_n(T) \leq \alpha_n(T^{**})$. ∎

6. Let $T: X \to X$ be an operator such that T^k is compact for some k. Such operators are called power-compact. For a power-compact operator T we define the sequence $(\lambda_n(T))_{n=1}^\infty$ which consists of all eigenvalues of T counted with multiplicities (cf. I. A. 18.) ordered in such a way that $|\lambda_1(T)| \geq |\lambda_2(T)| \geq |\lambda_3(T)| \geq \cdots$. Let us recall that $\lambda \in \mathbb{C}$ is an eigenvalue of an operator T if $Tx = \lambda x$ for some $x \in X, x \neq 0$. Since T is power compact, the Riesz theory holds for T (see I.A.16-19.) so the multiplicity of each eigenvalue is finite.

Let H and L be Hilbert spaces and let $T: H \to L$ be a compact operator. Then $|T| = \sqrt{T^*T}$ is a positive compact operator and there exists an isometry $U: \overline{|T|(H)} \to L$ such that $T = U|T|$. The spectral theorem for compact, positive operators shows that there exists an orthonormal system $(v_n)_{n=1}^\infty$ such that $|T|(x) = \sum_{n=1}^\infty \lambda_n(|T|)\langle x, v_n\rangle v_n$. Since $T = U|T|$ we have that for an arbitrary compact operator $T: H \to L$ there exist orthonormal systems $(v_n)_{n=1}^\infty$ and $(u_n)_{n=1}^\infty$ such that

$$T(x) = \sum_{n=1}^\infty \lambda_n(|T|)\langle x, v_n\rangle u_n. \tag{4}$$

Clearly $u_n = Uv_n$, $n = 1, 2, \ldots$. This is called the Schmidt decomposition and the numbers $\lambda_n(|T|)$ are called the singular numbers of T and are denoted $s_n(T)$.

If H and L are Hilbert spaces then $A^p(H,L)$ is called the p-th Schatten-von Neumann class and denoted $\sigma_p(H,L)$ and $a_p(T)$ is then denoted $\sigma_p(T)$. This may seem confusing but the hilbertian theory is so rich and special that it has its own traditional language.

7 Theorem. *Let H, L be Hilbert spaces and let $T: H \to L$ be a compact operator. Then $s_n(T) = \alpha_n(T)$, $n = 1, 2, \ldots$.*

Proof: Using (4) we get immediately

$$\alpha_k(T) \leq \left\| \sum_{n=k}^{\infty} \lambda_n(|T|)\langle \cdot, v_n \rangle u_n \right\| = s_k(T).$$

Conversely, given any operator $T_k: H \to L$ with rank $T_k < k$ let us take $x = \sum_{j=1}^{k} \beta_j v_j$ (where $(v_j)_{j=1}^{\infty}$ is given by (4)) such that $\|x\| = 1$ and $T_k x = 0$. From (4) we get

$$\|Tx\| = \left(\sum_{n=1}^{k} \lambda_n(|T|)^2 |\beta_j|^2 \right)^{\frac{1}{2}} \geq s_k(T)\|x\|$$

so

$$\|T - T_k\| \geq \|Tx - T_k x\| = \|Tx\| \geq s_k(T). \qquad \blacksquare$$

8.   Our basic tool for the study of Schatten-von Neumann classes is the following.

Theorem. (Weyl's inequality).   *Let $T: H \to H$ be a compact operator on a Hilbert space H. Then for every $n = 1, 2, \ldots$*

$$\prod_{k=1}^{n} |\lambda_k(T)| \leq \prod_{k=1}^{n} \alpha_k(T). \tag{5}$$

Proof: Clearly we can assume $\lambda_n(T) \neq 0$ since otherwise there is nothing to prove. Using the spectral theorem for compact operators and the Jordan decomposition in finite dimensional spaces we infer that for every n there exists a subspace $H_n \subset H$ such that

$$\dim H_n = n, \tag{6}$$

$$T(H_n) \subset H_n, \tag{7}$$

$$T \mid H_n: H_n \to H_n \quad \text{has eigenvalues } \lambda_1(T), \ldots, \lambda_n(T). \tag{8}$$

Let T_n denote $T|H_n\colon H_n \to H_n$. Let us fix an orthonormal basis $(x_j)_{j=1}^n$ in H_n, and for any operator $S\colon H_n \to H_n$ let $\det S = \det[\langle Sx_k, x_j\rangle]_{k,j=1}^n$. From well known properties of determinants of finite matrices we get $\det S = \prod_{k=1}^n \lambda_k(S)$. Thus (8) gives

$$\prod_{k=1}^n |\lambda_k(T)| = \prod_{k=1}^n |\lambda_k(T_n)| = |\det T_n|. \qquad (9)$$

Using the polar decomposition we get $T_n = U|T_n|$ where U is a unitary operator. Obviously $\alpha_k(T_n) \leq \alpha_k(T), k = 1, \ldots, n$ and Theorem 7 gives $\alpha_k(T_n) = \alpha_k(|T_n|) = \lambda_k(|T_n|), \ k = 1, 2, \ldots, n$. Thus we get

$$|\det T_n| = \det |T_n| = \prod_{k=1}^n |\lambda_k(|T_n|)| \leq \prod_{k=1}^n \alpha_k(T) \qquad (10)$$

and comparing (9) and (10) we get the claim. ∎

9. To use all the information contained in (5) we need a lemma about sequences of positive numbers.

Lemma. *Let $(\alpha_k)_{k=1}^N$ and $(\beta_k)_{k=1}^N$ be decreasing sequences of positive numbers such that $\sum_{k=1}^n \alpha_k \leq \sum_{k=1}^n \beta_k$ for $n = 1, 2, \ldots, N$ and let $\varphi\colon \mathbb{R} \to \mathbb{R}$ be a convex function such that $\varphi(x) \leq \varphi(|x|)$. Then*

$$\sum_{k=1}^N \varphi(\alpha_k) \leq \sum_{k=1}^N \varphi(\beta_k).$$

Proof: In $\mathbb{R}^N$ we define a convex set V by
$V = \operatorname{conv}\{(\varepsilon_k \beta_{\sigma(k)})_{k=1}^N \colon \varepsilon_k = \pm 1$ and σ is a permutation of the set $\{1, 2, \ldots, N\}\}$.
If $(\alpha_k)_{k=1}^N \notin V$ then by the Hahn-Banach theorem I.A.10 there exists a functional ϕ on $\mathbb{R}^N$ such that $\phi|V \leq 1$ and $\phi((\alpha_k)_{k=1}^N) > 1$. Since V is invariant under permutations of coordinates and changes of signs and $(\alpha_k)_{k=1}^\infty$ is a positive, decreasing sequence we can assume that

$$\phi((x_j)_{j=1}^N) = \sum_{j=1}^N c_j x_j \text{ with } c_1 \geq c_2 > \cdots \geq c_N \geq 0.$$

But

$$1 < \sum_{k=1}^{N} c_k \alpha_k = c_N \sum_{k=1}^{N} \alpha_k + \sum_{j=1}^{N-1} (c_j - c_{j+1}) \sum_{k=1}^{j} \alpha_k$$

$$\leq c_N \sum_{k=1}^{N} \beta_k + \sum_{j=1}^{N-1} (c_j - c_{j+1}) \sum_{k=1}^{j} \beta_k = \sum_{k=1}^{N} c_k \beta_k \leq 1.$$

This contradiction shows that $(\alpha_k)_{k=1}^{N} \in V$, so we have $(\alpha_k)_{k=1}^{N} = \sum_{j=1}^{s} \lambda_j (\varepsilon_k^j \beta_{\sigma_j(k)})_{k=1}^{N}$ for some λ_j's with $\sum \lambda_j = 1$ and $\lambda_j \geq 0$. This gives

$$\sum_{k=1}^{N} \varphi(\alpha_k) = \sum_{k=1}^{N} \varphi\left(\sum_{j=1}^{s} \lambda_j \varepsilon_k^j \beta_{\sigma_j(k)} \right)$$

$$\leq \sum_{k=1}^{N} \sum_{j=1}^{s} \lambda_j \varphi(\varepsilon_k^j \beta_{\sigma_j(k)}) \leq \sum_{j=1}^{s} \lambda_j \sum_{k=1}^{N} \varphi(\beta_{\sigma_j(k)})$$

$$= \sum_{k=1}^{N} \varphi(\beta_k). \qquad \blacksquare$$

10 Theorem. *For every compact operator $T: H \to H$ and every p, $0 < p < \infty$ and every $N = 1, 2, \ldots$ we have*

$$\left(\sum_{n=1}^{N} |\lambda_n(T)|^p \right)^{\frac{1}{p}} \leq \left(\sum_{n=1}^{N} \alpha_n(T)^p \right)^{\frac{1}{p}}. \tag{11}$$

In particular $\|(\lambda_n(T))\|_p \leq \sigma_p(T)$.

Proof: Without loss of generality we can assume $\lambda_N(T) \geq 1$ and $\alpha_N(T) \geq 1$. Applying Lemma 9 for $\alpha_n = p \log \lambda_n(T)$ and $\beta_n = p \log \alpha_n(T)$ and $\varphi(t) = \exp t$ we get the claim. $\qquad \blacksquare$

11. Weyl's inequality also allows us to show that $\sigma_p(H), 1 \leq p < \infty$, is actually a Banach space. As we know (see **3**), this is not true for general Banach spaces. We have

Proposition. *Let T and S be compact operators on a Hilbert space H. For every $n = 1, 2, \ldots$ and every $1 \leq p < \infty$ we have*

$$\left(\sum_{k=1}^{n} \alpha_k(S+T)^p \right)^{\frac{1}{p}} \leq \left(\sum_{k=1}^{n} \alpha_n(S)^p \right)^{\frac{1}{p}} + \left(\sum_{k=1}^{n} \alpha_k(T)^p \right)^{\frac{1}{p}}.$$

Proof: Using the Schmidt decomposition (4) and Theorem 7 we can write

$$(S + T)x = \sum_k \alpha_k (S + T) \langle x, v_k \rangle u_k.$$

Let U be a partial isometry defined by $U(v_k) = u_k$ and let P be the orthogonal projection onto $\mathrm{span}\{v_k\}_{k=1}^n$. We have

$$\sum_{k=1}^n \alpha_k(S + T) = \sum_{k=1}^n \langle U^*(S + T)v_k, v_k \rangle = tr PU^*(S + T)P$$

$$\leq |tr(PU^*SP)| + |tr(PU^*TP)|.$$

Since for finite rank operators the trace equals the sum of eigenvalues the last expression is majorized by

$$\sum_{k=1}^n |\lambda_k(PU^*SP)| + \sum_{k=1}^n |\lambda_k(PU^*TP)|.$$

Thus from (11) we get for $n = 1, 2, \ldots$

$$\sum_{k=1}^n \alpha_k(S + T) \leq \sum_{k=1}^n \alpha_k(PV^*SP) + \sum_{k=1}^n \alpha_k(PV^*TP)$$

$$\leq \sum_{k=1}^n \alpha_k(S) + \sum_{k=1}^n \alpha_k(T) \tag{12}$$

so our proposition is proved for $p = 1$. Applying Lemma 9 for $\alpha_k = \alpha_k(S + T)$ and $\beta_k = \alpha_k(S) + \alpha_k(T)$ and $\varphi(t) = t^p$ we get (use (12)) $\sum_{k=1}^n \alpha_k(S+T)^p \leq \sum_{k=1}^n (\alpha_k(S) + \alpha_k(T))^p$. Now the Hölder inequality gives the claim. ∎

12. The operators of class σ_2 are called Hilbert-Schmidt operators. Here are some equivalent characterizations.

Proposition. *Let H and L be Hilbert spaces and let $T: H \to L$. The following conditions are equivalent:*

(a) $T \in \sigma_2(H, L)$;

(b) *for every orthonormal basis $(h_j)_{j \in J}$ in the space H we have $\sum_{j \in J} \|Th_j\|^2 < \infty$;*

(c) *there exists an orthonormal basis $(h_j)_{j \in J}$ in the space H such that $\sum_{j \in J} \|Th_j\|^2 < \infty$;*

(d) T admits a factorization

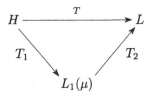

for some $L_1(\mu)$-space;

(e) T admits a factorization

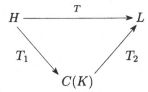

for some $C(K)$-space;

(f) $T \in \Pi_p(H, L)$ for every p, $1 \leq p < \infty$;

(g) $T \in \Pi_p(H, L)$ for some p, $1 \leq p < \infty$;

Proof: (a)$\Rightarrow$(b). Let $(h_j)_{j \in J}$ be any orthonormal basis in H. Using the Schmidt decomposition (4) we get

$$
\begin{aligned}
\sum_{j \in J} \|Th_j\|^2 &= \sum_{j \in J} \left\| \sum_k a_k(T)\langle h_j, v_k \rangle u_k \right\|^2 \\
&= \sum_{j \in J} \sum_k a_k(T)^2 |\langle h_j, v_k \rangle|^2 \\
&= \sum_k a_k(T)^2 \sum_{j \in J} |\langle h_j, v_k \rangle|^2 \qquad (13) \\
&= \sum_k a_k(T)^2 = \sigma_2(T)^2.
\end{aligned}
$$

(b)$\Rightarrow$(c). This is obvious.

(c)$\Rightarrow$(d). We define $v: H \to \ell_1(J)$ by

$$
v(x) = (\langle x, h_j \rangle \|Th_j\|)_{j \in J}
$$

and $\Sigma: \ell_1(J) \to L$ by

$$
\Sigma(\xi_j) = \sum_j \xi_j \frac{Th_j}{\|Th_j\|} \quad \text{with the convention } \frac{0}{0} = 0.
$$

Since $T = \Sigma \circ v$ we get the desired factorization.

(d)$\Rightarrow$(f). By Grothendieck's Theorem III.F.7 and Corollary III.F.9.

(f)$\Rightarrow$(g). Obvious.

(g)$\Rightarrow$(a). Since Hilbert space has cotype 2 (see III.A.23) we get from Lemma III.F.37 that $T \in \Pi_2(H, L)$. From Corollary III.F.9(b) we get that T is compact thus the Schmidt decomposition gives $T(x) = \sum_{k=1}^{\infty} \lambda_k \langle x, v_k \rangle u_k$. The definition of 2-absolutely summing map gives

$$\sum_k |\lambda_k|^2 = \sum_k \|Tv_k\|^2 \leq \pi_2(T)^2 \sup\{\sum_k |\langle x, v_k \rangle|^2 : \|x\| \leq 1\} = \pi_2(T)^2 \tag{14}$$

so $\sigma_2(T) \leq \pi_2(T)$.

Since (e) is a dual condition to (d) and (a) is a self dual condition (see Proposition 5) we infer that also (e) is equivalent to all the others.∎

Remark. If an operator $T: H \to L$ satisfies any of the conditions of the Proposition then

$$\pi_2(T) = \sigma_2(T) = \left(\sum_j \|Th_j\|^2 \right)^{\frac{1}{2}} \tag{15}$$

for any orthonormal basis $(h_j)_{j \in J}$ is H. We see from (13) and (14) that only $\pi_2(T) \leq \sigma_2(T)$ remains to be proved. If $(l_j)_{j \in J}$ is any orthonormal basis in L and $h \in H$ then

$$\|Th\|^2 = \sum_{j \in J} |\langle Th, l_j \rangle|^2 = \sum_{j \in J} |\langle h, T^* l_j \rangle|^2.$$

When we view this inequality as a special case of III.F.(9) we see that $\pi_2(T)^2 \leq \sum_{j \in J} \|T^* l_j\|^2$. From (13) we get $\pi_2(T) \leq \sigma_2(T^*) = \sigma_2(T)$.

13. One of the reasons why the Hilbert-Schmidt operators are important is that they admit a nice integral representation.

Proposition. *An operator $T: L_2(\Omega, \mu) \to L_2(\Sigma, \nu)$ is Hilbert-Schmidt if and only if there exists a function $K \in L_2(\Omega \times \Sigma, \mu \times \nu)$ such that*

$$Tf(\sigma) = \int_\Omega K(\omega, \sigma) f(\omega) d\mu(\omega) \tag{16}$$

and $\sigma_2(T) = \|K\|_{L_2(\Omega \times \Sigma, \mu \times \nu)}$.

Proof: From the Schmidt decomposition we get

$$Tf(\sigma) = \sum_k \lambda_k \int_\Omega f(\omega)\overline{v_k(\omega)}d\mu(\omega)u_k(\sigma)$$

$$= \int_\Omega \left[\sum_k \lambda_k \overline{v_k(\omega)} \cdot u_k(\sigma) \right] f(\omega)d\mu(\omega).$$

Since $\Sigma|\lambda_n|^2 < \infty$ and $\overline{v_k(\omega)} \cdot u_k(\sigma)$ is an orthonormal system in $L_2(\Omega \times \Sigma, \mu \times \nu)$ we get the desired function.

Conversely let $(f_j(\omega))_{j\in J}$ and $(h_s(\sigma))_{s\in S}$ be orthonormal bases in $L_2(\Omega, \mu)$ and $L_2(\Sigma, \nu)$ respectively. Then $(f_j(\omega) \cdot h_s(\sigma))_{(j,s)\in J\times S}$ is an orthonormal basis in the space $L_2(\Omega \times \Sigma, \mu \times \nu)$. We have

$$\sum_j \|Tf_j(\omega)\|^2 = \sum_j \sum_s |\langle Tf_j, h_s\rangle|^2 = \sum_j \sum_s |\langle K, f_j \otimes h_s\rangle|^2$$

$$= \|K\|^2_{L_2(\Omega\times\Sigma,\mu\times\nu)}.$$

So Proposition 12 implies that T is Hilbert-Schmidt and $\sigma_2(T) = \|K\|^2_{L_2(\Omega\times\Sigma,\mu\times\nu)}$.   ■

Example.   There exists an operator $T\colon L_2[0,1] \to L_2[0,1]$ such that $T \in \sigma_p$ for every $p > 2$ and there is no function $K(x,y)$ on $[0,1] \times [0,1]$ such that

$$Tf(x) = \int_0^1 K(x,y)f(y)dy \quad \text{a.e.}$$

where the integral is understood as a Lebesgue integral.

Observe that if such a representation exists then the function $\Omega(x) = \int |K(x,y)|dy$ is finite almost everywhere, so

$$|Tf(x)| \leq \|f\|_\infty \cdot \Omega(x). \tag{17}$$

We define our example as

$$T(e^{2\pi iny}) = \begin{cases} \lambda_n h_n, & n = 1, 2, \ldots, \\ 0, & n = 0, -1, -2, \ldots, \end{cases}$$

where h_n is the Haar system (see II.B.9) and $\lambda_n = n^{-\frac{1}{2}} \cdot \log n$. Obviously $T \in \sigma_p(L_2[0,1])$ for $p > 2$. Since $\sup_n |T(e^{2\pi iny})(x)| = \sup_n |\lambda_n h_n(x)| = \infty$ a.e. we see that (17) does not hold.

14.   Now we would like to apply these general notions to investigate the connection between the smoothness of functions and the size of Fourier coefficients.

Theorem. *The Fourier coefficients of every function $f \in Lip_\alpha(\mathbb{T})$, $0 < \alpha \leq 1$ belong to ℓ_p for every $p > \frac{2}{(2\alpha+1)}$.*

Proof:

Given $f \in Lip_\alpha(\mathbb{T})$ let us consider the operator $H_f\colon L_2(\mathbb{T}) \to L_2(\mathbb{T})$ given by $H_f(g) = f * g$. Since the eigenvalues of H_f are $(\hat{f}(n))_{n=-\infty}^{+\infty}$ by Theorem 10 it is enough to show that $H_f \in \sigma_p$. Since $f \in Lip_\alpha(\mathbb{T})$ the operator H_f actually maps $L_1(\mathbb{T})$ into $Lip_\alpha(\mathbb{T})$. Let $(f_n)_{n=0}^\infty$ be the Franklin system (see III.D.20) and let $F(f) = (\langle f, f_n\rangle(n+1)^{\alpha+\frac{1}{2}})_{n=0}^\infty$. From III.D.27 we get that $F\colon Lip_\alpha(\mathbb{T}) \to \ell_\infty$. Let $\Sigma\colon \ell_\infty \to L_2(\mathbb{T})$ be given by $\Sigma(\xi_n) = \sum_{n=0}^\infty \xi_n(n+1)^{-\alpha-\frac{1}{2}}f_n$. We have the commutative diagram

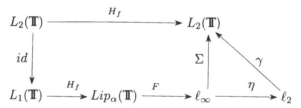

where $\eta(\xi_n) = ((n+1)^{-\frac{1}{2}-\varepsilon}\xi_n)_{n=0}^\infty$ and $\gamma(\xi_n) = \sum_{n=0}^\infty \xi_n(n+1)^{-\alpha+\varepsilon}f_n$. Clearly $\gamma \in \sigma_s$ with $s > \frac{1}{(\alpha-\varepsilon)}$ and $\eta F H_f$ id $\in \sigma_2$ by Proposition 12. Thus by Proposition 4 $H_f \in \sigma_p$ with $\frac{1}{p} = \frac{1}{2} + \frac{1}{s}$. Since ε was arbitrary we get the theorem. ∎

15. Now we want to discuss eigenvalues of operators on general Banach spaces. Estimating eigenvalues of an operator on a Banach space is more difficult than on a Hilbert space since the geometry is more complicated. The main idea is to reduce the problem to a related problem on a Hilbert space. One way to do it is to use the concept of related operators.

Two operators $S\colon X \to X$ and $T\colon Y \to Y$ are related if there exist operators $A\colon X \to Y$ and $B\colon Y \to X$ such that

$$S = BA \text{ and } T = AB.$$

The following simple lemma is very useful when estimating eigenvalues.

Lemma. *Let S and T be related operators. If one is power-compact then the other is and $(\lambda_n(T))_{n=1}^\infty = (\lambda_n(S))_{n=1}^\infty$.*

Proof: Since $T^{k+1} = AS^kB$ and $B^{k+1} = BT^kA$ the first claim follows. Let $\lambda \neq 0$ be an eigenvalue of S with multiplicity k and let V be the corresponding eigenspace, so $\dim V = k$ and $\ker(\lambda - S)^s = V$ for some s. Suppose that for some $v \in V$ with $v \neq 0$ we have $A(v) = 0$. Then

$$0 = (\lambda - S)^s(v) = \sum_{j=1}^{s}(-1)^j \binom{s}{j}\lambda^{s-j}(BA)^j(v) = \lambda^s v$$

so $\lambda = 0$. But this is impossible, so for $W = A(V)$ we have $\dim W = k$. For $w = Av \in W$ we have

$$(\lambda - T)^s w = \sum_{j=1}^{s}(-1)^j \binom{s}{j}\lambda^{s-j}(AB)^j w = \sum_{j=1}^{s}(-1)^j \binom{s}{j}\lambda^{s-j}(AB)^j Av$$

$$= A\left(\sum_{j=1}^{s}(-1)^j \binom{s}{j}\lambda^{s-j}(BA)^j v\right) = A(\lambda - S)^s v = 0.$$

This shows that W is contained in the eigenspace of T corresponding to λ. Repeating the same argument with S and T interchanged we get the lemma.   ∎

16.   Directly from Lemma 15 we get

Theorem.   *If* $T: X \to X$ *is a p-absolutely summing operator with* $p \leq 2$ *then* $(\lambda_n(T))_{n=1}^{\infty} \in \ell_2$.

Proof: It is enough to assume $\pi_2(T) = 1$ (see Corollary III.F.9(a)). Then we have a factorization

$$
\begin{array}{ccc}
X & \xrightarrow{\ \ T\ \ } & X \\[2pt]
{\scriptstyle i}\downarrow & & \uparrow{\scriptstyle B} \\[2pt]
C(K) & \xrightarrow[\ \ id\ \ ]{} & L_2(\mu)
\end{array}
$$

The operator $S: L_2(\mu) \to L_2(\mu)$ defined as $S = id \circ i \circ B$ is similar to T and from Proposition 12 we get $\sigma_2(S) < \infty$. Theorem 10 completes the proof.   ∎

17.   Surprisingly this is an optimal result. Namely we have

Proposition.   *There exists a nuclear operator* T *such that* $(\lambda_n(T))_{n=1}^{\infty}$ $\notin \ell_p$ *for any* $p < 2$.

This follows directly from III.A.25 and the following.

18 Proposition. *Let G be a compact abelian group with Haar measure m, and let $T: C(G) \to C(G)$ be given by $Tf = f * h$ for some $h \in C(G)$. Then T is nuclear.*

Proof: It is clearly sufficient to show that $n_1(T) \leq \|h\|_\infty$ for h a finite combination of characters. Such T being finite dimensional is nuclear. It follows from II.E.5(e) and the definition of the nuclear norm that T can be approximated in the nuclear norm by operators of the form $P_1 T P_2$ where P_1 and P_2 are projections onto finite dimensional subspaces in $C(G)$ with $d(ImP_1, \ell_\infty^N) \leq 1+\varepsilon$ and $d(ImP_2, \ell_\infty^M) \leq 1+\varepsilon$. From III.F.12 we infer that $\pi_1(T) \leq \|h\|_\infty$ so it is enough to show that for $S: \ell_\infty^N \to \ell_\infty^M$ we have $n_1(S) \leq \pi_1(S)$. This follows directly from Corollary III.F.24 since $\pi_1(S) = i_1(S)$ (see remarks after III.F.22). ∎

Remark. The above Proposition 17 contrasts with the situation in Hilbert spaces. For $T: H \to H$ we have $n_1(T) = \sigma_1(T)$. If T has a nuclear representation $T(x) = \sum_j \langle x, x_j \rangle y_j$ then from Proposition 11 we get $\sigma_1(T) \leq \sum_j \sigma_1(\langle \cdot, x_j \rangle y_j) = \sum_j \|x_j\| \, \|y_j\|$ so $\sigma_1(T) \leq n_1(T)$. The converse follows directly from the Schmidt decomposition. Thus Theorem 10 shows that any nuclear operator on a Hilbert space has absolutely summable eigenvalues.

19. The behaviour of eigenvalues of p-absolutely summing maps for $p > 2$ is given in

Theorem. *If $T: X \to X$ is a p-absolutely summing operator, $p \geq 2$, then*

$$\left(\sum_{n=1}^\infty |\lambda_n(T)|^p \right)^{\frac{1}{p}} \leq \pi_p(T). \tag{18}$$

The proof of this theorem follows from the following two facts.

20 Lemma. *There exists a constant C_p such that for every operator $T: X \to X$ we have*

$$\lambda_n(T) \leq C_p n^{-\frac{1}{p}} \pi_p(T), \quad p \geq 2, n = 1, 2, \ldots . \tag{19}$$

21 Proposition. *Let p be given. Suppose that for some q there exists a constant C_q such that for every operator $T: X \to X$ and any Banach* space X we have $(\sum_{n=1}^{\infty} |\lambda_n(T)|^q)^{\frac{1}{q}} \leq C_q \pi_p(T)$. Then we can take $C_q = 1$.

Proof of Theorem 19.   It follows from Lemma 20 that for every $q > p$ we have $(\sum |\lambda_n(T)|^q)^{\frac{1}{q}} \leq C_q \pi_p(T)$. Applying Proposition 21 and passing to the limit as $q \to p$ we get (18). ∎

Proof of Lemma 20.   From finite dimensional linear algebra we infer that for every $n = 1, 2, 3, \ldots$ there exists a subspace $X_n \subset X$, $\dim X_n = n$ such that $T(X_n) \subset X_n$ and $\lambda_j(T|X_n) = \lambda_j(T)$, $j = 1, 2, \ldots, n$. Clearly $\pi_p(T|X_n) \leq \pi_p(T)$ so applying the Pietsch factorization Theorem III.F.8 we have the factorization

$$
\begin{array}{ccc}
X_n & \xrightarrow{\;\;T|X_n\;\;} & X_n \\[2mm]
i\Big\downarrow & & \Big\uparrow\alpha \\[2mm]
X_n^{\infty} & \xrightarrow{\quad id \quad} & X_n^p
\end{array}
$$

where $\pi_p(id) \leq 1$, $\|\alpha\| \leq \pi_p(T)$ and X_n^p is an n-dimensional subspace of some $L_p(\mu)$. From Corollary III.B.9 we get operators A and B such that $X_n^p \xrightarrow{A} \ell_2^n \xrightarrow{B} X_n^p$ and such that $BA = id_{X_n^p}$ and $\|A\| \cdot \|B\| \leq n^{\frac{1}{2} - \frac{1}{p}}$. The operator $T|X_n \colon X_n \to X_n$ is related to the following composition which we will call S:

$$
\ell_2^n \xrightarrow{B} X_n^p \xrightarrow{\alpha} X_n \xrightarrow{i} X_n^{\infty} \xrightarrow{id} X_n^p \xrightarrow{A} \ell_2^n.
$$

Since $\pi_p(S) \leq \pi_p(id)\|A\| \cdot \|i\| \cdot \|\alpha\| \cdot \|B\| \leq n^{\frac{1}{2} - \frac{1}{p}} \pi_p(T)$ we get from Proposition 12 that there exists a constant C_p such that

$$
\sigma_2(S) \leq C_p \pi_p(S) \leq C_p n^{\frac{1}{2} - \frac{1}{p}} \pi_p(T). \tag{20}
$$

From Theorem 10 and (20) we get

$$
\sqrt{n} \cdot |\lambda_n(T)| = \sqrt{n}|\lambda_n(S)| \leq \left(\sum_{j=1}^{n} |\lambda_j(S)|^2 \right)^{\frac{1}{2}} \leq \sigma_2(S) \leq C_p n^{\frac{1}{2} - \frac{1}{p}} \pi_p(T)
$$

so (19) follows. ∎

Proof of Proposition 21.   Let us denote the smallest possible C_q by K. If $K > 1$ then there exists an operator $T \colon X \to X$ such that

$\pi_p(T) = 1$ and $(\sum_{n=1}^{\infty} |\lambda_n(T)|^q)^{\frac{1}{q}} > \sqrt{K}$. Without loss of generality we can treat X as a subspace of $C(\Omega)$ for some compact space Ω. From III.F.8 we get a probability measure μ on Ω and an operator $\widetilde{T}: X_p \to X$ with $\|\widetilde{T}\| = 1$ where X_p is the closure of X in $L_p(\Omega, \mu)$. Let $Y \subset C(\Omega \otimes \Omega)$ be the closure of the set of functions of the form $\sum_{j=1}^{n} x_j(\omega_1) \cdot z_j(\omega_2)$ where $x_j, z_j \in X, j = 1, \ldots, n$ and let Y_p be the closure in $L_p(\Omega \times \Omega, \mu \times \mu)$ of Y. We define an operator $T \otimes T: Y \to Y$ by the formula $T \otimes T(\sum x_j(\omega_1) y_j(\omega_2)) = \sum T(x_j)(\omega_1) \cdot T(y_j)(\omega_2)$. One easily checks that T is continuous and that

$$\sum_{n=1}^{\infty} |\lambda_n(T \otimes T)|^q \geq \sum_{n=1}^{\infty} |\lambda_n(T)|^q \cdot \sum_{n=1}^{\infty} |\lambda_n(T)|^q. \tag{21}$$

Let $\widetilde{T} \otimes \widetilde{T}: Y_p \to Y$ be defined by

$$(\widetilde{T} \otimes \widetilde{T})(\sum f_j(\omega_1) \cdot g_j(\omega_2)) = \sum \widetilde{T}(f_j)(\omega_1) \cdot \widetilde{T}(g_j)(\omega_2).$$

Since the formal identity from Y into Y_p has norm at most 1 we get $\pi_p(T \otimes T) \leq \|\widetilde{T} \otimes \widetilde{T}\|$. But for $F = \sum_{j=1}^{n} f_j(\omega_1) g_j(\omega_2) \in Y_p$ we have

$$|(\widetilde{T} \otimes \widetilde{T})(F)(\omega_1, \omega_2)| = \left| \widetilde{T} \left(\sum_{j=1}^{n} \widetilde{T}(g_j)(\omega_2) f_j \right)(\omega_1) \right|$$

$$\leq \|\widetilde{T}\| \left(\int_{\Omega} \left| \sum_{j=1}^{n} \widetilde{T}(g_j)(\omega_2) f_j(\omega_1) \right|^p d\mu(\omega_1) \right)^{\frac{1}{p}}$$

$$= \|\widetilde{T}\| \left(\int_{\Omega} \left| \widetilde{T} \left(\sum_{J=1}^{n} f_j(\omega_1) g_j \right)(\omega_2) \right|^p d\mu(\omega_1) \right)^{\frac{1}{p}}$$

$$\leq \|\widetilde{T}\|^2 \left(\int_{\Omega} \left\| \sum_{j=1}^{n} f_j(\omega_1) g_j \right\|_p^p d\mu(\omega_1) \right)^{\frac{1}{p}}$$

$$\leq \|\widetilde{T}\|^2 \left(\int_{\Omega \times \Omega} \left| \sum_{j=1}^{n} f_j(\omega_1) g_j(\omega_2) \right|^p d(\mu \times \mu)(\omega_1, \omega_2) \right)^{\frac{1}{p}}$$

$$\leq \|\widetilde{T}\|^2 \|F\|_{Y_p}.$$

Thus $\pi_p(T \otimes T) \leq 1$, so (21) together with the choice of T gives

$$\left(\sum_{n=1}^{\infty} |\lambda_n(T \otimes T)|^q \right)^{\frac{1}{q}} \geq (\sum |\lambda_n(T)|^q)^{\frac{2}{q}} > K.$$

This contradicts the definition of K.   ∎

Remark. The reader familiar with tensor products will easily see that the above argument gives that an ε-tensor product of p-summing maps is p-summing.

22. As an example of the applicability of previous results let us consider Hille-Tamarkin integral operators. Let (Ω, μ) be a probability measure space and let $K(\omega_1, \omega_2)$ be a function on $\Omega \times \Omega$ such that

$$\int_\Omega \left(\int_\Omega |K(\omega_1, \omega_2)|^{p'} d\mu(\omega_2) \right)^{\frac{p}{p'}} d\mu(\omega_1) < \infty, \qquad 2 \le p < \infty. \quad (22)$$

Then the formula $T_K(f)(\omega) = \int_\Omega K(\omega, \omega_2) f(\omega_2) d\mu(\omega_2)$ defines, as is easily seen, a linear operator $T_K \colon L_p(\Omega, \mu) \to L_p(\Omega, \mu)$. For such an operator we have $\sum_{n=1}^\infty |\lambda_n(T_K)|^p < \infty$. This follows from Theorem 19 and the fact that T_K is p-absolutely summing. To see this put $\varphi(\omega_1) = (\int_\Omega |K(\omega_1, \omega_2)|^{p'} d\mu(\omega_2))^{\frac{1}{p'}}$. One checks that $S(f) = T_K(f) \cdot \varphi^{-1}$ is a linear map from $L_p(\Omega, \mu)$ into $L_\infty(\Omega, \mu)$ so we have to check that for $\varphi \in L_p$ the map $f \mapsto f \cdot \varphi$ is p-absolutely summing from $L_\infty(\Omega)$ into $L_p(\Omega, \mu)$. This was observed in III.F.4.

Note that Hille-Tamarkin integral operators are direct generalizations of Hilbert-Schmidt operators (see Proposition 13).

Let G be an abelian compact group with normalized Haar measure m and dual group Γ. For $f \in L_{p'}(G, m), p \ge 2$ we define a kernel $K(g_1, g_2) = f(g_1 - g_2)$. This kernel clearly satisfies (22). Since $T_K(g) = f * g$ we see that eigenvalues of T_K coincide with $\hat{f}(\gamma)$ so we get the Hausdorff-Young inequality

$$\left(\sum_{\gamma \in \Gamma} |\hat{f}(\gamma)|^p \right)^{\frac{1}{p}} \le \|f\|_{p'}, \quad p \ge 2. \quad (23)$$

Notes and Remarks.
There are two excellent books which treat the matters explained in this chapter, and much more. They are Pietsch [1987] and König [1986]. The concept of approximation number is so natural that we have been unable to trace proper historical references. Nowadays it is an example of the general notion of s-numbers (see Pietsch [1987]). *Proposition 5* is due to Hutton [1974]. It is a quantitative version of the classical Schauder Theorem asserting that an operator is compact if and only if its adjoint is compact. Our material on Schatten-von Neumann classes is classical and can be found in many places. The above mentioned books contain nice

presentations but also Gohberg-Krein [1969] and Simon [1979] should be mentioned.

Proposition 12 showing the connection between Hilbert-Schmidt and p-absolutely summing operators on a Hilbert space was proved by Pełczyński [1967]. *Theorem 14* is the classical result of Fredholm [1903] but in the theory of Fourier series it is usually associated with Bernstein [1914] and Szasz [1922]. Our proof is taken from Wojtaszczyk [1988].

Theorem 16 and *Proposition 17* and *18* are basically due to Grothendieck [1955]. In the present generality *Theorem 16* was proved by Pietsch [1963].

The fact that the eigenvalues of a nuclear operator on Hilbert space are absolutely summable actually characterizes spaces isomorphic to Hilbert space among all Banach spaces (see Johnson-König-Maurey-Retherford [1979]). This paper contains also the first proof of *Theorem 19*. Our proof of *Theorem 19* is a modification of a proof given in Pietsch [1986]. The application of *Corollary III.B.9* allows us to avoid the use of the general theory of Weyl's numbers. One should be aware that the subject of eigenvalue estimates of operators on X is related to best projections on finite dimensional subspaces of X. This is made clear in König [1986] 4.b where the estimates for eigenvalues are used to prove *Corollary III.B.9*. Our discussion of Hille-Tamarkin integral operators is taken from Johnson-König-Maurey-Retherford [1979].

Exercises

1. Show that $\sigma_p(\ell_2)^* = \sigma_q(\ell_2)$ for $1 < p < \infty$ and $\frac{1}{p} + \frac{1}{q} = 1$, and also $\sigma_\infty(\ell_2)^* = \sigma_1(\ell_2)$ and $\sigma_1(\ell_2)^* = L(\ell_2)$. The duality is given by $\langle T, S \rangle = trTS$.

2. Show that the space $\sigma_1(\ell_2)$ has cotype 2.

3. For $A \subset \mathbb{N}$ let $P_A \colon \ell_2 \to \ell_2$ denote the natural coordinate projection defined by $P_A(\sum_{j=1}^\infty a_j e_j) = \sum_{j \in A} a_j e_j$.

   (a) Show that maps $T \mapsto TP_A$ and $T \mapsto P_A T$ are contractions on $\sigma_p(\ell_2), 1 \le p \le \infty$.

   (b) Show that if $p \ne 2$ then the operators $T_{ij}, i, j = 1, 2, \ldots$ defined by $T_{ij}(\sum_{k=1}^\infty a_k e_k) = a_i e_j$ do not form an unconditional basis in $\sigma_p(\ell_2)$.

   (c) Show that $\sigma_p, p \ne 2$ and $p \ne \infty$, is not isomorphic to any subspace of $L_p(\mu)$.

4. If X is an infinite dimensional subspace of $\sigma_\infty(\ell_2)$, then X contains an infinite dimensional subspace X_1 complemented in $\sigma_\infty(\ell_2)$ such that either $X_1 \sim c_0$ or $X_1 \sim \ell_2$.

5. Show that $\sigma_p \sim (\sum \sigma_p)_p$ for $1 \leq p < \infty$ and $\sigma_\infty \sim (\sum \sigma_\infty)_0$.

6. Show that there is no norm on $A^1(\ell_\infty, \ell_1)$ which is equivalent to the quasi-norm $a_1(\cdot)$.

7. Show that for every $(\lambda_n)_{n=1}^\infty \in \ell_2$ there exists a nuclear operator $T: \ell_1 \oplus \ell_\infty \longrightarrow \ell_1 \oplus \ell_\infty$ such that the eigenvalues of T are precisely $\pm\lambda_1, \pm\lambda_2, \pm\lambda_3, \ldots$.

8. Suppose that $K(x,y)$ is integrable on $[0,1] \times [0,1]$ and that

    $$\int_0^1 |K(x_1,y) - K(x_2,y)|^2 dy \to 0 \text{ as } |x_1 - x_2| \to 0.$$

    Show that the operator $Tf(x) = \int_0^1 K(x,y)f(y)dy$ acts from $C[0,1]$ into $C[0,1]$ and has square summable eigenvalues.

9. Suppose (Ω, μ) is a probability measure space and $K(\xi, \eta)$ is a function on $\Omega \times \Omega$ such that

    $$\int_\Omega \left(\int_\Omega |K(\xi,\eta)|^q d\mu(\eta) \right)^{p/q} d\mu(\xi) < \infty,$$

    where p and q are positive numbers such that $\frac{1}{p} + \frac{1}{q} \leq 1$. Show that the operator T_K given as $T_K(f)(\xi) = \int_\Omega K(\xi,\eta)f(\eta)d\mu(\eta)$ maps $L_{q'}(\Omega, \mu)$ into $L_{q'}(\Omega, \mu)$, $\frac{1}{q} + \frac{1}{q'} = 1$, and $(\lambda_n(T_k))_{n=1}^\infty \in \ell_{q^+}$ where $q^+ = \max(q', 2)$.

10. (a) Let $I_n: \sigma_\infty(\ell_2^n) \to \sigma_2(\ell_2^n)$ be the formal identity. Show that there exist constants $0 < c < C$ such that $cn \leq \pi_1(I_n) \leq Cn$ for $n = 1, 2, \ldots$.

    (b) Let $J_n: \sigma_1(\ell_2^n) \to \sigma_2(\ell_2^n)$ be the formal identity. Show that $\pi_1(J_n) \leq c\sqrt{n}$ for some $c > 0$.

    (c) Show that $\gamma_1(J_n) \geq cn$ for some constant $c > 0$, where $\gamma_1(J_n) = \inf\{\|\alpha\|\|\beta\| : \alpha: \sigma_1(\ell_2^n) \to \ell_1$ and $\beta: \ell_1 \to \sigma_2(\ell_2^n)$ and $\beta\alpha = J_n\}$.

    (d) Use the above to show that there exist a 1-absolutely summing operator which does not factor through any $L_1(\mu)$-space. Another such example is given in Exercise III.I.2.

(e) Show that, if X is a complemented subspace of a Banach space with an unconditional basis, then every 1-absolutely summing operator from X into any Banach space factors through ℓ_1.

(f) Show that $\sigma_1(\ell_2)$ and $\sigma_\infty(\ell_2)$ are not isomorphic to any complemented subspace of a space with an unconditional basis.

11. Suppose that $T: \ell_2^n \to \ell_2^n$ is such that $\|Te_j\| \geq 1$ for $j = 1, 2, \ldots, n$. Show that rank $T \geq \frac{n}{\|T\|^2}$.

12. Let $u: \ell_2^n \to \ell_2^n$ be a linear operator given by the matrix $(u(i,j))_{i,j=1}^n$. Show that

$$\sum_{i,j} |u(i,j)| \leq n\sigma_1(u).$$

13. (a) Show that every 1-absolutely summing operator $T: C(K) \to \ell_2$ is 1-nuclear. Note that $A(\mathbb{D})$ does not have this property (see III.F.6(a)).

(b) Show that the space $C^1(\mathbb{T}^2)$ is not isomorphic to a quotient space of any $C(S)$-space.

(c) Show that every linear operator $T: L_1(\mu) \to L_2(\nu)$ maps order-bounded sets into order bounded sets, i.e. sets of the form $V_g = \{f \in L_1(\mu) : |f| \leq g \text{ for } g \in L_1(\mu)\}$ are mapped into sets $\{f \in L_2(\nu) : |f| \leq g, \ g \in L_2(\nu)\}$.

14. (a) Let E be any n-dimensional Banach space and let id_E be the identity operator on E. Show that $\pi_2(id_E) = \sqrt{n}$.

(b) Let (Ω, μ) be a probability measure space and let $E \subset L_\infty(\Omega, \mu)$ be an n-dimensional subspace. Show that there exists $e \in E$ such that $\|e\|_\infty = 1$ and $\|e\|_2 \leq n^{-\frac{1}{2}}$.

(c) Let E be an n-dimensional Banach space. Show using (a) that $d(E, \ell_2^n) \leq \sqrt{n}$. This gives an alternative proof of an important special case of III.B.9.

(d) Let E be an n-dimensional Banach space. Show using (a) that $\lambda(E) \leq \sqrt{n}$. This gives an alternative proof of an important special case of III.B.10.

15. Suppose that $\|\cdot\|^1$ and $\|\cdot\|^2$ are two Hilbertian norms on an n-dimensional space X. Show that there exists a subspace $X_1 \subset X$, $\dim X_1 \geq \frac{n}{2}$ and a constant α such that

$$\|x\|^1 = \alpha\|x\|^2 \quad \text{for all} \quad x \in X_1.$$

16. Let E be a subspace of ℓ_∞^N.

   (a) Show that there exists an $x \in E$ such that $\|x\| = 1$ and $|\{i\colon |x(i)| = 1\}| \geq \dim E$.

   (b) Show that $d(E, \ell_2^n) \geq \frac{1}{2}\frac{n}{\sqrt{N}}$ where $n = \dim E$.

17. (a) Let T_n^∞ be the space of trigonometric polynomials on $\mathbb{T}$, of degree at most n, with the sup-norm. Suppose that E is a subspace of T_n^∞ and $\dim E \geq \alpha(2n+1)$. Show that there exists a polynomial $p(\theta) = \sum_{k=-n}^n a_k e^{ik\theta} \in E$ such that $\|p\|_\infty = 1$ but $(\sum_{k=-n}^n |a_k|^2)^{\frac{1}{2}} \geq C(\alpha)$ where $C(\alpha) > 0$, does not depend on n.

   (b) Let $W_n^\infty(\mathbf{S}_d)$ be the space of polynomials homogeneous of degree n restricted to $\mathbf{S}_d$ and equipped with the sup norm. Suppose that E is a subspace of $W_n^\infty(\mathbf{S}_d)$ with $\dim E \geq \alpha \dim W_n^\infty(\mathbf{S}_d)$. Show that there exists $p \in E$ such that $\|p\|_\infty \leq C(\alpha, d)\|p\|_2$ where $C(\alpha, d) > 0$ does not depend on n (for more information about the spaces $W_n^p(\mathbf{S}_d)$ see III.B.14).

III.H. Factorization Theorems

This chapter discusses some very important and powerful theorems about factorization of operators with values in L_p-spaces, $0 \leq p \leq \infty$. These theorems express the fact that sometimes the map (not necessarily linear) from a Banach space X into $L_p(\Omega, \mu)$ is 'essentially' a map into some smaller function space, notably L_q or $L_{q,\infty}$ for some $q > p$. Contrary to our general custom we will discuss operators more general than linear. We start with the Nikishin theorems which discuss a factorization through weak L_p. As a preparation we discuss some properties of operators from a Banach space X into $L_0(\Omega, \mu)$. Then we discuss factorization through L_p spaces. For linear operators we show the connection between factorization and p-absolutely summing operators and prove that every linear operator from L_p, $2 \leq p \leq \infty$, into L_q, $0 < q \leq 2$, factors strongly through L_2. We apply these results to reflexive subspaces of $L_1(\mu)$, to the structure of series unconditionally convergent in measure, to the Menchoff-Rademacher theorem about almost everywhere convergence of orthogonal series or series unconditionally convergent in measure, to Fourier coefficients of Hölder functions and to multipliers of some spaces of analytic functions.

1. Let us start with the precise definition of sublinear operators, for which we will prove our basic factorization results.

Definition. An operator $T: X \to L_p(\Omega, \mu)$, $0 \leq p \leq \infty$, where X is a Banach space is called sublinear if

(a) $|T(x + y)| \leq |T(x)| + |T(y)|$ for $x, y \in X$,

(b) $|T(\lambda x)| = |\lambda| \cdot |T(x)|$ for $x \in X$ and all scalars λ,
where the above inequalities are understood pointwise μ-almost everywhere.

Clearly every linear operator is sublinear. If $S: X \to L_0$ is a linear operator then $Tx(\omega) = |S(x)(\omega)|$ is a sublinear operator. More generally if $S: X \to L_0(Y)$ is a linear operator, where X, Y are Banach spaces and $L_0(Y)$ is the space of strongly measurable Y-valued functions, then

$T(x) = \|S(x)\|_Y$ is a sublinear operator. In particular if we have a sequence $S_n: X \to L_0$ of linear operators then

$$M(x)(\omega) = \max_n |S_n(x)(\omega)| \quad \text{and} \quad Q(x)(\omega) = \sqrt{\sum_{n=1}^{\infty} |S_n(x)(\omega)|^2}$$

are sublinear operators (provided they do exist), because we can consider the operator

$$S: X \to L_0(\ell_\infty) \quad \text{or} \quad S: X \to L_0(\ell_2)$$

defined by $S(x) = (S_n(x))_{n=1}^{\infty}$. This covers a wide variety of maximal operators and square function type operators. It is such examples that convinced us to consider sublinear operators.

2 Proposition. Let $T: X \to L_0(\Omega, \mu)$, with μ a finite measure, be a sublinear operator. The following conditions are equivalent:

(a) T is continuous at 0;

(b) $T(B_X)$ is a bounded set in $L_0(\Omega, \mu)$;

(c) there exists a function $C(\lambda)$ defined for $\lambda \geq 0$ such that

$$\mu\{\omega \in \Omega: |Tx(\omega)| > \lambda \cdot \|x\|\} \leq C(\lambda)$$

and $\lim C(\lambda) = 0$ as $\lambda \to \infty$.

Proof: $a \Rightarrow b$ and $c \Rightarrow a$ are obvious from the definition. Let us prove $b \Rightarrow c$. Since $T(B_X)$ is absorbed by every neighbourhood of zero in $L_0(\Omega, \mu)$ we get that for every $\lambda > 0$ there exists an N_λ such that

$$T(B_X) \subseteq N_\lambda \cdot \{f \in L_0(\Omega, \mu): \mu\{\omega \in \Omega: |f(\omega)| > \lambda\} < 1/\lambda\}.$$

This gives $\mu\{\omega: |Tx(\omega)| \geq \lambda \cdot N_\lambda \cdot \|x\|\} \leq 1/\lambda$, so the desired $C(\lambda)$ exists.   ■

3 Proposition. Let $T_n: X \to L_0(\Omega, \mu)$, $n = 1, 2, \ldots$, be a sequence of continuous linear operators. Assume that $T^\bullet(x) = \sup_n |T_n(x)(\omega)| < \infty$, μ-almost everywhere for every $x \in X$. Then $T^\bullet$ is a continuous sublinear operator from X into $L_0(\Omega, \mu)$.

Proof: Let us put $L_0(\ell_\infty) = \{(f_n)_{n=1}^{\infty}: \|f_n(\omega)\|_\infty \in L_0(\Omega, \mu)\}$. With the natural structure it is an F-space. It follows from the closed graph theorem I.A.6 that the operator $U: X \rightarrow L_0(\ell_\infty)$ defined by $U(x) = (T_n(x))_{n=1}^\infty$ is continuous. Thus $T^\bullet(x)(\omega) = \|U(x)(\omega)\|_\infty$ is also continuous. ∎

A word about notation. It is customary in harmonic analysis to denote the maximal operators such as our $T^\bullet = \sup_n |T_n(x)(\omega)|$ as T^*. This, however contradicts the functional analysis usage of T^* as an adjoint operator.

4 Definition. *Let X be a Banach space and let $T: X \rightarrow L_0(\Omega, \mu)$ be a sublinear operator. We say that T factors strongly through $L_{p,\infty}$ if there exists a function g on Ω such that*

$$\mu\left\{\omega: \left|\frac{T(x)(\omega)}{g(\omega)}\right| > \lambda\right\} \leq \left(\frac{\|x\|}{\lambda}\right)^p \quad \text{for all} \quad x \in X.$$

In more operator-theoretical terms we can express this as a commutative diagram

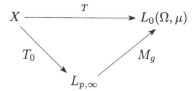

where $M_g(f) = g \cdot f$.

5 Proposition. *Let X be a Banach space and let $T: X \rightarrow L_0(\Omega, \mu)$ be a sublinear operator continuous at 0. Let us assume that $\mu(\Omega) = 1$ and $0 < p < \infty$. The following conditions are equivalent:*

(a) *T factors strongly through $L_{p,\infty}$;*

(b) *there exists a function $C(\lambda)$ with $\lim_{\lambda \to \infty} C(\lambda) = 0$ such that*

$$\mu\left\{\omega \in \Omega: \sup |T(x_j)(\omega)| \geq \lambda\left(\sum_j \|x_j\|^p\right)^{1/p}\right\} \leq C(\lambda)$$

for all finite sequences (x_j) in X;

(c) *for every positive ε there exists a constant $C_\varepsilon > 0$ and a set $E_\varepsilon \subset \Omega$ with $\mu(\Omega \backslash E_\varepsilon) < \varepsilon$ such that*

$$\mu\{\omega \in E_\varepsilon: |T(x)(\omega)| > \lambda\} \leq C_\varepsilon \left(\frac{\|x\|}{\lambda}\right)^p$$

for all x in X and all positive λ.

Remark.
The condition (b) means that the operator $\widetilde{T}\colon (\Sigma X)_p \to L_0(\ell_\infty)$ defined by $\widetilde{T}(x_j) = (T(x_j))$ is continuous.

Proof: (a)$\Rightarrow$(b). Let us take $(x_n)_{n=1}^\infty$ with $\sum \|x_n\|^p = 1$. Using the notation from **4** we have

$$\mu\{\omega\colon \sup |T(x_j)| \geq \lambda\} = \mu\{\omega\colon g(\omega) \cdot \sup |T_0(x_j)| \geq \lambda\}$$
$$\leq \mu\{\omega\colon g(\omega) \geq \sqrt{\lambda}\} + \mu\{\omega\colon \sup |T_0(x_j)| \geq \sqrt{\lambda}\}$$
$$\leq \mu\{\omega\colon g(\omega) \geq \sqrt{\lambda}\} + \sum_j \mu\{\omega\colon |T_0(x_j)| \geq \sqrt{\lambda}\}$$
$$\leq \mu\{\omega\colon g(\omega) \geq \sqrt{\lambda}\} + \sum_j \left(\frac{\|x_j\|}{\sqrt{\lambda}}\right)^p$$
$$\leq \mu\{\omega\colon g(\omega) \geq \sqrt{\lambda}\} + \lambda^{-p/2} = C(\lambda).$$

One checks that this function $C(\lambda)$ satisfies the desired conditions so we have (b).

(b)$\Rightarrow$(c). Let us fix a function $C(\lambda)$ as in (b) and for a given $\varepsilon > 0$ let us fix a number R such that $C(R) < \varepsilon$. Let us consider the following condition on a subset $F \subset \Omega$

$$\exists x \in X, \|x\| \leq 1 \text{ such that } \mu(F) \cdot |Tx(\omega)|^p > R^p \text{ for all } \omega \in F. \quad (*)$$

If no subset $F \subset \Omega$ satisfies $(*)$ then for every $x \in X$ with $\|x\| = 1$ we have
$$\mu\{\omega \in \Omega\colon |Tx(\omega)| > \lambda\} \leq \left(\frac{R}{\lambda}\right)^p.$$

If there are sets in Ω satisfying $(*)$ let us fix a maximal family of disjoint sets (F_j) satisfying $(*)$ with corresponding $x_j \in X, \|x_j\| \leq 1$. For $c_j = \mu(F_j)^{1/p}$ we have $\Sigma\|c_j x_j\|^p \leq 1$ and $\sup_j |T(c_j x_j)(\omega)| > R$ a.e. on $F = \bigcup_{j=1}^\infty F_j$. Condition (b) yields $\mu(F) \leq C(R) \leq \varepsilon$. We will show that $E = \Omega\backslash F$ satisfies (c) with $C_\varepsilon = R^p$. If not, there exist $x \in X$ with $\|x\| \leq 1$ and a number $\lambda > 0$ such that $\mu\{\omega \in E\colon |Tx(\omega)| \geq \lambda\} > (R/\lambda)^p$. Thus $\{\omega \in E\colon |Tx(\omega)| \geq \lambda\}$ satisfies $(*)$ and is disjoint with F. Since (F_j) was a maximal family we get (c).

(c)$\Rightarrow$(a). For $\varepsilon_n = 1/n$, $n = 1, 2, \ldots$ let E_n and $C_n = C_{\varepsilon_n}$ be given by (c). Let us fix a sequence of positive numbers a_n such that $\sum C_n a_n^{-p} = 1$ and for $F_n = E_n\backslash\bigcup_{j<n} E_j$, $n = 1, 2, \ldots$ let us define

$g(\omega) = \sum_{n=1}^{\infty} a_n \chi_{F_n}$. One sees that $\bigcup_{n=1}^{\infty} E_n = \bigcup_{n=1}^{\infty} F_n = \Omega$ and F_n's are disjont. For $x \in X$ and $\lambda > 0$ we have

$$\mu\left\{\omega \in \Omega : \frac{|T(x(\omega))|}{g(\omega)} > \lambda\right\} = \sum_{n=1}^{\infty} \mu\{\omega \in F_n : |Tx(\omega)| > a_n\lambda\}$$

$$\leq \sum_{n=1}^{\infty} \mu\{\omega \in E_n : |Tx(\omega)| > a_n\lambda\}$$

$$\leq \sum_{n=1}^{\infty} C_n \left(\frac{\|x\|}{a_n\lambda}\right)^p$$

$$= \left(\sum_{n=1}^{\infty} C_n a_n^{-p}\right) \cdot (\|x\|/\lambda)^p = (\|x\|/\lambda)^p. \ \blacksquare$$

The really important implication in Proposition 5 is (b)$\Rightarrow$(a). It allows us to establish that some operators do factor strongly through $L_{p,\infty}$. The main example of it is the following

6 Theorem. (Nikishin). *Let X be a Banach space of type p and let (Ω, μ) be a σ-finite measure space. Then every sublinear operator T from X into $L_0(\Omega, \mu)$ continuous at zero, factors strongly through $L_{p,\infty}$.*

Proof: As is well known (see II.B.2) there exist a probability measure ν on Ω and a measurable function ϕ such that the map $f \mapsto f \cdot \phi$ is a topological isomorphism between $L_0(\Omega, \mu)$ and $L_0(\Omega, \nu)$. Thus without loss of generality we can assume $\mu(\Omega) = 1$. In order to check condition (b) of Proposition 5 let us fix $x_1, x_2, \ldots, x_n \in X$ with $\sum \|x_j\|^p \leq 1$. Let us put $g_t = \sum_j r_j(t)x_j \in X$ (the r_j's are Rademacher functions; see I.B.8). For every k, $1 \leq k \leq n$, let $g_t^k = \sum_j \varepsilon_j r_j(t)x_j$ where $\varepsilon_k = 1$ and $\varepsilon_j = -1$ for $j \neq k$. Clearly $2|T(x_k)| = |T(g_t + g_t^k)| \leq |T(g_t)| + |T(g_t^k)|$. For almost all $\omega \in \Omega$, the functions $|T(g_t)(\omega)|$ and $|T(g_t^k)(\omega)|$ have the same distribution as a function of t, so

$$|\{t \in [0,1] : |Tg_t(\omega)| \geq |Tx_k(\omega)|\}| \geq 1/2.$$

Since $|Tx_k(\omega)|$ does not depend on t we get that for almost all $\omega \in \Omega$

$$|\{t \in [0,1] : |Tg_t(\omega)| \geq \max_k |Tx_k(\omega)|\}| \geq 1/2.$$

Thus for every $\lambda > 0$

$$\mu\{\omega: \sup_k |Tx_k(\omega)| \geq \lambda\} \leq 2 \cdot \int |\{t \in [0,1]: |Tg_t(\omega)| \geq \lambda\}| d\mu(\omega)$$

$$\leq 2 \cdot \int \left[|\{t \in [0,1]: \|g_t\| \geq \sqrt{\lambda}\}| \right. \tag{1}$$

$$+ |\{t \in [0,1]: |Tg_t(\omega)| \geq \sqrt{\lambda}\|g_t\|\}| \Big] d\mu(\omega)$$

$$= 2 \cdot |\{t \in [0,1]: \|g_t\| \geq \sqrt{\lambda}\}|$$

$$+ 2 \cdot \int \mu\{\omega \in \Omega: |Tg_t(\omega)| \geq \sqrt{\lambda}\|g_t\|\} dt.$$

Since X has type p, the Markov inequality, sometimes also called Chebyshev's inequality, gives

$$|\{t \in [0,1]: \|g_t\| \geq \sqrt{\lambda}\}| \leq T_p(X)^p \cdot \lambda^{-p/2}.$$

Since T is continuous at zero, Proposition 2 shows that the last integral in (1) tends to zero as $\lambda \to \infty$. This verifies condition (b) of Proposition 5 and proves the theorem.   ∎

7 Corollary. *Let G be a compact abelian group with Haar measure m. Assume that X is a Banach space of type p and that we have a representation $g \mapsto I_g$ of G into the isomorphisms of X. Let $T_g f(h) = f(h+g)$ for $f \in L_0(G,m)$. Assume that $T: X \to L_0(G,m)$ is a continuous sublinear operator such that $T_g T I_{g^{-1}} = T$ for all $g \in G$. Then T is of weak type q with $q = \min(p,2)$.*

Proof: Using the Nikishin theorem we see that condition (c) of Proposition 5 gives a set $E \subset G$ with $m(E) > \frac{1}{2}$ such that

$$m(\{g \in G: |Tx(g)| > \lambda\} \cap E) \leq C\left(\frac{\|x\|}{\lambda}\right)^p \quad \text{for all} \quad x \in X. \tag{2}$$

Since $T_g T I_{g^{-1}} = T$ we get

$$\{g \in G: |TI_h(x)(g)| > \lambda\} = h + \{g \in G: |Tx(g)| > \lambda\}. \tag{3}$$

Comparing (2) and (3) we see that for every $h \in G$

$$m\big(\{g \in G: |Tx(g)| > \lambda\} \cap (E - h)\big) \leq C\left(\frac{\|I_h(x)\|}{\lambda}\right)^p. \tag{4}$$

Integrating (4) with respect to h and using the fact that $\|I_h\| \leq C$ for $h \in G$ we get

$$m\{g \in G : |Tx(g)| > \lambda\} \leq m(E)^{-1} C \left(\frac{\|x\|}{\lambda}\right)^p. \qquad \blacksquare$$

8 Comment. The Nikishin theorem and Corollary 7 formalize several well known equivalences between existence almost everywhere and weak type (1-1) for certain operators. In particular we have

Example 1. Let $f \in L_1(\mathbb{T})$ and let $f(re^{i\theta})$ denote its harmonic extension. Let $\tilde{f}(re^{i\theta})$ denote the harmonic function conjugate to $f(re^{i\theta})$. For $f \geq 0$ the function $G(re^{i\theta}) = (1 + f(re^{i\theta}) + i\tilde{f}(re^{i\theta}))^{-1}$ is bounded and analytic. The Fatou theorem shows that $\lim_{r \to 1} G(re^{i\theta})$ exists a.e.. This implies that $\lim_{r \to 1} \tilde{f}(re^{i\theta})$ exists a.e.. This implies that such a limit exists a.e. for arbitrary f. For the background on this see I.B.19 and the references given there. So (see Proposition 3) $Mf(\theta) = \max_{0 < r < 1} |\tilde{f}(re^{i\theta})|$ is a sublinear continuous operator from $L_1(\mathbb{T})$ into $L_0(\mathbb{T})$. Clearly M commutes with translations so by Proposition 7 M is of weak type (1-1).

Example 2. The following are equivalent for $1 \leq p \leq 2$.

(a) For every $f \in L_p(\mathbb{T})$ the series $\sum_{-\infty}^{\infty} \hat{f}(n) \cdot e^{in\theta}$ converges to f a.e.

(b) For every $f \in L_p(\mathbb{T})$ we have $\max_N |\sum_{-N}^{N} \hat{f}(n)e^{in\theta}| < \infty$ a.e.

(c) The map $f \mapsto \max_N |\sum_{-N}^{N} \hat{f}(n)e^{in\theta}|$ is of weak type (p, p).
The only non-trivial (given Nikishin's theorem) implication is (c)$\Rightarrow$(a). This is however a standard approximation argument. We have almost everywhere convergence on the dense set of smooth functions and (c) allows us to get the same for all $f \in L_p(\mathbb{T})$.

Actually the well known Carleson-Hunt theorem (see Carleson [1966] and Hunt [1968]) says that (a),(b),(c) hold for $p > 1$ and the classical Kolmogorov [1926] example shows that they are false for $p = 1$.

9. The following definition is clearly analogous to Definition 4.

Definition. Let X be a Banach space and let $T : X \to L_p(\Omega, \mu)$, $0 < p < \infty$, be a sublinear operator. We say that T factors strongly through L_q, $q > p$ if there exists a positive function $g \in L_r(\Omega, \mu)$, with $1/q + 1/r = 1/p$, and a continuous sublinear operator $u : X \to L_q(\Omega, \mu)$ such that $Tx = g \cdot u(x)$ for $x \in X$.

Note that if $T: X \to L_p(\Omega, \mu)$ factors strongly through L_q, $0 \le p < q \le \infty$, then $T(X) \subset L_p(E, \mu)$ where $E = \operatorname{supp} g$ is a σ-finite set. Also, there exists a function φ on E and a probability measure ν on E such that the map $f \mapsto f \cdot \varphi$ is an isometry between $L_p(E, \mu)$ and $L_p(E, \nu)$; see III.A.2. These remarks show that we can always assume that (Ω, μ) is a probability measure space.

10.   Our next important result will be the following proposition which gives conditions on an individual operator to factor strongly through $L_p(\Omega, \mu)$. It is analogous to Proposition 5.

Proposition.   *Let T be a sublinear operator from X into $L_p(\Omega, \mu)$, $0 < p < \infty$. Then, the following conditions are equivalent:*

(a)   *T factors strongly through L_q, $q > p$;*

(b)   *there exists a constant K such that for every finite sequence $(x_j)_{j=1}^n \subset X$*

$$\left(\int_\Omega \left(\sum_{j=1}^n |T(x_j)|^q \right)^{p/q} d\mu \right)^{1/p} \le K \left(\sum_{j=1}^n \|x_j\|^q \right)^{1/q}.$$

Proof: Observe that condition (b) implies that $T(X) \subset L_p(E, \mu)$ for some σ-finite set $E \subset \Omega$. If this is not the case then there exist a sequence $(x_j)_{j=1}^\infty \subset B_X$ and disjoint subsets of Ω, $(A_j)_{j=1}^\infty$ such that $\int_{A_j} |T(x_j)|^p d\mu \ge c$ for some $c > 0$ and all j's. But then for all N we have

$$KN^{1/q} \ge \left(\int_\Omega \left(\sum_{j=1}^N |T(x_j)|^q \right)^{p/q} d\mu \right)^{1/p}$$

$$\ge \left(\sum_{j=1}^N \int_{A_j} |T(x_j)|^p d\mu \right)^{1/p} \ge (cN)^{1/p}$$

which is impossible for large N. Thus we can assume (see **9**) that (Ω, μ) is a probability measure space.

(a)$\Rightarrow$(b). This is easy. We have

$$\left(\int_\Omega \left(\sum_{j=1}^n |T(x_j)|^q \right)^{p/q} d\mu \right)^{1/p}$$

$$= \left(\int_\Omega g^p \left(\sum_{j=1}^n |u(x_j)|^q \right)^{p/q} d\mu \right)^{1/p}$$

$$\leq \left(\int_\Omega g^r d\mu \right)^{1/r} \left(\int_\Omega \sum_{j=1}^n |u(x_j)|^q d\mu \right)^{1/q}$$

$$\leq C(\sum_j \|u(x_j)\|_q^q)^{1/q} \leq C\|u\|(\sum_j \|x_j\|^q)^{1/q}.$$

The implication (b)⇒(a) is more difficult. Let us observe that actually we are looking for a positive function g in $L_r(\Omega, \mu)$ such that $\|g^{-1}T(x)\|_q \leq K\|x\|$. The case $q = \infty$ is rather simple. In this case (b) reads

$$\left(\int_\Omega \sup_j |Tx_j|^p d\mu \right)^{1/p} \leq K \sup \|x_j\|.$$

This implies that

$$g = \sup\{|T(x)|: x \in X, \|x\| \leq 1\}$$

exists in $L_p(\Omega, \mu)$ and $\|g\|_p \leq 1$ (see Exercise III.A.1). This is the desired g (note that for $q = \infty$ we get $p = r$).

In the case $q < \infty$ let us define

$$K_n = \sup \left\{ \left(\int_\Omega \left(\sum_{j=1}^n |Tx_j|^q \right)^{p/q} d\mu \right)^{1/p} : \sum_{j=1}^n \|x_j\|^q \leq 1 \right\}. \quad (5)$$

We put $K = \lim K_n$. Fix $x_1^n, \ldots, x_n^n$ with $\sum_{j=1}^n \|x_j^n\|^q \leq K_n^{-q}(1 + 1/n)$ such that $\int_\Omega (\sum_j |Tx_j^n|^q)^{p/q} d\mu = 1$. Let us define functions $f_n = (\sum_j |Tx_j^n|^q)^{p/q}$. Clearly $f_n \geq 0$ and $\int f_n = 1$ for $n = 1, 2, \ldots$. For $(A_k)_{k=1}^N$ any sequence of disjoint subsets of Ω and $(n_k)_{k=1}^N$ any sequence of integers, we have using (5)

$$\sum_{k=1}^N \int_{A_k} f_{n_k} = \int_\Omega \left(\sum_{k=1}^N (f_{n_k} \chi_{A_k})^{q/p} \right)^{p/q} d\mu$$

$$\leq \int_\Omega \left(\sum_{k=1}^N f_{n_k}^{q/p} \right)^{p/q} d\mu$$

$$= \int_\Omega \left(\sum_{k=1}^N \sum_{j=1}^{n_k} |Tx_j^{n_k}|^q \right)^{p/q} d\mu$$

$$\leq \left(\sum_{k=1}^N \sum_{j=1}^{n_k} \int_\Omega |Tx_j^{n_k}|^q d\mu \right)^{p/q}$$

$$\leq K^p \left(\sum_{k=1}^N \sum_{j=1}^{n_k} \|x_j^{n_k}\|^q \right)^{p/q} \leq CN^{p/q}.$$

This shows (see III.C.12) that the sequence $(f_n)_{n=1}^{\infty}$ is uniformly integrable in $L_1(\Omega, \mu)$. Let h be any $\sigma(L_1(\Omega, \mu), L_\infty(\Omega, \mu))$ cluster point of this sequence. Clearly h is nonnegative and $\int h d\mu = 1$. For any element $x \in X$ with $\|x\| = 1$ and any number $t \geq 0$ and any $n = 1, 2, \ldots$ we have from (5)

$$\int_\Omega (f_n^{q/p} + t^q |Tx|^q)^{p/q} d\mu = \int_\Omega \left(\sum_{j=1}^n |Tx_j^n|^q + |T(tx)|^q \right)^{p/q} d\mu$$

$$\leq K_{n+1}^p \left(\sum_{j=1}^n \|x_j^n\|^q + t^q \right)^{p/q}$$

$$\leq K_{n+1}^p (K_n^{-q}(1 + 1/n) + t^q)^{p/q}.$$

Passing to the limit with n we get

$$\int_\Omega (h^{q/p} + t^q |Tx|^q)^{p/q} d\mu \leq (1 + K^q t^q)^{p/q}.$$

Since for $t = 0$ this inequality becomes an equality we see that for small s we have

$$\frac{d}{dt} \left(\int_\Omega (h^{q/p} + t^q |Tx|^q)^{p/q} d\mu \right) \Big|_{t=s} \leq \frac{d}{dt} (1 + K^q t^q)^{p/q} \big|_{t=s}.$$

Routine differentiation gives

$$\int h^{(\frac{p}{q}-1)q/p} |Tx|^q d\mu \leq K^q \quad \text{for} \quad x \in X, \|x\| = 1.$$

Thus we can put $g = h^{\frac{1}{p} - \frac{1}{q}}$. $\blacksquare$

11. It is possible to present a result analogous to Nikishin's theorem 6 for strong factorization through $L_p(\Omega, \mu)$ of linear operators. It requires, however, a different notion of type (see Notes and remarks). Only the case of type 2, the most important for applications, can be easily handled with our concept of type.

Corollary. *Assume that X is a Banach space of type 2 and that $T: X \rightarrow L_p(\Omega, \mu), 0 < p < 2$, be a continuous linear operator. Then T factors strongly through $L_2(\Omega, \mu)$.*

Proof: We will check condition (b) of Proposition 10 assuming that (Ω, μ) is a probability measure space. As we know (see **9** and the beginning of the proof of **10**) this is enough. For a finite sequence $(x_j)_{j=1}^n \subset X$ we have from the Khintchine inequality (see I.B.8)

$$\left(\int \left(\sum_j |Tx_j|^2 \right)^{p/2} d\mu \right)^{1/p} \leq C_p \left(\int \int \left| \sum_j r_j(t) Tx_j(\omega) \right|^p dt d\mu(\omega) \right)^{1/p}$$

$$= C_p \left(\int \int \left| T\left(\sum_j r_j(t) x_j \right)(\omega) \right|^p d\mu(\omega) dt \right)^{1/p}$$

$$\leq C_p \|T\| \left(\int \left\| \sum_j r_j(t) x_j \right\|^p dt \right)^{1/p}$$

$$\leq C_p \|T\| \, T_2(X) \left(\sum_j \|x_j\|^2 \right)^{1/2}. \qquad \blacksquare$$

12. For spaces of type s, with $s < 2$ we have the following result which is weaker than one might expect looking at Corollary 11. Unfortunately it cannot be improved.

Theorem. If X is a Banach space of type $s, 1 < s < 2$ and $T : X \rightarrow L_p(\Omega, \mu)$ is a linear operator, $0 < p < s$, then for every q, with $p < q < s$, T factorizes strongly through $L_q(\Omega, \mu)$.

Proof: We will check that condition (b) of Proposition 10 holds. Let $(x_j)_{j=1}^n$ be a finite sequence in X. Let $(\eta_j(\tau))_{j=1}^n$ be a sequence of q-stable independent random variables (see III.A.13). We have (see III.A.16)

$$\left(\int \left(\sum_{j=1}^n |T(x_j)(\omega)|^q \right)^{\frac{p}{q}} d\mu(\omega) \right)^{\frac{1}{p}}$$

$$= C \left(\int \int \left| \sum_{j=1}^n \eta_j(\tau) T(x_j)(\omega) \right|^p d\tau d\mu(\omega) \right)^{\frac{1}{p}}$$

$$= C \left(\int \int \left| T\left(\sum_{j=1}^n \eta_j(\tau) x_j \right)(\omega) \right|^p d\mu(\omega) d\tau \right)^{\frac{1}{p}} \qquad (6)$$

$$\leq C \|T\| \left(\int \left\| \sum_{j=1}^n \eta_j(\tau) x_j \right\|^p d\tau \right)^{\frac{1}{p}}.$$

Because $\eta_j(\tau)$ are symmetric variables and X has type s we have

$$\int \left\| \sum_{j=1}^{n} \eta_j(\tau) x_j \right\|^p d\tau = \int \int \left\| \sum_{j=1}^{n} r_j(t)\eta_j(\tau) x_j \right\|^p d\tau dt \qquad (7)$$

$$\leq T_s(X)^p \int \left(\sum_{j=1}^{n} |\eta_j(\tau)|^s \|x_j\|^s \right)^{\frac{p}{s}} d\tau.$$

Now we take a sequence $(\gamma_j(\theta))_{j=1}^{n}$ of s-stable, independent random variables. Like previously we get

$$\int \left(\sum_{j=1}^{n} |\eta_j(\tau)|^s \|x_j\|^s \right)^{\frac{p}{s}} d\tau$$

$$= \int \int \left| \sum_{j=1}^{n} \gamma_j(\theta)\eta_j(\tau)\|x_j\| \right|^p d\theta d\tau$$

$$= \int \left(\sum_{j=1}^{n} |\gamma_j(\theta)|^q \|x_j\|^q \right)^{\frac{p}{q}} d\theta \qquad (8)$$

$$\leq \left(\int \sum_{j=1}^{n} |\gamma_j(\theta)|^q \|x_j\|^q d\theta \right)^{\frac{p}{q}}$$

$$= \left(\int |\gamma_1(\theta)|^q d\theta \right)^{\frac{p}{q}} \left(\sum_{j=1}^{n} \|x_j\|^q \right)^{\frac{p}{q}}.$$

Since $q < s$ we get from III.A.15 that $\int |\gamma_1(\theta)|^q d\theta < \infty$ so putting together (6), (7), (8) we see that condition (b) of Proposition 10 holds.∎

13.   Now we would like to discuss an application of the previous theorem. Let X be a reflexive subspace of $L_1(\mu)$, μ a probability measure space. From III.C.18 we see that X has type s for some $s > 1$. Applying Theorem 12 to the identity operator $i\colon X \to L_1(\mu)$ we get a factorization through $L_q(\mu)$ for any $1 < q < s$. Explicitly for every $1 < q < s$ there exists a function $g(\omega) \geq 0$, $\|g\|_r = 1$, $\frac{1}{r} + \frac{1}{q} = 1$ and a constant K such that

$$\left(\int \left| \frac{x(\omega)}{g(\omega)} \right|^q d\mu \right)^{\frac{1}{q}} \leq K\|x\|.$$

If we write $\Delta(\omega) = g^r(\omega)$ we have

Corollary.   If X is a reflexive subspace of $L_1(\mu)$, μ a probability measure , then there exist $q > 1$, and K and a positive function Δ with

$\|\Delta\|_1 = 1$ *such that for every* $x \in X$

$$\left(\int \left| \frac{x(\omega)}{\Delta(\omega)} \right|^q \Delta(\omega) d\mu \right)^{\frac{1}{q}} \leq K \|x\|_1.$$

Note that this means in particular that every reflexive subspace of an $L_1(\mu)$ space is isomorphic to a subspace of some $L_p(\mu)$ space for $p > 1$.

14. There is a connection between absolutely summing operators and the factorization problems discussed here. We do not intend to present here a full picture, so we will limit ourselves to the following remarks. Suppose we know that for some X and some p, $L(C(K), X) = \Pi_p(C(K), X)$. This gives us for every operator $T: C(K) \to X$ a factorization

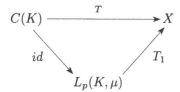

Dualizing we get

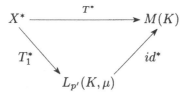

Note that if we treat $M(K)$ as $L_1(\nu)$ then $id^*: L_{p'}(K, \mu) \to L_1(\nu)$ is a multiplication operator. Since both properties '$\Pi_p(C(K), X) = L(C(K), X)$' and 'every operator from X^* into $L_1(\mu)$ factorizes strongly through $L_p(\mu)$' are local (for the last one see Proposition 10) there is no problem with taking duals. This is rather imprecise but for future reference we will prove

Proposition. *If* X^* *has type* p *for some* $p > 1$ *then* $L(C(K), X) = \Pi_q(C(K), X)$ *for every* $q > p'$, *where* $\frac{1}{p} + \frac{1}{p'} = 1$.

Proof: It is enough to check that $L(c_0, X) = \Pi_q(c_0, X)$. If $T: c_0 \to X$ then $T^*: X^* \to \ell_1$. Theorem 12 gives a factorization

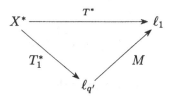

Thus $T^{**}: \ell_\infty \to X^{**}$ has a factorization $M^* T_1^{**}$. Observe that $M(\xi_n) = (m_n \xi_n)$ for some $(m_n) \in \ell_q$, thus $M^*: \ell_\infty \to \ell_q$ is given by $M^*(\xi_n) = (m_n \xi_n)$. It is immediate (see III.F.4) that M^* is q-absolutely summing. So T^{**} is q-absolutely summing and we get that T is q-absolutely summing. ∎

15. The following proposition, going in the other direction, is more involved. It gives another condition for a space X ensuring that every linear operator from X into $L_p(\Omega, \mu)$ factors strongly through $L_q(\Omega, \mu)$.

Proposition. *Let X be a Banach space and let $0 < p < q, q \geq 1$. Let us assume that for some $C < \infty$ and for every linear operator $u: X^* \to \ell_q$ we have $\pi_p(u) \leq C \pi_q(u)$. Let us assume moreover that if $p < 1$ then X has the bounded approximation property. Then every linear operator $T: X \to L_p(\Omega, \mu)$ factorizes strongly through $L_q(\Omega, \mu)$.*

Proof. We will check condition (b) of Proposition 10. Given a finite sequence $(x_j)_{j=1}^n$ in X we can find (using the b.a.p. of X if $p < 1$ or the b.a.p. of $L_p(\Omega, \mu)$ if $p \geq 1$) an operator $\tilde{T}: X \to L_p(\Omega, \mu)$ such that $\|\tilde{T}\| \leq K\|T\|$, $\tilde{T}(x) = \sum_{k=1}^\infty x_k^*(x) \chi_{A_k}$ for some disjoint sets $A_k \subset \Omega$ and

$$2\left(\int (\sum_{j=1}^n |\tilde{T}(x_j)|^q)^{p/q} d\mu \right)^{1/p} \geq \left(\int (\sum_{j=1}^n |T(x_j)|^q)^{p/q} d\mu \right)^{1/p}. \quad (9)$$

Let us define $u: X^* \to \ell_q^n$ by the formula $u(x^*) = (x_j(x^*))_{j=1}^n$. One can easily check that $\pi_q(u) \leq (\sum_{j=1}^n \|x_j\|^q)^{1/q}$ so $\pi_p(u) \leq C(\sum_{j=1}^n \|x_j\|^q)^{1/q}$. Let us define a function $\varphi: \Omega \to X^*$ by the formula $\varphi(\omega)(x) = \tilde{T}(x)(\omega)$. The properties of $\tilde{T}$ ensure that φ is well defined.

We have (see III.F.33 and (9))

$$\left(\int(\sum_{j=1}^{n}|T(x_j)|^q)^{p/q}d\mu\right)^{1/p}$$

$$\leq 2\left(\int(\sum_{j=1}^{n}|\tilde{T}(x_j)|^q)^{p/q}d\mu\right)^{1/p}$$

$$\leq 2\left(\int\sum_{j=1}^{n}|\varphi(\omega)(x_j)|^q\right)^{p/q}d\mu(\omega)\right)^{1/p}$$

$$= 2\left(\int\|u(\varphi(\omega))\|^pd\mu(\omega)\right)^{1/p}$$

$$\leq 2\cdot\pi_p(u)\sup_{\|x^{**}\|\leq 1}\left(\int|x^{**}(\varphi(\omega))|^pd\mu(\omega)\right)^{1/p}$$

$$\leq 2\cdot\pi_p(u)\sup_{\|x^{**}\|\leq 1}\left(\int|S(x^{**})(\omega)|^pd\mu(\omega)\right)^{1/p}$$

$$\leq 2\cdot C\cdot(\sum_{j=1}^{n}\|x_j^*\|^q)^{1/q}\|\tilde{T}\|\leq 2KC\|T\|(\sum_{j=1}^{n}\|x_j^*\|^q)^{1/q}.$$

In the above $S\colon X^{**}\to L_p(\Omega,\mu)$ is given by $S(x^{**})=\sum_{k=1}^{\infty}x^{**}(x_k^*)\chi_{A_k}$. Clearly $\|S\|=\|\tilde{T}\|$. ∎

16 Corollary. *Every linear operator $T\colon L_p\to L_q(\Omega,\mu)$, $0<q\leq 2\leq p\leq\infty$, factors strongly through $L_2(\Omega,\mu)$.*

Proof: The case $p<\infty$ follows immediately from Corollary 11 and Theorem III.A.23. For the case $p=\infty$ let us note that by Corollary III.F.35 we have $B(L_1,L_2)=\Pi_p(L_1,L_2)$ for $0<p\leq 2$, so we can apply Proposition 15.

Remark. Corollary 16 for $q\geq 1$ and $p=\infty$ can be derived directly from Corollary 11 and Theorem III.F.29.

17. We say that the series $\sum f_n$ is unconditionally convergent in measure if $\sum c_n f_n$ is convergent in measure for every $(c_n)\in\ell_\infty$. Let us discuss now the structure of such series.

Theorem. (Ørno). *If $\sum_{n=1}^{\infty}f_n$ is unconditionally convergent in measure on $[0,1]$, then there exist a function $g(t)$ on $[0,1]$, a sequence*

$(\alpha_n)_{n=1}^{\infty} \in \ell_2$ and an orthonormal system $(w_n)_{n=1}^{\infty}$ in $L_2[0,2]$ such that

$$f_n(t) = \alpha_n \cdot g(t) \cdot w_n(t) \text{ for } t \in [0,1] \text{ and } n = 1,2,\ldots .$$

Proof: By the closed graph theorem I.A.6. the map $T: \ell_{\infty} \to L_0[0,1]$ defined by $T(c_n) = \sum_{n=1}^{\infty} c_n f_n$ is continuous. We have the following commuting diagram:

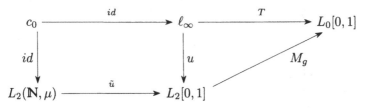

Operators u and M_g (recall that $M_g(f) = g \cdot f$) are obtained by applying first Theorem 6 and next Corollary 16 to the operator T. The operator u is 2-absolutely summing by Theorem III.F.29, so by the Pietsch theorem III.F.8 there exists a probability measure μ on the set $\mathbf{N}$ of natural numbers and an operator $\tilde{u}$ such that the above diagram commutes. Let us put $h_n = f_n(\|\tilde{u}\| \sqrt{\mu(\{n\})}g)^{-1}$. Since $\tilde{u}(e_n) = u(e_n) = f_n \cdot g^{-1}$, we get for an arbitrary sequence of scalars $(\alpha_n)_{n=1}^{\infty}$

$$\left\| \sum_{n=1}^{\infty} \alpha_n h_n \right\|_{L_2[0,1]} = \left\| \sum_{n=1}^{\infty} \frac{\alpha_n}{\|\tilde{u}\| \sqrt{\mu(\{n\})}} \tilde{u}(e_n) \right\|_{L_2[0,1]}$$

$$\leq \left\| \sum_{n=1}^{\infty} \frac{\alpha_n}{\sqrt{\mu(\{n\})}} e_n \right\|_{L_2(\mathbf{N},\mu)} = \left(\sum |\alpha_n|^2 \right)^{1/2} .$$

In order to complete the proof we appeal to the following.

18 Lemma.   If functions $(h_n)_{n=1}^{\infty} \subset L_2[0,1]$ are such that $\|\sum_{n=1}^{\infty} \alpha_n h_n\| \leq (\sum_{n=1}^{\infty} |\alpha_n|^2)^{1/2}$ for every finite sequence of scalars (α_n), then there exists an orthonormal system w_n in $L_2[0,2]$ such that $w_n|[0,1] = h_n$ for $n = 1,2,\ldots .$

This lemma follows (see below) from the following general result about operators on Hilbert spaces:

19 Theorem. (Dilation Theorem).   Let H be a Hilbert space and let $T: H \to H$ be a linear operator with $\|T\| \leq 1$. Then there exist a Hilbert space $\mathcal{H} \supset H$ and an isometry $\mathcal{U}: \mathcal{H} \to \mathcal{H}$ such that $T = P\mathcal{U}|H$ where P is the orthogonal projection from $\mathcal{H}$ onto H.

Proof of Theorem 19.   Let us define $\mathcal{H} = \left(\sum_{n=1}^{\infty} H\right)_2$ and embed H into $\mathcal{H}$ onto the first coordinate. Let $\mathcal{U}(h_1, h_2, \ldots) = (Th_1, Sh_1, h_2, h_3, \ldots)$ where $S = (id - T^*T)^{1/2}$. Clearly $P\mathcal{U}|H = T$. In order to check that $\mathcal{U}$ is an isometry it is enough to check that $\|Th\|^2 + \|Sh\|^2 = \|h\|^2$ for all $h \in H$. But

$$\|Th\|^2 + \|Sh\|^2 = \langle Th, Th \rangle + \langle (id - T^*T)^{1/2}h, (id - T^*T)^{1/2}h \rangle$$
$$= \langle T^*Th, h \rangle + \langle (id - T^*T)h, h \rangle = \langle h, h \rangle = \|h\|^2. \quad \blacksquare$$

Proof of Lemma 18.   Let $(\varphi_n)_{n=1}^{\infty}$ be any complete, orthonormal system in $L_2[0, 1]$. The operator T, defined by $T(f) = \sum \langle f, \varphi_n \rangle h_n$, has norm at most 1. Apply Theorem 19 with $H = L_2[0, 1]$ and identify $\mathcal{H}$ with $L_2[0, 2]$. Put $w_n = \mathcal{U}\varphi_n$ for $n = 1, 2, \ldots$. $\quad \blacksquare$

20 Corollary.   *If the series $\sum f_n$ converges unconditionally in measure on [0,1] then for every $\varepsilon > 0$ there exists a set $E_\varepsilon \subset [0, 1]$ such that $|E_\varepsilon| > 1 - \varepsilon$ and $\sum f_n | E_\varepsilon$ converges unconditionally in $L_2(E_\varepsilon)$.*

Proof: Let $f_n = \alpha_n g w_n$ as in Theorem 17. Let us define $E_\varepsilon = \{t : |g(t)| < M\}$ where M is such that $|E_\varepsilon| > 1 - \varepsilon$. Since $(\alpha_n) \in \ell_2$, the series $\sum_{n=1}^{\infty} \alpha_n w_n$ is unconditionally convergent in $L_2[0, 2]$, thus $\sum_{n=1}^{\infty} \alpha_n \cdot g \cdot w_n \chi_{E_\varepsilon} = \sum_{n=1}^{\infty} f_n \chi_{E_\varepsilon}$ is also unconditionally convergent. $\blacksquare$

21.   Ørno's theorem 17 can be very useful in transferring properties of general orthonormal series to series unconditionally convergent in measure. There are several instances known when this is possible. One such case is the extension of the Menchoff-Rademacher theorem on almost everywhere convergence to general series unconditionally convergent in measure (see Corollary 25). The proof of the Menchoff-Rademacher theorem which we will present now uses the theory of p-absolutely summing operators. Let us state the following proposition whose proof will be given later.

Proposition.   *Let $(a_n)_{n=1}^{\infty}$ be such that $\sum_{n=1}^{\infty} |a_n|^2 \log^2(n + 1) < \infty$. Then the operator $\sigma : \ell_2 \to \ell_\infty$ defined by $\sigma(\xi_j) = \left(\sum_{j=1}^{n} \xi_j a_j\right)_{n=1}^{\infty}$ is 2-absolutely summing and $\pi_2(\sigma) \le C\sqrt{\sum_{n=1}^{\infty} |a_n|^2 \log^2(n + 1)}$.*

22.   Using the above proposition we will show the following classical result:

Theorem. (Menchoff-Rademacher). Let $(f_n)_{n=1}^\infty$ be an orthonormal system in $L_2(\Omega, \mu)$ and let $\sum_{n=1}^\infty |a_n|^2 \log^2(n+1) < \infty$. Then the series $\sum_{n=1}^\infty a_n f_n$ converges μ-almost everywhere.

Proof: Let us define functions $\Phi_s: \Omega \to \ell_2$, $s = 1, 2, \ldots$ by the formula

$$\Phi_s(\omega) = (f_1(\omega), f_2(\omega), \ldots, f_s(\omega), 0, 0, \ldots).$$

From Proposition 21 and III.F.33(b) we get

$$\int \|\sigma \Phi_s(\omega)\|_\infty^2 d\mu(\omega)$$

$$\leq C\left(\sum_{n=1}^\infty |a_n|^2 \log^2(n+1) \right) \cdot$$

$$\sup_{\xi \in \ell_2, \|\xi\| \leq 1} \int |\xi(\Phi_s(\omega))|^2 d\mu(\omega). \tag{11}$$

Using the definitions of σ and Φ_s and (11) we get

$$\int \sup_{n \leq s} \left| \sum_{j=1}^n a_j f_j(\omega) \right|^2 d\mu(\omega) \leq C\left(\sum_{n=1}^\infty |a_n|^2 \log^2(n+1) \right). \tag{12}$$

Since s was arbitrary from (12) we infer that

$$\int \sup_n \left| \sum_{j=1}^n a_j f_j(\omega) \right|^2 d\mu(\omega) \leq C\left(\sum_{n=1}^\infty |a_n|^2 \log^2(n+1) \right). \tag{13}$$

For a fixed sequence (a_n), (13) obviously yields

$$\int \sup_{K \geq N} \left| \sum_{j=N}^K a_j f_j(\omega) \right|^2 d\mu(\omega) \leq C\left(\sum_{n=N}^\infty |a_n|^2 \log^2(n+1) \right).$$

This easily implies that $\sum_{n=1}^\infty a_n f_n(\omega)$ converges μ-almost everywhere. ∎

In the proof of Proposition 21 we will use the following two lemmas.

23 Lemma. Let $T_n: X_n \to Y_n$ be a sequence of operators between Banach spaces and let us define $T: (\Sigma X_n)_2 \to (\Sigma Y_n)_2$ by $T((x_n)) = (T_n(x_n))$. Then $\pi_2(T) \leq (\sum \pi_2(T_n)^2)^{1/2}$.

The proof of this lemma easily follows from Theorem III.F.8 and is left as an exercise.

24 Lemma. *Let $\sigma: \ell_2^N \to \ell_\infty^N$ be defined as $\sigma(\xi_j) = \left(\sum_{j=1}^n a_j \xi_j \right)_{n=1}^N$. Then $\pi_2(\sigma) \leq C \left(\sum_{j=1}^N |a_j|^2 \right)^{1/2} \log(N+1)$.*

Proof: One can easily compute that σ is given by a matrix $[a_{n,m}]_{n,m=1}^N$ where

$$a_{n,m} = \begin{cases} a_m & \text{if} \quad m \leq n, \\ 0 & \text{if} \quad m > n. \end{cases}$$

Let us consider the matrix valued function $A(\Theta) = \left(\sum_{k=0}^N e^{ik\Theta} \right)$ $[e^{i(n-m)\Theta} a_m]_{n,m=1}^N$. We easily see that for every Θ, $A(\Theta)$ represents a one-dimensional operator. When we treat $A(\Theta)$ as an operator from ℓ_2^N into ℓ_∞^N we get

$$\pi_2(A(\Theta)) = \|A(\Theta)\| = \left| \sum_{k=1}^N e^{ik\Theta} \right| \cdot \left(\sum_{m=1}^N |a_m|^2 \right)^{1/2}.$$

This implies

$$\pi_2 \left(\frac{1}{2\pi} \int_0^{2\pi} A(\Theta)d\Theta \right) \leq \left(\sum_{M=1}^N |a_m|^2 \right)^{1/2} \frac{1}{2\pi} \int_0^{2\pi} \left| \sum_{k=1}^N e^{ik\Theta} \right| d\Theta$$

$$\tag{14}$$

$$\leq C \log(N+1) \left(\sum_{m=1}^N |a_m|^2 \right)^{1/2}.$$

Integrating coordinatewise we check that $(2\pi)^{-1} \int_0^{2\pi} A(\Theta)d\Theta = \sigma$ so (14) gives the lemma. ∎

Proof of Proposition 21. Let us fix a sequence $(a_n)_{n=1}^\infty$. Let

$$I_N = \{n: 2^N < n \leq 2^{N+1}\}$$

and let $\ell_2(I_N)$ (resp. $\ell_\infty(I_N)$) denote the subspace of ℓ_2 (resp. ℓ_∞) consisting of vectors supported on I_N. Let $v_N = (a_n)_{n=1}^\infty | I_N$. Let us write $\ell_2 = X_1 \oplus X_0$ where $X_1 = \operatorname{span}\{v_N\}_{N=0}^\infty$. Clearly $X_0 = \left(\sum_{N=0}^\infty \ell_2^0(I_N) \right)_2$ where $\ell_2^0(I_N) = \{x \in \ell_2(I_N): x \perp v_N\}$. We will show the desired estimates for $\pi_2(\sigma|X_0)$ and $\pi_2(\sigma|X_1)$. We have

$$\|\sigma(v_N \|v_N\|^{-1})\| \leq \left(\sum_{j \in I_N} |a_j|^2 \right)^{\frac{1}{2}} \leq 1/N \left(\sum_{j \in I_N} |a_j|^2 \log^2(j+1) \right)^{\frac{1}{2}}.$$

Thus $\sigma|X_1$ admits a natural factorization

$$
\begin{array}{ccc}
X_1 & \xrightarrow{\quad\sigma\quad} & \ell_\infty \\[2pt]
\Big\downarrow{\scriptstyle\alpha} & & \Big\uparrow{\scriptstyle v} \\[2pt]
\ell_2 & \xrightarrow{\quad\beta\quad} & \ell_1
\end{array}
$$

where

$$
\alpha(v_N\|v_N\|^{-1}) = \frac{1}{N}e_N,
$$
$$
\beta(e_N) = N\cdot\|v_N\|^{-1}\|\sigma(v_N)\|\cdot e_N,
$$
$$
v(e_N) = \sigma(v_N)\cdot\|\sigma(v_N)\|_\infty^{-1}.
$$

Since α is a diagonal operator between Hilbert spaces with the diagonal $(1/N)_{N=1}^\infty$ it is Hilbert-Schmidt, so by III.G.12 we have $\pi_2(\alpha) \le C$. Since $\|\beta\| \le \big(\sum_{n=1}^\infty |a_n|^2\log^2(n+1)\big)^{1/2}$ the desired estimate for $\pi_2(\sigma|X_1)$ follows. In order to estimate $\pi_2(\sigma|X_0)$ let us note that $\sigma(\ell_2^0(I_N)) \subset \ell_\infty^0(I_N)$. From Lemma 24 we get

$$
\pi_2(\sigma|\ell_2^0(I_N)) \le \log 2^N \bigg(\sum_{j\in I_N}|a_j|^2\bigg)^{\frac{1}{2}} \le C\bigg(\sum_{j\in I_N}|a_j|^2\log^2(j+1)\bigg)^{\frac{1}{2}}.
$$

This implies (use Lemma 23) that $\sigma|X_0\colon X_0 \to \big(\sum_N \ell_\infty^0(I_N)\big)_2$ has 2-summing norm at most $C\big(\sum_{n=1}^\infty |a_n|^2\log^2(n+1)\big)^{1/2}$. A fortiori, the same holds for $\sigma\colon X_0 \to \ell_\infty$. ∎

25 Corollary. *Let $\sum_{n=1}^\infty f_n$ be a series unconditionally convergent in measure. Then $\sum_{n=1}^\infty (f_n/\log(n+1))$ converges almost everywhere.*

Proof: Use Theorem 17 and Theorem 22. ∎

26.   We will prove a result already announced in the previous chapter, namely we want to show that Theorem III.G.14 is best possible in the following very strong sense.

Theorem.   *Let $0 < \alpha \le 1$ and let $(g_n)_{n=1}^\infty$ be any complete orthonormal system. Then there exists a function $f \in Lip_\alpha(\mathbb{T})$ such that $\sum |\langle f, g_n\rangle|^p = \infty$ for $p = \frac{2}{(2\alpha+1)}$.*

Proof: If this is not the case then we have a continuous map $G: Lip_\alpha \to \ell_p$ defined as $G(f) = (\langle f, g_n \rangle)_{n \geq 1}$. Let us consider the case $0 < \alpha < 1$. In this case we have the commutative diagram

$$
\begin{array}{ccccccc}
\ell_2 & \xrightarrow{\;id\;} & \ell_\infty & \xrightarrow{\;F\;} & Lip_\alpha & \xrightarrow{\;id\;} & L_2 \\
& & \Big\downarrow{\scriptstyle F_1} & & \Big\downarrow{\scriptstyle G} & & \Big\uparrow{\scriptstyle \Sigma} \\
& & \ell_2 & \xrightarrow{\;\Delta\;} & \ell_p & \xrightarrow{\;id\;} & \ell_2
\end{array}
$$

where $F(\xi_n) = \sum_{n=-1}^{\infty} (n+2)^{-\alpha-1/2} \xi_n f_n$ and $(f_n)_{n=-1}^{\infty}$ denotes the orthonormal Franklin system. Theorem III.C.27 shows that F is a continuous map. The operators F_1 and Δ form a factorization given by Corollary 16 applied to the operator GF. Thus Δ is a diagonal operator with $\Delta = (\delta_n) \in \ell_r$ where $1/2 + 1/r = 1/p$. The operator Σ is defined as $\Sigma(\xi_n) = \Sigma \xi_n g_n$.

The operator $id \circ F \circ id$ is a diagonal operator on ℓ_2 given by the sequence $(n+2)^{-\alpha-1/2}$, thus $id \circ F \circ id \notin \sigma_p$ (see III.G.10).

On the other hand $F_1 \circ id \in \sigma_2$ (see III.G.12) and $id \circ \Delta \in \sigma_r$. Proposition III.G.4 gives that $id \circ F \circ id = \Sigma \circ (id \circ \Delta) \circ (F_1 \circ id) \in \sigma_p$. This contradiction shows the claim.

In order to show the Theorem for $\alpha = 1$ we consider the diagram

$$
\begin{array}{ccccccc}
\ell_2 & \xrightarrow{\;D\;} & L_\infty^0(\mathbb{T}) & \xrightarrow{\;v\;} & Lip_1 & \xrightarrow{\;id\;} & L_2 \\
& & \Big\downarrow{\scriptstyle v_1} & & \Big\downarrow{\scriptstyle G} & & \Big\uparrow{\scriptstyle \Sigma} \\
& & \ell_2 & \xrightarrow{\;\Delta\;} & \ell_{2/3} & \xrightarrow{\;id\;} & \ell_2
\end{array}
$$

where G and Σ have the same meaning as before; $L_\infty^0(\mathbb{T})$ is the subspace of $L_\infty(\mathbb{T})$ consisting of all functions whose integral is zero; $v(f)(s) = \int_0^s f(t)dt$ and v_1 and Δ are the factorization given by Corollary 16 of the operator $G \circ v$. The operator D is given by the formula $D(\xi_n) = \sum_{n \neq 0} d_n \xi_n e^{2\pi int}$ where (d_n) is a sequence of positive numbers such that $\sum_{n \neq 0} d_n^2 = 1$ and $\sum_{n \neq 0} (d_n n^{-1})^{2/3} = \infty$. Using the orthonormal system $(e^{2\pi int})_{n \neq 1}$ one checks that $id \circ v \circ D \notin \sigma_{2/3}$ but analogously as in the previous case we have $\Sigma \circ (id \circ \Delta) \circ (v_1 \circ D) \in \sigma_{2/3}$. This contradiction completes the proof of the theorem. ∎

27. As one more example of an application of our results we would like to present a description of certain multipliers. For $s > 0$ we define

$$X_s = \{f(z): f \text{ is analytic for } |z| < 1$$
$$\text{and } |||f|||_s = \sup_{|z|<1} |f(z)|(1 - |z|)^s < \infty\}.$$

We have encountered these spaces in the remark after the proof of Theorem III.A.11 and we know that each of them is isomorphic to ℓ_∞. This fact will be important in our considerations. We are interested in coefficient multipliers from X_s into the Bergman space $B_p(\mathbb{D}), 0 < p \leq 2$, i.e. we look for sequences of complex numbers $(\lambda_n)_{n=0}^\infty$ such that $\sum_{n=0}^\infty \lambda_n a_n z^n$ is in $B_p(\mathbb{D})$ for every $\sum_{n=0}^\infty a_n z^n \in X_s$.

Proposition. *Suppose* $\Lambda = (\lambda_n)_{n=0}^\infty$ *is a multiplier from* X_s *into* $B_p(\mathbb{D})$, $0 < s < \infty, 0 < p < 2$. *Then we can write* $\lambda_n = \mu_n \cdot v_n, n = 0, 1, 2, \ldots,$ *in such a way that* $(\mu_n)_{n=0}^\infty$ *is a multiplier from* X_s *into* ℓ_2 *and* $(v_n)_{n=0}^\infty$ *is a multiplier from* ℓ_2 *into* $B_p(\mathbb{D})$.

Proof: Since $X_s \sim \ell_\infty$ and $B_p(\mathbb{D}) \subset L_p(\mathbb{D})$ we can apply Corollary 16, so there exists a function $g_1(z) \geq 0, |z| < 1$ such that $\left\|\frac{\Lambda(f)(z)}{g_1(z)}\right\|_{L_2(\mathbb{D})} \leq C|||f|||_s$. Since the operator Λ commutes with rotations of the disc and the norms involved are rotation invariant, one can choose $g_1(z) = g(|z|)$ (simply average over rotations). Since the sequence $\left(\frac{z^n}{g(|z|)}\right)_{n=0}^\infty$ is orthogonal in $L_2(\mathbb{D})$ we obtain

$$\left(\sum_{n=0}^\infty |\lambda_n a_n|^2 |\beta_n|^2\right)^{\frac{1}{2}} \leq C|||f|||_s \qquad (15)$$

for every $f = \sum_{n=0}^\infty a_n z^n \in X_s$ and $\beta_n = \left\|\frac{z^n}{g(|z|)}\right\|_{L_2(\mathbb{D})}$. This means that the sequence $(\lambda_n \beta_n)_{n=0}^\infty = (\mu_n)_{n=0}^\infty$ is a multiplier from X_s into ℓ_2. One easily checks that $v_n = \beta_n^{-1}, n = 0, 1, 2, \ldots,$ determines a multiplier from ℓ_2 into $B_p(\mathbb{D})$. ∎

This proposition splits the original problem into two. We will address those two in the next two propositions.

28 Proposition. *The sequence* $\Lambda = (\lambda_n)_{n=0}^\infty$ *is a multiplier from* X_s *into* ℓ_2 *if and only if*

$$\sum_{k=0}^\infty [\sup_{2^k \leq n < 2^{k+1}} n^s |\lambda_n|]^2 = K < \infty. \qquad (16)$$

Proof: Let $X_s^0 = \{f(z): \lim_{|z| \to 1} |f(z)|(1-|z|)^s = 0\}$. This space X_s^0 is clearly a closed subspace of X_s containing all polynomials. The first observation we need is that Λ is a multiplier from X_s^0 into ℓ_2 if and only if it is a multiplier from X_s into ℓ_2. Since convolution with the Fejér kernel $\mathcal{F}_n$ has norm 1 in sup-norm (see I.B.16) we see that for $f \in X_s$ we have $|||f * \mathcal{F}_n|||_s \leq |||f|||_s$ and $f * \mathcal{F}_n \in X_s^0$. Thus we see that if $\Lambda: X_s^0 \to \ell_2$ then Λ also maps X_s into ℓ_2. The converse is obvious.

Since $X_s \sim \ell_\infty$ we infer from Theorem III.F.29 that $\Lambda: X_s \to \ell_2$ is 2-absolutely summing, so $\Lambda: X_s^0 \to \ell_2$ is also 2-absolutely summing. Let us consider the isometric embedding $i: X_s^0 \to C(\overline{\mathbb{D}})$ defined as $i(f) = f(z) \cdot (1 - |z|)^s$. The Pietsch theorem III.F.8 and III.F.(13) give the factorization

$$
\begin{array}{ccc}
X_s^0 & \xrightarrow{\quad \Lambda \quad} & \ell_2 \\
{\scriptstyle i} \downarrow & & \uparrow {\scriptstyle \widetilde{\Lambda}} \\
C(\overline{\mathbb{D}}) & \xrightarrow[\quad id \quad]{} & L_2(\overline{\mathbb{D}}, \mu)
\end{array}
\qquad (17)
$$

where μ is a rotation invariant measure on $\overline{\mathbb{D}}$. Since $i(X_s^0)$ vanishes on $\partial \mathbb{D}$, we can additionally assume $\mu(\{z: |z| = 1\}) = 0$. So there exists a probability measure μ_1 on $[0,1)$ such that for $\varphi \in C(\overline{\mathbb{D}})$ we have

$$
\int_{\mathbb{D}} \varphi(z) d\mu(z) = \frac{1}{2\pi} \int_0^1 \int_0^{2\pi} \varphi(re^{i\theta}) d\theta d\mu_1(r).
$$

Now we see that $(i(z^n))_{n=0}^\infty$ is an orthogonal sequence in $L_2(\mathbb{D}, \mu)$ so $\widetilde{\Lambda}$ is bounded if and only if

$$
|\lambda_n| \leq C \|i(z^n)\|_{L_2(\overline{\mathbb{D}}, \mu)}
$$
$$
= C \left(\int_0^1 r^{2n}(1-r)^{2s} d\mu_1(r) \right)^{\frac{1}{2}}, \quad n = 0, 1, 2, \ldots \quad (18)
$$

The above reasoning is clearly reversible so we see that Λ is a multiplier from X_s into ℓ_2 if and only if (18) holds for some probability measure μ_1.

Now assume that (16) holds with $K = 1$. Let us fix n_k, $2^k \leq n_k < 2^{k+1}$ such that $\sup_{2^k \leq n < 2^{k+1}} n^s |\lambda_n| = n_k^s |\lambda_{n_k}|$, for $k = 0, 1, 2, \ldots$. Let $a_k = (1 - n_k^{-1})$ and let us define $\nu = \sum_{k=0}^\infty n_k^{2s} |\lambda_{n_k}|^2 \delta_{a_k}$, where δ_{a_k} is the Dirac measure concentrated at a_k. This is clearly a probability measure on $[0,1)$.

Take arbitrary n and let k_0 be such that $2^{k_0} \leq n < 2^{k_0+1}$. Then we have

$$\int_0^1 r^{2n}(1-r)^{2s}d\nu(r) = \sum_{k=0}^{\infty} a_k^{2n}(1-a_k)^{2s}n_k^{2s}|\lambda_{n_k}|^2$$

$$= \sum_{k=0}^{\infty}(1-n_k^{-1})^{2n}|\lambda_{n_k}|^2$$

$$\geq (1-n_{k_0}^{-1})^{2n}|\lambda_{n_{k_0}}|^2 \geq C|\lambda_n|^2.$$

so (18) holds.

Conversely, assuming (18) we have

$$\sum_{k=0}^{\infty}[\sup_{2^k \leq n < 2^{k+1}} n^s|\lambda_n|]^2$$

$$\leq C\sum_{k=0}^{\infty} \sup_{2^k \leq n < 2^{k+1}} n^{2s}\int_0^1 r^{2n}(1-r)^{2s}d\mu_1(r)$$

$$\leq C\sum_{k=0}^{\infty} 2^{2ks}\int_0^1 r^{2\cdot 2^k}(1-r)^{2s}d\mu_1(r) \qquad (19)$$

$$= C\int_0^1 (1-r)^{2s}\left(\sum_{k=0}^{\infty} 2^{2ks}r^{2\cdot 2^k}\right)d\mu_1(r).$$

Now we want to show that the integrand is a bounded function. For $1-2^{-N} \leq r \leq 1-2^{-N-1}$ we have for some q independent of $N, 0 < q < 1$

$$\sum_{k=0}^{\infty} 2^{2ks}\cdot r^{2\cdot 2^k} \leq \sum_{k=0}^{N} 2^{2sk}(1-2^{-N-1})^{2\cdot 2^k} + \sum_{k=N+1}^{\infty} 2^{2sk}(1-2^{-N-1})^{2\cdot 2^k}$$

$$\leq \sum_{k=0}^{N} 2^{2sk} + \sum_{k=N+1}^{\infty} 2^{2sk}q^{2^{k-(N+1)}}$$

$$\leq C2^{2sN} + C2^{2sN}\sum_{k=1}^{\infty} 2^{2sk}q^{2^k} \leq C2^{2sN} \leq C(1-r)^{-2s}$$

so the integrand is really bounded and so (16) holds. ∎

29 Proposition.   *The sequence* $\Lambda = (\lambda_n)_{n=0}^{\infty}$ *is a multiplier from* ℓ_2 *into* $B_p(\mathbb{D}), 0 < p \leq 2$ *if and only if*

$$\left(\sum_k (\sup_{2^k \leq n < 2^{k+1}} n^{-\frac{1}{p}}|\lambda_n|)^r\right)^{\frac{1}{r}} < \infty \quad \text{for} \quad \frac{1}{r}+\frac{1}{2}=\frac{1}{p}. \qquad (20)$$

Proof: Let $(n_k)_{k=0}^{\infty}$ be any sequence such that $2^k \leq n_k < 2^{k+1}$. Passing to polar coordinates and using Khintchine's inequality I.B.8 (see also Exercise III.A.9) we get that there exists a constant C such that for any such sequence $(n_k)_{k=0}^{\infty}$ and any sequence of scalars $(a_k)_{k=0}^{\infty}$

$$\left\| \sum_k a_k z^{n_k} \right\|_p \geq C \left(\sum_k |a_k|^p n_k^{-1} \right)^{\frac{1}{p}}. \tag{21}$$

From (21) we see that for a multiplier $\Lambda\colon \ell_2 \to B_p(\mathbb{D})$ we have

$$\left(\sum_{k=1}^{\infty} |a_{n_k}|^p |\lambda_{n_k}|^p n_k^{-1} \right)^{\frac{1}{p}} \leq C \left(\sum_k |a_{n_k}|^2 \right)^{\frac{1}{2}}. \tag{22}$$

Since (22) holds with the same constant for all scalars $(a_n)_{n=0}^{\infty}$ and all sequences $(n_k)_{k=0}^{\infty}$ as above we get (20).

In order to prove the other implication we use the following inequality

$$\left\| \sum_{n=0}^{\infty} a_n z^n \right\|_p \leq \left(\sum_k \left\| \sum_{2^k}^{2^{k+1}-1} a_n z^n \right\|_p^p \right)^{\frac{1}{p}}. \tag{23}$$

For $p \leq 1$ this is just the p-convexity of the space $L_p(\mathbb{D})$ (see I.B.2). For $p = 2$ it follows directly from orthogonality of $z^n, n = 0, 1, 2, \ldots,$ in $L_2(\mathbb{D})$. The remaining cases follow by standard interpolation.

For a sequence $(a_n)_{n=0}^{\infty}$ such that $\sum_{n=0}^{\infty} |a_n|^2 = 1$ and $(\lambda_n)_{n=0}^{\infty}$ satisfying (20) we obtain from (23) and the Hölder inequality

$$\left\| \sum_{n=0}^{\infty} a_n \lambda_n z^n \right\|_p^p \leq \sum_k \left\| \sum_{n=2^k}^{2^{k+1}-1} a_n \lambda_n z^n \right\|_p^p$$

$$= \sum_k \int_0^1 \frac{1}{2\pi} \int_0^{2\pi} \left| \sum_{n=2^k}^{2^{k+1}-1} a_n \lambda_n r^n e^{in\theta} \right|^p d\theta r \, dr$$

$$\leq \sum_k \int_0^1 \left(\frac{1}{2\pi} \int_0^{2\pi} \left| \sum_{n=2^k}^{2^{k+1}-1} a_n \lambda_n r^n e^{in\theta} \right|^2 d\theta \right)^{\frac{p}{2}} r \, dr$$

$$= \sum_k \int_0^1 \left(\sum_{n=2^k}^{2^{k+1}-1} |a_n|^2 |\lambda_n|^2 r^{2n} \right)^{\frac{p}{2}} r \, dr$$

$$\leq \sum_k \int_0^1 r^{p2^k} \left(\sum_{n=2^k}^{2^{k+1}-1} |a_n|^2 |\lambda_n|^2 \right)^{\frac{p}{2}} r \, dr$$

$$\leq C \sum_k 2^{-k} \left(\sum_{n=2^k}^{2^{k+1}-1} |a_n|^2 |\lambda_n|^2 \right)^{\frac{p}{2}}$$

$$\leq C \sum_k \left(\max_{2^k \leq n < 2^{k+1}} n^{-1} |\lambda_n|^p \right) \left(\sum_{2^k}^{2^{k+1}-1} |a_n|^2 \right)^{\frac{p}{2}}$$

$$\leq C \left(\sum_k \left(\max_{2^k \leq n < 2^{k+1}} n^{-1} |\lambda_n|^p \right)^{\frac{2}{2-p}} \right)^{\frac{(2-p)}{2}} \left(\sum_{n=0}^{\infty} |a_n|^2 \right)^{\frac{p}{2}}.$$

If Λ satisfies (20) this shows that Λ is a multiplier.   ∎

Now we are able to prove the description of multipliers from X_s into $B_p(\mathbb{D})$. From Propositions 27 and 28 and 29 and Hölder's inequality we get

30 Theorem.   *The sequence* $\Lambda = (\lambda_n)_{n=0}^{\infty}$ *is a multiplier from* X_s *into* $B_p(\mathbb{D}), 0 < s, 0 < p \leq 2$ *if and only if*

$$\sum_{k=0}^{\infty} \left(\sup_{2^k \leq n < 2^{k+1}} n^{s-\frac{1}{p}} |\lambda_n| \right)^p < \infty.$$

Notes and Remarks.

Various sublinear operators are of paramount importance in modern harmonic analysis; they include a variety of maximal operators, square functions or area functions etc. Texts such as de Guzman [1981], Garcia-Cuerva–Rubio de Francia [1985], Folland-Stein [1982], Torchinsky [1986] etc. make their importance absolutely clear. In our presentation we give only the most general results which fall naturally into the scope of Banach space theory. Kolmogorov [1925] has shown the weak type (1,1) of the trigonometric conjugation operator (see **8** *Example 1*). This was probably the first paper where finiteness almost everywhere was shown to imply weak type (1,1). The principle was generalized in Stein [1961] where a version of our *Corollary 7* was proved. E.M. Nikishin was led to consider his general theorems by problems connected with the structure of systems of convergence in measure for ℓ_2. His main results in this area are published in Nikishin [1970]. This paper basically contains *Theorem 6*. Later Maurey [1974] gave a more abstract presentation. It is his approach that we follow in this book.

A sequence of functions $(\phi_n)_{n=1}^{\infty} \subset L_0[0,1]$ is called a system of convergence in measure for ℓ_2 if every series $\sum_{n=1}^{\infty} a_n \phi_n$ with $(a_n) \in \ell_2$ converges in measure. Clearly there is a correspondence between systems of convergence in measure for ℓ_2 and continuous linear operators $T: \ell_2 \to L_0$. Some other similar notions have been investigated (see Exercises 13, 14 and 16). **8** *Example 2* is an old theorem of Calderon (see Zygmund [1968] XIII.1.22). The fundamental theorem due to Carleson [1966] (for $p = 2$) and extended by Hunt [1968] to $1 < p < \infty$ asserts that for every $f \in L_p(\mathbb{T})$ its Fourier series converges almost everywhere. An example of an L_1-function whose Fourier series diverges a.e. was given by Kolmogorov [1923]. The example was improved in Kolmogorov [1926] to yield an L_1-function with everywhere divergent Fourier series. Very recently the Armenian mathematician Kheladze gave a remarkbly simple construction of an L_1-function for which condition (c) of 8.Example 2 fails.

Inspired by Nikishin [1970] and Rosenthal [1973] B. Maurey under-took his study of operators from X into $L_p(\Omega, \mu)$ which factors strongly through $L_q(\Omega, \mu)$. His results are presented in Maurey [1974]. Our presentation of 10-12 and 15 follows that monograph, with the excep-tion that the proof of (b)⇒(a) in *Proposition 10* is taken from Pisier [1986a]. We recommend the reader to consult this paper. It contains many additional results, also in the setting of C^*-algebras. Its main interest is to present necessary and sufficient conditions for the operator $T: X \to L_p(\Omega, \mu)$ to factorize strongly through $L_{q,\infty}(\Omega, \mu)$, that is, the topic between Nikishin's and Maurey's theorems.

Corollary 13 is one of the main results of Rosenthal [1973]. This paper played a very important role in the development of the theory.

We would like to mention that the notion of type used in Maurey [1974] is different from the one used in this book. We use the type and cotype which is sometimes called in the literature 'Rademacher type' and 'Rademacher cotype', while Maurey uses the so-called stable type.

Let $1 \le p \le 2$ and let Θ_j be a sequence of independent, identically distributed standard p-stable random variables. A Banach space X is called of stable type p if there exists a constant C such that for any finite sequence (x_j) in X we have

$$\left(\int \| \sum x_j \Theta_j \|^{1/2} \right)^2 \le C \left(\sum \| x_j \|^p \right)^{1/p}.$$

Actually the use of the exponent $1/2$ in the left hand side integral is irrelevant. It can be replaced by any number $q < p$. With this, one checks that $L_p, 1 \le p < 2$, is not of type p but is of type s for any $s < p$. The following fact analogous to *Theorem 6* holds.

Theorem A. *If X is a Banach space of stable type $p, 1 \leq p \leq 2$, then every linear operator from X into $L_r(\Omega, \mu), r < p$ factors strongly through $L_p(\Omega, \mu)$.*

We have practically proved this theorem during the proof of *Theorem 12*. The usefulness of the notion of stable type can be seen from this proof.

There is an obvious analogy between *Definitions 4* and *9* and between *Proposition 5* and *Proposition 10*. Let us note that conditions (b) of those propositions can be interpreted as vector valued inequalities (see the remark after *Proposition 5*). This point of view is explained in detail in Garcia-Cuerva–Rubio de Francia [1985], as is the equivalence between factorization and weighted norm inequalities. We do not discuss this important subject here.

Our informal discussion in 14 is more or less folklore. It can be found in full detail in Maurey [1974]. *Proposition 14* is a special case of the following result due to Maurey [1974].

Theorem B. *The following conditions on the Banach space X are equivalent:*

(a) *c_0 is not finitely representable in X;*

(b) *there exists a $q < \infty$ such that $\Pi_q(C(K), X) = L(C(K), X)$.*

This result and its consequences for Banach space theory are discussed in great detail in Rosenthal [1976].

The Menchoff-Rademacher *Theorem 22* was proved in Menchoff [1923] and Rademacher [1922] improving many earlier results. This is the best result. Menchoff [1923] constructed an orthonormal system $(\Psi_n)_{n=1}^\infty$ on [0,1] such that for every sequence $(\omega_n)_{n=1}^\infty$ with $1 = \omega_1 \leq \omega_2 \leq \dots$ and $\omega_n = o(\log^2 n)$ there exists a series $\sum_{n=1}^\infty a_n \Psi_n$ divergent almost everywhere and such that $\sum_{n=1}^\infty |a_n|^2 \omega_n < \infty$. The connection between the theory of p-summing (or radonifying) operators and the Menchoff-Rademacher theorem was noted in Schwartz [1970] and Kwapień-Pełczyński [1970]. This last paper also contains some generalizations of the Menchoff-Rademacher theorem in the spirit of *Corollary 25* which was proved in Maurey [1974]. Later Bennett [1976] gave another, more elementary, but in fact closely related, treatment of such generalizations. *Theorem 17*, proved in Ørno [1976], shows that series unconditionally convergent in measure (in particular unconditionally convergent in L_p) are closely related to orthogonal series. Our *Lemma*

18 is a classical result of I. Schur, published first in Rademacher [1922] (see also Kashin-Saakian [1984]). The *dilation Theorem 19* is classical by now and is a basis of a large part of the theory of operators on Hilbert spaces (see Nagy-Foias [1967]). Our proof of the Menchoff-Rademacher theorem is a mixture of various published proofs like Kwapień-Pełczyński [1970], Bennett [1976], Nahoum [1973], Schwartz [1970]. Our *Corollary 20* was proved in Ørno [1976]. It improves an earlier result of Kashin [1974]. *Theorem 26* for $\alpha = 1/2$ was proved by Mitiagin [1964] and the general case was shown by Bočkariov [1978]. Our proof follows Wojtaszczyk [1988]. Actually it is possible to obtain analogous results for systems more general than orthonormal and for more general moduli of smoothness. We refer the interested reader to Wojtaszczyk [1988] for formulations, proofs and the history of the subject. *Theorem 30* and its proof are taken from Wojtaszczyk [P].

We would like to mention also the paper Bichteler [1981] where factorization theorems are applied to the theory of stochastic integration.

The factorization theorems are basically a type of Tauberian theorem; they assert that the operator is actually better that it seems to be. This is useful both ways; we get stronger information once we prove something weaker or conversely we show the 'very' bad behaviour once we show a 'moderately' bad one.

Exercises

1. Let $G(f)(x) = [M(|f|^2)]^{1/2}$ where M is the Hardy-Littlewood maximal operator. Show that G is a sublinear operator on $L_2[0,1]$ which is of weak type (2,2) but not continuous on $L_2[0,1]$.

2. Show that if every operator $T: X^* \to L_p(\Omega, \mu)$ factors strongly through $L_q(\Omega, \mu)$ (and X^* has b.a.p. if $p < 1$) then $\Pi_p(X, \ell_q) = \Pi_q(X, \ell_q), p < q, q \geq 1$.

3. Let $(\varphi_j)_{j=1}^\infty$ be a sequence of independent, p-stable random variables in $L_1(\mu)$ and let $X = \text{span}(\varphi_j)_{j=1}^\infty \cong \ell_p$. Show that $i^*: L_\infty(\mu) \xrightarrow{\text{onto}} X^*$, where i is the identity embedding of X into $L_1(\mu)$, is not p'-absolutely summing, $\frac{1}{p} + \frac{1}{p'} = 1$.

4. Show, without using *Proposition 15*, that every operator from L_∞ into $L_q, 1 \leq q \leq 2$, factors strongly through L_2.

5. Let $T: L_p \to L_0$, $1 \leq p < \infty$, be a continuous sublinear operator. Assume moreover that T is monotone (i.e. if $|f| \leq |g|$ then also $|Tf| \leq |Tg|$). Then T factors strongly through $L_{p,\infty}$.

6. Let $T: L_p \to L_0(\Omega, \mu)$ be a positive (i.e. if $f \leq g$ then $Tf \leq Tg$) linear operator and assume $\mu(\Omega) = 1$ and $p \geq 1$. Then T factors strongly through $L_p(\Omega, \mu)$.

7. Show that there exists a positive linear operator (see Exercise 6) $T: \ell_p \to L_q(\Omega, \mu)$, with $1 > p > q > 0$, and (Ω, μ) a probability measure, which does not factor strongly through $L_p(\Omega, \mu)$.

8. Let X be a Banach space and $0 < r < p < \infty$. Show that the following properties of the bounded linear operator $T: X \to L_r(\Omega, \mu)$ are equivalent:

    (a) there exists a constant C such that for all finite sequences $(x_i) \subset X$ we have
    $$\| \sup_i |T(x_i)| \|_p \leq C \left(\sum_i \|x_i\|^p \right)^{\frac{1}{p}};$$

    (b) there exists a constant C' such that there exists a function $f \in L_1(\Omega, \mu), f \geq 0, \int_\Omega f d\mu = 1$ such that for all $x \in X$ and all measurable subsets $E \subset \Omega$ we have
    $$\|T(x) \cdot \chi_E\|_r \leq C'\|x\| \left(\int_E f d\mu \right)^{\frac{1}{r} - \frac{1}{p}};$$

(c) the operator T admits a factorization of the form

$$X \xrightarrow{\widetilde{T}} L_{p,\infty}(\Omega, f d\mu) \xrightarrow{M} L_r(\mu)$$

with $\widetilde{T}$ bounded, $f \in L_1(\Omega, \mu)$, $f \geq 0$, $\int_\Omega f d\mu = 1$ where M is an operator of multiplication by $f^{\frac{1}{r}}$.

9. Every operator from $L_p[0,1]$ into ℓ_q, $1 \leq q < 2 < p \leq \infty$, is compact.

10. Let $(f_n)_{n=1}^\infty$ be an orthonormal system in $L_2[0,1]$. Show that $N^{-1} \sum_{n=1}^N f_n \to 0$ almost everywhere. For a given number $x, 0 < x < 1$ let $(\varepsilon_j(x))_{j\geq 1}$ be its dyadic expansion ($\varepsilon_j = 0$ or $\varepsilon_j = 1$). Show that for almost all $x \in [0,1]$ we have $N^{-1} \sum_{n=1}^N \varepsilon_j(x) \to 1/2$, i.e. almost every number has asymptotically equal number of 0's and 1's in its dyadic expansion.

11. Let $(f_n)_{n=1}^\infty$ be a complete orthonormal system in $L_2[0,1]$. Show that $\sum_{n=1}^\infty |f_n| = \infty$ on a set of positive measure.

12. Let $\mathcal{A}_\alpha$, $0 < \alpha < 1$, be the space of all functions in the disc algebra such that $|f(e^{i\theta}) - f(e^{i(\theta-h)})| \leq C|h|^\alpha$. Show that there exists a $f \in \mathcal{A}_\alpha$ such that $f' \notin N$ (N denotes the Nevanlinna class).

13. A system of functions $(f_n)_{n\geq 1} \subset L_0[0,1]$ is called a system of convergence in measure for ℓ_2 if every series $\sum_{n\geq 1} a_n f_n$ with $(a_n) \in \ell_2$ converges in measure. Show that $(f_n)_{n\geq 1}$ is a system of convergence in measure for ℓ_2 if and only if for every $\varepsilon > 0$ there exist a set $E_\varepsilon \subset [0,1]$ with $|E_\varepsilon| > 1 - \varepsilon$, a constant C_ε and an orthonormal system $(\varphi_n)_{n\geq 1}$ on $[0,1]$ such that $f_n|E_\varepsilon = C_\varepsilon \varphi_n|E_\varepsilon$ for $n = 1, 2, \ldots$.

14. Show that the following conditions on a system of functions $(f_n)_{n\geq 1} \subset L_0[0,1]$ are equivalent.

(a) For every $(a_n) \in \ell_1$, we have $\sum |a_n f_n| < \infty$ a.e.

(b) For every $\varepsilon > 0$ there exist $E_\varepsilon \subset [0,1]$ with $|E_\varepsilon| > 1 - \varepsilon$ and a constant C_ε such that $\sup_n \int_{E_\varepsilon} |f_n| \leq C_\varepsilon$.

15. Show that there exists a function $f \in L_1(\mathbb{T}^2)$ such that

$$\lim_{r \to 1} \sum_{n,m\geq 1} r^{n+m} \hat{f}(n,m) e^{in\theta} e^{im\varphi}$$

does not exist on a set of positive measure.

16. Let $\varphi \in L_2(\mathbb{T})$ with $\int \varphi(t)dt = 0$. Show that the following conditions are equivalent:

    (a) $f_n(t) = \varphi(nt), n = 1, 2, 3, \ldots$, is a system of convergence in measure for ℓ_2;

    (b) $\int \left|\sum_{n\geq 1} a_n f_n\right|^2 \leq C \sum_{n\geq 1} |a_n|^2$ for some constant C and all sequences of scalars.

    Use this to show that $f_n(x) = \mathrm{sgn}\, \sin(nx), n = 1, 2, \ldots$ is not a system of convergence in measure of ℓ_2. This means that the smoothness of trigonometric functions, and not only the distribution of signs, plays a role in the almost everywhere convergence of trigonometric series of L_2-functions.

17. If $(\varphi_n)_{n\geq 1}$ is a uniformly bounded orthonormal system on $[0, 1]$ and $(d_n)_{n\geq 1}$ is a sequence of positive numbers such that $\sum_{n\geq 1} d_n^q = \infty$ where $q = 2p(2 - p)^{-1}$ and $0 < p < 2$, then there exists a function $f \in C[0, 1]$ such that $\sum_{n\geq 1} |d_n \langle f, \varphi_n \rangle|^p = \infty$.

18. Show that the map $f = \sum_{-\infty}^{\infty} \hat{f}(n)e^{in\theta} \mapsto \left(\lambda_n \cdot \hat{f}(n)\right)_{n=-\infty}^{\infty}$ maps $C(\mathbb{T})$ into ℓ_p, $0 < p < 2$, if and only if $\sum_{-\infty}^{\infty} |\lambda_n|^q < \infty$ where $q = 2p(2 - p)^{-1}$.

19. Show the von Neumann inequality using *Theorem 19*. Let us recall (see Exercise III.B.8) that the von Neumann inequality says that for a contraction T on a Hilbert space (i.e. $\|T\| \leq 1$) and any polynomial $p(z) = \sum_{n=0} a_n z^n$ we have $\|\sum_{n=0} a_n T^n\| \leq \|p\|_A$.

20. (a) Suppose $Y \supset X$ and assume that both X and Y/X have some type $p > 1$. Show that Y has some type $q > 1$.

    (b) Let X, Y, Z be Banach spaces with $Y \supset X$ and let $T: X \to Z$. Show that there exist a space V, an isometric embedding $j: Z \to V$ and an operator $T_1: Y \to V$ such that $jT = T_1|X$ and the spaces Y/X and V/Z are isometric.

    (c) Suppose that $X \subset C(S)$ is a subspace such that $C(S)/X$ is reflexive. Show that every operator from X into $\ell_p, 1 \leq p \leq 2$, is 2-absolutely summing.

    (d) Suppose $\Lambda \subset \mathbb{Z}$ is a Λ_p-set for some $p > 1$ (see I.B.14) and suppose that $(\varphi_n)_{n=1}^{\infty}$ is a complete orthonormal system in $L_2(\mathbb{T})$. Show that there exists $f \in C(\mathbb{T})$ such that $\hat{f}(n) = 0$ for $n \in \Lambda$ and $\sum_{n=1}^{\infty} |\langle f, \varphi_n \rangle|^s = \infty$ for all $s < 2$.

21. (a) Describe the coefficient multipliers from X_s into ℓ_p, for $s > 0$, and $0 < p \le 2$.

    (b) Describe the coefficient multipliers from $B_p(\mathbb{D})$ into $B_q(\mathbb{D})$ for $0 < q \le 2 \le p < \infty$.

III.I. Absolutely Summing Operators On The Disc Algebra

We start this chapter with the construction of a non-compact, 1-absolutely summing operator from any proper uniform algebra into ℓ_2. This shows in particular that such an algebra is never complemented in $C(K)$. Then we study p-absolutely summing operators on the disc algebra A. We construct an 'analytic projection' which maps some weighted $L_p(\mathbb{T}, \Delta d\lambda)$ spaces onto the closure of analytic polynomials and has properties analogous to the properties of the classical Riesz projection $\mathcal{R}$. Then we show that every p-absolutely summing operator from the disc algebra is p-integral, $p > 1$. We also show that A^* has cotype 2 and derive some corollaries of these results. Next we study reflexive subspaces $Y \subset L_1/H_1$ and show that any linear operator $T: Y \to H_\infty$ extends to an operator $\tilde{T}: L_1/H_1 \to H_\infty$. This is applied to some interpolation problems on $\mathbb{D} \times \mathbb{D}$.

1. In this chapter we present the detailed study of p-summing and related operators defined on the disc algebra A. Such a study is motivated both by the intrinsic beauty of the problems and by important applications. Actually we have already seen one application. In Proposition III.F.6(a) we have exhibited an absolutely summing operator $P: A \to \ell_2$ which was later used to prove the Grothendieck theorem III.F.7. This example suggests that absolutely summing operators on A may have some rather unexpected properties. Actually this phenomenon is not restricted to the disc algebra, but is shared (to a certain degree) by all proper uniform algebras. Let $B \subset C(S)$ be a proper, point-separating subalgebra of $C(S)$, with $1 \in B$.

2 Proposition. *There exists a non-compact, 1-absolutely summing operator* $T: B \to \ell_2$.

Proof: Take $\mu \in M(S)$ such that $\mu \in B^\perp$ and $\mu \notin \overline{B}^\perp$ where $\overline{B} = \{f \in C(S): \bar{f} \in B\}$. It follows from the Stone-Weierstrass theorem that $B \neq \overline{B}$ so such a μ exists. Let ν denote the Hahn-Banach extension of $\mu|\overline{B}$ to $C(S)$. We assume that $\|\nu\| = 1$. Thus we can take a sequence $(f_n)_{n=1}^\infty \subset B$ with $\|f_n\|_\infty \leq 1$ and such that $\int \bar{f}_n d\nu \to 1$ for $n \to \infty$. Considering this sequence in $L_2(|\nu| + |\mu|)$ we can pass to a subsequence and convex combinations to get a new sequence (still denoted by $(f_n)_{n=1}^{\infty}$) such that $f_n \in B$ for $n = 1, 2, \ldots$ and $\|f_n\|_{\infty} \leq 1$ and $\lim_{n \to \infty} \int \bar{f}_n d\nu = 1$ and $f_n \to F$, $(|\nu| + |\mu|)$-almost everywhere for some F (see III.A.29 but in fact easier). Clearly we get $\int \bar{F} d\nu = 1 = \|\nu\|$ so $|F| = 1$, $|\nu|$-a.e. Now let us consider $V = B^{\perp} \cap L_1(|\nu| + |\mu|)$ and define $T_1 : V \to \ell_2$ by $T_1(\sigma) = (\int \bar{F}^{2^k} d\sigma)_{k=1}^{\infty}$. In order to show that T_1 is continuous it is enough to show that

$$\sum_k \left| \int \bar{f}_n^{2^k} d\sigma \right|^2 \leq C \|\sigma\|^2$$

for every n with the constant independent of n. Let $R_n : C(\mathbb{T}) \to C(S)$ be defined by $R_n(\varphi) = H(\varphi) \circ f_n$ where $H(\varphi)$ is the harmonic extension of φ. Clearly $\|R_n\| \leq 1$. Since $R_n(A) \subset B$ we get that $R_n^* : B^{\perp} \to A^{\perp} = H_1^0$ so by the Paley theorem (see I.B.24)

$$\sum_k \left| \int \bar{f}_n^{2^k} d\sigma \right|^2 = \sum_k \left| \int R_n(\bar{z}^{2^k}) d\sigma \right|^2 = \sum_k |\langle R_n^*(\sigma), \bar{z}^{2^k} \rangle|^2$$

$$\leq C \|R_n^*(\sigma)\|^2 \leq C \|\sigma\|^2.$$

Now we look for the operator $i : B \to V$ given by $i(f) = f \cdot \rho$ for some $\rho \in V$, such that $T = T_1 \circ i$ will be non-compact. Clearly such T is 1-absolutely summing (see III.F.4). We take $\rho = \mu + \overline{(\nu - \mu)} \cdot F^2$. Since for $f \in B$

$$\int f d\rho = \int f d\mu + \int F^2 f d\bar{\nu} - \int F^2 f d\bar{\mu} = 0 + \int \overline{F^2 f} d(\nu - \mu) = 0$$

we see that $\rho \in V$. Moreover

$$\lim_k \sup_{\|f\| \leq 1} |\langle Tf, e_k \rangle| = \lim_k \sup_{\|f\| \leq 1} \left| \int \bar{F}^{2k} f d\rho \right|$$

$$\geq \lim_k \sup_n \left| \int \bar{F}^{2^k} f_n^{2^k - 1} d\rho \right|$$

$$\geq \lim_k \left| \int \bar{F}^{2^k} F^{2^k - 1} d\rho \right|$$

$$= \left| \int_{\{|F|=1\}} \bar{F} d\rho \right|$$

$$= \left| \int_{\{|F|=1\}} \bar{F} d\mu + \int_{\{|F|=1\}} F d\bar{\nu} - \int_{\{|F|=1\}} F d\bar{\mu} \right|$$

$$= \left| 1 + \left(\int_{\{|F|=1\}} \bar{F} d\mu - \int_{\{|F|=1|\}} \bar{F} d\mu \right) \right|.$$

Since the bracket above is purely imaginary we see that

$$\lim_{k} \sup_{\|f\|\leq 1} |\langle Tf, e_k \rangle| \geq 1$$

so T is not compact.   ■

3.   From this proposition we get

Corollary.   *Let B be a proper, closed subalgebra of $C(S)$ separating points and with 1. Then B is not isomorphic (as a Banach space) to any quotient of any $C(K)$-space. In particular B is not complemented in $C(S)$.*

Proof: Suppose that there exists a map $q\colon C(K) \xrightarrow{\text{onto}} B$. Let $T\colon B \to \ell_2$ be a 1-absolutely summing, non-compact operator. Then $T \circ q\colon C(K) \to \ell_2$ is a 1-absolutely summing, non-compact operator. From III.F.8 we infer that we have the factorization

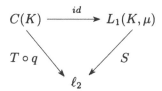

Since $L_1(K,\mu)$ has the Dunford-Pettis property (see III.D.33,34) and both S and *id* are obviously weakly compact we get that $T \circ q$ is compact. This contradiction finishes the proof.   ■

Remark.   The above corollary shows that some algebraic properties of a uniform algebra are determined by its Banach space structure. In particular there does not exist a multiplication on the disc algebra A which makes it into a commutative C^*-algebra.

4.   There are some very natural limitations to what can be proved about the Banach space structure of a general uniform algebra. We have the following.

Proposition.   *For every complex separable Banach space X there exists a uniform algebra $\mathcal{U}_X$ such that X is isomorphic to a 1-complemented subspace of $\mathcal{U}_X$.*

Proof: Let K be the unit ball in X^* equipped with the $\sigma(X^*, X)$-topology. We define $\mathcal{U}_X$ to be the smallest closed subalgebra of $C(K)$ containing all functions $\varphi_x(k) = \langle k, x \rangle$ for $x \in X$ and constants. For a function $f \in \mathcal{U}_X$ we define $Pf(k) = (2\pi)^{-1} \int e^{i\theta} f(e^{-i\theta}k)d\theta$. Since the elements $\sum_{r=1}^{N} \langle \cdot, x_r \rangle^{j_r}$ for $x_r \in X$ and $j_r \in \mathbb{N}$ are dense in $\mathcal{U}_X$ we easily see that for every $f \in \mathcal{U}_X, Pf(k) = \langle k, x \rangle$ for some $x \in X$. Since clearly $\|P\| \le 1$ we get the claim (see II.A.10).   ∎

5.   Now we return to the disc algebra A. Suppose we have a p-absolutely summing operator $T: A \to X$, $1 \le p < \infty$. The Pietsch theorem III.F.8 gives that there exists a measure μ on the circle $\mathbb{T}$ such that $\|Tf\| \le C(\int |f|^p d\mu)^{\frac{1}{p}}$. In other words T extends to the space $H_p(\mu)$ which is the closure of the polynomials in $L_p(\mathbb{T}, \mu)$. Thus it is natural that we start the detailed investigation of p-absolutely summing operators on A with some observations about spaces $H_p(\mu)$.

6 Proposition.   *Let μ be a probability measure on $\mathbb{T}$, with the Lebesgue decomposition $\mu = \mu_s + fd\lambda$ where λ is the normalised Lebesgue measure. Then*

$$H_p(\mu) = L_p(\mu_s) \oplus H_p(fd\lambda), \ 1 \le p < \infty.$$

Proof: What we really have to show is the following: given $g_1 \in L_p(\mu_s)$ and $g_2 \in H_p(fd\lambda)$ we have to find $h \in A(\mathbb{T})$ such that $\|g_1 - h\|_{L_p(\mu_s)} \le \varepsilon$ and $\|g_2 - h\|_{H_p(fd\lambda)} \le \varepsilon$. We can obviously assume that g_1 is continuous and $g_2 \in A$. Since $\mu_s \perp \lambda$ there exists a closed set $\Delta \subset \mathbb{T}$ with $\lambda(\Delta) = 0$ such that $(\int_{\mathbb{T} \setminus \Delta} |g_1|^p d\mu_s)^{\frac{1}{p}} < \frac{\varepsilon}{10}$ and $(\int_{\mathbb{T} \setminus \Delta} |g_2|^p d\mu_s)^{\frac{1}{p}} < \frac{\varepsilon}{10}$. Using Proposition III.E.2 we find a $\varphi \in A$ such that $\varphi|\Delta = g_1|\Delta$ and $(\int_{\mathbb{T}} |\varphi|^p fd\lambda)^{\frac{1}{p}} \le \frac{\varepsilon}{10}$ and $\int_{\mathbb{T} \setminus \Delta} |\varphi|^p d\mu_s \le \varepsilon/10$. Let us use Proposition III.E.2 once more to get a function $\psi \in A$ such that $\|\psi\| = 1$, $\psi|\Delta = 1$ and $(\int |\psi g_2|^p fd\lambda)^{\frac{1}{p}} < \varepsilon/10$. We put $h = \varphi + (1 - \psi)g_2$. One checks that we have imposed enough conditions to make it the right h.   ∎

Note that Proposition 6 in particular implies that a p-summing operator $T: A \to X$ whose Pietsch measure is singular with respect to the Lebesgue measure extends to a p-summing operator $\tilde{T}: C(\mathbb{T}) \to X$.

7.   Now we want to give a heuristic indication of what will be done in the subsequent sections **8, 9** and **10**. We would like to reduce the investigation of p-summing operators on A to the study of certain operators on $C(\mathbb{T})$ or $L_p(\mathbb{T})$. From Proposition 6 we see that this requires investigating a projection from $L_p(f d\lambda)$ onto $H_p(f d\lambda)$, where f is a positive function and we impose the normalization $\int f d\lambda = 1$. In general we cannot assume any properties of f. The only freedom which we retain is that we can replace f by any $f_1 \geq f$ (because if $f d\lambda$ was a Pietsch measure for an operator T, then $f_1 d\lambda$ also is). However in order to keep things under control we have to control $\int f_1 d\lambda$. The classical case when $f = 1$, i.e. $H_p(f d\lambda) = H_p(\mathbb{T})$, correspond, to operators which are rotation invariant (see Exercise 2), and in this case there is the Riesz projection $\mathcal{R}$ whose properties are well known (see I.B.20).

The main result of this chapter, Theorem 10, will give a very useful analogue of the Riesz projection in our general case. Technically, our projection P will be built by patching up pieces of the Riesz projection $\mathcal{R}$. More precisely (but not really precisely) we will construct a sequence of H_∞-functions $(\varphi_j)_{j=1}^{\infty}$ (depending on f) such that $\sum_{j=1}^{\infty} \varphi_j^2 = 1$ and we put $Pg = \sum_{j=1}^{\infty} \varphi_j \mathcal{R}(\varphi_j g)$. Formally P is a projection onto analytic functions. In Theorem 9 we will construct such a sequence of functions having very elaborate properties, and the projection P will be constructed in Theorem 10.

8 Lemma. *There exist constants C and M such that for every $f \in L_1(\mathbb{T})$ such that $f \geq 0$ and $\int f d\lambda = 1$ we have a function $G \in L_1(\mathbb{T})$ and a sequence $(\varphi_j)_{j=0}^{\infty}$ of H_∞-functions such that*

$$\sum_{j=0}^{\infty} \varphi_j = 1, \tag{1}$$

$$\left\| \sum_{j=0}^{\infty} |\varphi_j|^p \right\|_{\infty} < \infty \quad \text{for every} \quad p, \ 0 < p < \infty, \tag{2}$$

$$f \leq G, \tag{3}$$

$$\sum_{j=0}^{\infty} M^j |\varphi_j|^3 \leq G, \tag{4}$$

$$|\varphi_j| G \leq C M^j, \tag{5}$$

$$\|G\|_1 \leq C. \tag{6}$$

Comment. We are trying to get analytic functions to imitate the situation when $\varphi_j = \chi_{\{M^j \leq f \leq M^{j+1}\}}$. In order to be sure that (1) holds we define first auxiliary functions ψ_j and then the functions φ_j. The complicated formula (8) defining ψ_j is needed to make sure that (2) holds. Note also that formulas like (9)-(11) to define a sequence satisfying (2) have already been used in the proof of Theorem III.D.31 formulas (17) and (27).

Proof: Let us fix numbers M, ε, δ in such a way that $0 < \varepsilon < \frac{1}{2}$, $0 < \delta < \frac{1}{2}$, $M > 2$, $M\delta^2 > 16$, $M\delta^3 < 1$. Assume additionally that $\|f\|_\infty < \infty$ and put

$$J = \max\{j : \|f\|_\infty > M^j\}.$$

We define sets $A_j = \{t \in \mathbb{T} : f(t) > M^j\}$, $j = 0, 1, \ldots, J$. We define outer functions (see I.B.23) $(\psi_j)_{j=0}^J$ by the conditions

$$|\psi_J| = 1 - (1 - \varepsilon)\chi_{A_J}, \tag{7}$$

$$|\psi_j| = (1 - (1 - \varepsilon)\chi_{A_j})$$
$$\left[\sup\left\{1, \delta^{-1}|1 - \psi_{j+1}|, \delta^{-2}|1 - \psi_{j+2}|, \ldots, \delta^{j-J}|1 - \psi_J|\right\}\right]^{-1}$$
$$\text{for} \quad j = 0, 1, 2, \ldots, J - 1. \tag{8}$$

The desired functions $(\varphi_j)_{j=0}^{J+1}$ are defined as

$$\varphi_{J+1} = 1 - \psi_J, \tag{9}$$

$$\varphi_k = \psi_J \cdot \ldots \cdot \psi_k(1 - \psi_{k-1}) \qquad k = 1, 2, \ldots, J, \tag{10}$$

$$\varphi_0 = \psi_J \cdot \psi_{J-1} \cdot \ldots \cdot \psi_0. \tag{11}$$

We define G to be

$$G = (1 - \varepsilon)^{-3} M \sum_{j=0} M^j |1 - \psi_j|^3 + 1. \tag{12}$$

With the above definitions we have to check conditions (1)-(6). This is mostly routine.

Ad (1). Follows directly from (9)-(11).

Ad (2). Observe that (8) implies $|\psi_j| \le (\delta^{-s}|1 - \psi_{j+s}|)^{-1}$ so

$$|\psi_j|\,|1 - \psi_{j+s}| \le \delta^s \text{ for } j = 0, 1, \ldots, J - 1 \text{ and } s = 1, 2, \ldots, J - j. \tag{13}$$

For a given $t \in \mathbb{T}$ let n be the smallest integer such that $|\psi_n(t)| > \varepsilon$. We see from (10) and (11) that $|\varphi_j(t)| \le 2\varepsilon^{n-j}$ for $j = 0, 1, \ldots, n$, so

$$\sum_{j=0}^n |\varphi_j(t)|^p \le C(\varepsilon). \tag{14}$$

For $j > n$ we get from (13) that $|1 - \psi_{n+s}(t)| \leq \frac{\delta^s}{\varepsilon}$ so (9) and (10) yield

$$\sum_{j=n+1}^{J+1} |\varphi_j(t)|^p \leq C(\varepsilon, \delta).$$

Ad (3).   We see from (7) and (8) that $|\psi_j(t)| \leq \varepsilon$ for $t \in A_j$ so $M^j|1 - \psi_j|^3 \geq M^j(1 - \varepsilon)^3$ on A_j. In particular on $A_j - A_{j+1}$ we have (see (12)) $G \geq M^{j+1} \geq f$.

Ad (4).   Since $|\psi_j| \leq 1$ (see (7), (8)) we have $|\varphi_j| \leq |1 - \psi_{j-1}|$ for $j = 1, 2, \ldots, J + 1$, so $\sum_{j=0}^{J+1} M^j|\varphi_j|^3 \leq 1 + \sum_{j=0}^{J} M^{j+1}|1 - \psi_j|^3 \leq G$.

Ad (5).   From (10), (12) and (13) we have

$$|\varphi_k|G \leq C \sum_{j=0}^{J} M^j|1 - \psi_j|^3|\varphi_k|$$

$$\leq C \sum_{j=0}^{k+2} M^j|1 - \psi_j|^3|\varphi_k|$$

$$+ \sum_{s \geq 3} M^{k+s}|1 - \psi_{k+s}|^3|\psi_k| \cdot |\psi_{k+1}||\psi_{k+2}|$$

$$\leq CM^k + M^k \sum_{s \geq 3}(M\delta^3)^s \leq CM^k.$$

Ad (6).   This condition is the most difficult to check. Let us recall (see I.B.23) that $\psi_j = |\psi_j| \cdot \exp i\log |\psi_j|$. Integrating the obvious numerical inequality $|1 - \alpha e^{i\beta}| \leq |1 - \alpha| + |\beta|$ we get

$$\int |1 - \psi_j|^2 \leq 2 \left(\int |1 - |\psi_j||^2 + \int \widetilde{|\log |\psi_j||}^2 \right).$$

Since $|\psi_j| \leq 1$ we get $1 - |\psi_j| \leq |\log |\psi_j||$ and using the fact that the trigonometric conjugation operator has norm 1 in $L_2(\mathbb{T})$ we get

$$\int |1 - \psi_j|^2 \leq 4 \int |\log |\psi_j||^2.$$

Using (8) we estimate it further as

$$\int |1 - \psi_j|^2 \tag{15}$$

$$\leq 8 \left(\int |\log(1 - (1 - \varepsilon)\chi_{A_j})|^2 \right.$$

$$+ \int \log \sup\{1, \delta^{-1}|1 - \psi_{j+1}|, \ldots, \delta^{j-J}|1 - \psi_J|\}^2 \Big)$$

$$\leq 8(\log \varepsilon)^2 |A_j| + 8 \sum_{k=j+1}^{J} \int \delta^{2(j-k)} |1 - \psi_k|^2.$$

Summing (15) over j we get

$$\sum_{j=0} M^j \int |1 - \psi_j|^2 \tag{16}$$

$$\leq 8(\log \varepsilon)^2 \sum_{j=0}^{J} M^j |A_j| + 8 \sum_{j=0}^{J} M^j \sum_{k=j+1}^{J} \delta^{2(j-k)} \int |1 - \psi_k|^2$$

$$\leq C + 8 \sum_{k=1}^{\infty} M^k \sum_{j=0}^{k-1} \left(M\delta^2\right)^{j-k} \int |1 - \psi_k|^2.$$

Since $8 \sum_{j=1}^{\infty} (M\delta^2)^{-j} < 1$, (16) gives

$$\sum_{j=0} M^j \int |1 - \psi_j|^2 \leq C. \tag{17}$$

But $\int |G| \leq M(1 - \varepsilon)^{-3} 2 \sum_{j=0}^{J} M^j \int |1 - \psi_j|^2 + 1$ so (6) follows from (17).

Now we have to remove the assumption $\|f\|_\infty < \infty$. This can be done as follows. Given f with $\|f\|_\infty = \infty$ we apply the previous construction to the functions $f_J = \min(f, M^J)$. Since no constant in the previous construction depends on J we see that for each j the functions ψ_j and φ_j (dependent on J) will tend to a limit (almost uniformly in $\mathbb{D}$ or $\sigma(L_\infty, L_1/H_1)$) as $J \to \infty$. Obviously the limits will also satisfy (1)-(6).   ∎

9.   Now we will improve this decomposition.

Theorem.   Given a function $\Delta \in L_1(\mathbb{T})$ with $\Delta \geq 0$, $\int \Delta = 1$ and an integer $r \geq 2$ there exist a sequence of scalars $(c_j)_{j=0}^{\infty}$ and two sequences of H_∞-functions $(\theta_j)_{j=0}^{\infty}$ and $(\tau_j)_{j=0}^{\infty}$ such that

$$\|\theta_j\|_\infty \leq 1 \quad \text{for} \quad j = 0, 1, 2, \ldots, \tag{18}$$

$$\left\| \sum_{j=0}^{\infty} |\tau_j| \right\|_\infty \leq C, \tag{19}$$

$$\sum_{j=0}^{\infty} \theta_j \tau_j^r = 1, \tag{20}$$

and for $F = \sum_{j=0}^{\infty} c_j |\tau_j|$ we have

$$\Delta \leq F, \tag{21}$$

$$|\tau_j| F \leq C c_j, \tag{22}$$

$$\|F\|_1 \leq C. \tag{23}$$

for some constant C independent of Δ.

Comment.   Ideally we would like to have $|\tau_j| = |\varphi_j|^{\frac{1}{r}}$. In this way, however, we lose control over $|\tau_j| \cdot F$ and $\|F\|_1$.

Proof: Let us take $(\varphi_j)_{j=0}^{\infty}$ as given by Lemma 8 and write

$$1 = \left(\sum_{j=0}^{\infty} \varphi_j \right)^{3r} = \sum_{j=0}^{\infty} \varphi_j \sum_{j_1,\ldots,j_{3r-1} \geq j} a(j_1,\ldots,j_{3r-1}) \varphi_{j_1} \cdot \varphi_{j_2} \cdots \varphi_{j_{3r-1}}$$

$$= \sum_j \varphi_j \cdot \phi_j = \sum_j \varphi'_j \tag{24}$$

where the above equalities actually define functions $a(j_1,\ldots,j_{3r-1})$; ϕ_j and φ'_j. Since $|a(j_1,\ldots,j_{3r-1})| \leq C(r)$ and $|\phi_j| \leq C(r)$ we see that conditions (1), (2), (5), hold with φ_j replaced by φ'_j. We also have (using (4))

$$\sum_{j=0}^{\infty} M^j |\varphi'_j|^{\frac{1}{r}} \leq C \sum_{j=0}^{\infty} M^j \left(\sum_{i \geq j} |\varphi_i| \right)^3$$

$$\leq C \sum_{j=0}^{\infty} \left(\sum_{i \geq j} M^{\frac{(j-i)}{3}} \cdot M^{\frac{i}{3}} |\varphi_i| \right)^3 \tag{4'}$$

$$\leq C \sum_{j=0}^{\infty} \sum_{i \geq j} M^{\frac{(j-i)}{3}} M^i |\varphi_i|^3$$

$$= C \left(\sum_{i=0}^{\infty} M^i |\varphi_i|^3 \right) \left(\sum_{j \leq i} M^{\frac{(j-i)}{3}} \right)$$

$$\leq C \cdot G.$$

Once more we write

$$1 = \left(\sum_{j=0}^{\infty} \varphi'_j \right)^r = \sum_{j=0}^{\infty} \varphi'_j \sum_{j_1,\ldots,j_{r-1} \leq j} b(j_1,\ldots,j_{r-1}) \varphi_{j_1},\ldots,\varphi_{j_{r-1}}$$

$$= \sum_{j=0}^{\infty} \varphi'_j \cdot \phi'_j = \sum_{j=0}^{\infty} \varphi''_j.$$

where the above equalities actually define functions $b(j_1, \ldots, j_{r-1})$, ϕ'_j and φ''_j. Since $|b(j_1, \ldots, j_{r-1})| \leq C(r)$ one gets $|\phi'_j| \leq C(r)$ so conditions (1), (2), (4') hold for $(\varphi''_j)_{j=0}^{\infty}$.

We also have (using 5)

$$|\varphi''_j|^{\frac{1}{r}} G \leq |\varphi'_j|^{\frac{1}{r}} |\phi'_j|^{\frac{1}{r}} G \leq C |\varphi'_j|^{\frac{1}{r}} \left(\sum_{i=0}^{j} |\varphi'_i| \right)^{\frac{(r-1)}{r}} G \qquad (5')$$

$$\leq C \left(\sum_{i=0}^{j} |\varphi'_i| \right) G \leq C \sum_{i=0}^{j} M^i \leq C M^j.$$

Now we use the inner-outer factorization to write

$$\varphi''_j = \theta_j \cdot \tau'_j, \qquad j = 0, 1, 2, \ldots$$

with θ_j inner (so (18) holds) and τ'_j outer. Since τ'_j is outer $\tau_j = (\tau'_j)^{\frac{1}{r}}$ makes sense. Since (φ''_j) satisfies (1) and (2) we get (19) and (20). Since $1 \leq \sum_{j=0}^{\infty} |\theta_j| |\tau_j|^r = \sum_{j=0}^{\infty} |\tau_j|^r \leq C$ we get also $1 \leq \sum_{j=0}^{\infty} |\tau_j|^2 \leq C$ so by (5') and (4') we have

$$G \leq \sum_{j=0}^{\infty} |\tau_j|^2 G \leq \sum_{j=0}^{\infty} C M^j |\tau_j| \leq C G.$$

This shows that for $c_j = C M^j$ we have $G \leq F \leq C G$ so (21) follows from (3), (22) follows from (5') and (23) follows from (6). ∎

10. Now we are ready to state the main result of this chapter.

Theorem. *There exists a C such that given $\Delta \in L_1(\mathbb{T})$ such that $\Delta \geq 0$ and $\int \Delta \leq 1$ there exists $\overline{\Delta} \geq \Delta$ with $\int \overline{\Delta} \leq C$ and a projection P such that*

(a) *P is a continuous projection in $L_p(\mathbb{T}, \overline{\Delta} d\lambda)$, $1 < p < \infty$ and the image of P is $H_p(\overline{\Delta} d\lambda)$ and $\|P\| \leq C \max(p, \frac{1}{(p-1)})$,*

(b) *P is of weak type (1-1) with respect to the measure $\overline{\Delta} \lambda$,*

(c) *the dual projection P^* is also of weak type (1-1) with respect to the measure $\overline{\Delta} \lambda$.*

First we reformulate what it means for an operator T to be of weak type (1-1). We have

11 Lemma. *An operator $T: L_1(\mu) \to L_0(\mu)$, μ a probability measure, is of weak type (1-1) if and only if for every $\alpha \in L_1(\mu)$ and every $\beta \in L_\infty(\mu)$ we have*

$$\int |T(\alpha)|^{\frac{1}{2}} |\beta| d\mu \le C \|\alpha\|_1^{\frac{1}{2}} \|\beta\|_\infty^{\frac{1}{2}} \|\beta\|_1^{\frac{1}{2}}. \tag{25}$$

Proof: Assume that T is of weak type (1-1) and take β with $\|\beta\|_\infty = 1$ and put $\|\beta\|_1 = \xi$. This is enough since (25) is homogenous in β. Under these restrictions $\int |T(\alpha)|^{\frac{1}{2}} |\beta| d\mu$ is maximal if $\beta = \chi_A$ where A is a set such that $\{|T(\alpha)| \ge \eta\} \supset A \supset \{|T(\alpha)| > \eta\}$ for some η and $\mu(A) = \xi$. So

$$\int |T(\alpha)|^{\frac{1}{2}} |\beta| d\mu = \int_A |T(\alpha)|^{\frac{1}{2}} \le C \|\alpha\|_1^{\frac{1}{2}} \int_0^\xi \lambda^{-\frac{1}{2}} d\lambda$$

$$\le C \|\alpha\|_1^{\frac{1}{2}} \xi^{\frac{1}{2}} = C \|\alpha\|_1^{\frac{1}{2}} \|\beta\|_\infty^{\frac{1}{2}} \|\beta\|_1^{\frac{1}{2}}.$$

Conversely, let us define $\beta = \chi_{\{|T(\alpha)| > \lambda\}}$ and we obtain from (25)

$$\mu\{|T(\alpha)| > \lambda\} \le \int \left| T\left(\frac{\alpha}{\lambda}\right) \right|^{\frac{1}{2}} \beta d\mu \le C\lambda^{-\frac{1}{2}} \|\alpha\|_1^{\frac{1}{2}} \|\beta\|_\infty^{\frac{1}{2}} \|\beta\|_1^{\frac{1}{2}}$$

$$= C\lambda^{-\frac{1}{2}} \|\alpha\|_1^{\frac{1}{2}} \mu\{|T(\alpha)| > \lambda\}^{\frac{1}{2}}$$

so T is of weak type (1-1). ∎

Proof of Theorem 10. Obviously it is enough to assume $\int \Delta = 1$ so let us apply Theorem 9 to Δ with $r = 8$ and put $\overline{\Delta} = F$. We define the desired projection by

$$P(f) = \sum_{j=0}^\infty \theta_j \tau_j^4 \mathcal{R}(\tau_j^4 \cdot f)$$

where $\mathcal{R}$ is the classical Riesz projection (see I.B.20.). Clearly P is algebraically a projection onto analytic functions.

For $1 < p < \infty$ we have

$$\int |Pf|^p \overline{\Delta} d\lambda = \int \left| \sum_{j=0}^\infty \theta_j \tau_j^4 \mathcal{R}(\tau_j^4 f) \right|^p \overline{\Delta} d\lambda$$

$$\le C \int \sum_{j=0}^\infty |\theta_j| |\tau_j|^4 |\mathcal{R}(\tau_j^4 f)|^p \overline{\Delta} d\lambda \qquad \text{by (19)}$$

$$\le C \sum_{j=0}^\infty \int |\mathcal{R}(\tau_j^4 f)|^p |\tau_j| \overline{\Delta} d\lambda \qquad \text{by (18)}$$

$$\leq C \sum_{j=0}^{\infty} c_j \int |\mathcal{R}(\tau_j^4 f)|^p d\lambda \qquad \text{by (22)}$$

$$\leq C \|\mathcal{R}\|_p^p \sum_{j=0}^{\infty} c_j \int |\tau_j^4 f|^p d\lambda \qquad \text{by (I.B.20)}$$

$$\leq C \|\mathcal{R}\|_p^p \int |f|^p \cdot \sum_j c_j |\tau_j|^{4p} d\lambda$$

$$\leq C \|\mathcal{R}\|_p^p \int |f|^p \sum_j c_j |\tau_j| d\lambda \qquad \text{by (19)}$$

$$\leq C \|\mathcal{R}\|_p^p \int |f|^p \overline{\Delta} d\lambda.$$

This shows that P is continuous in $L_p(\overline{\Delta} d\lambda)$ and the estimate for the norm follows from the well known estimates for $\|\mathcal{R}\|$ (see I.B.20).

Let us take $\beta \in L_\infty(\mathbf{T}, \lambda)$ with $\beta \geq 0$ and $\|\beta\|_\infty = 1$. Then for $f \subset L_1(\mathbf{T}, \overline{\Delta} d\lambda)$ we have

$$\int |Pf|^{\frac{1}{2}} \beta \overline{\Delta} d\lambda = \int \left| \sum_{j=0}^{\infty} \theta_j \tau_j^4 \mathcal{R}(\tau_j^4 f) \right|^{\frac{1}{2}} \beta \overline{\Delta} d\lambda$$

$$\leq \int \sum_{j=0}^{\infty} |\tau_j|^2 |\mathcal{R}(\tau_j^4 f)|^{\frac{1}{2}} \beta \overline{\Delta} d\lambda \leq \qquad \text{by (18)}$$

$$\leq C \sum_{j=0}^{\infty} c_j \int |\tau_j| |\mathcal{R}(\tau_j^4 f)|^{\frac{1}{2}} \beta d\lambda \qquad \text{by (22)}$$

$$\leq C \sum_{j=0}^{\infty} c_j \left(\int |\tau_j^4 f| d\lambda \right)^{\frac{1}{2}} \left(\int |\tau_j| \beta d\lambda \right)^{\frac{1}{2}} \qquad \text{by (25)}$$

$$\leq C \left(\sum_{j=0}^{\infty} c_j \int |\tau_j^4 f| d\lambda \right)^{\frac{1}{2}} \left(\sum_{j=0}^{\infty} c_j \int |\tau_j| \beta d\lambda \right)^{\frac{1}{2}}$$

$$\leq C \left(\int |f| \left(\sum_{j=0}^{\infty} c_j |\tau_j|^4 \right) d\lambda \right)^{\frac{1}{2}} \left(\int \beta \left(\sum_{j=0}^{\infty} c_j |\tau_j| \right) d\lambda \right)^{\frac{1}{2}}$$

$$\leq C \left(\int |f| \overline{\Delta} d\lambda \right)^{\frac{1}{2}} \left(\int \beta \overline{\Delta} d\lambda \right)^{\frac{1}{2}}.$$

Since this holds for arbitrary β we infer from Lemma 11 that P is of weak type (1-1) with respect to the measure $\overline{\Delta} d\lambda$.

Given $f \in L_p(\overline{\Delta}d\lambda)$ and $h \in L_{p'}(\overline{\Delta}d\lambda)$ we have

$$\int hP(f)\overline{\Delta}d\lambda = \sum_{j=0}^{\infty} \int h\theta_j \tau_j^4 \mathcal{R}(\tau_j^4 f)\overline{\Delta}d\lambda$$

$$= \sum_{j=0}^{\infty} \int \tau_j^4 f\mathcal{R}(h\theta_j \tau_j^4 \overline{\Delta})d\lambda$$

$$= \int f \sum_{j=0}^{\infty} \frac{\tau_j^4}{\overline{\Delta}}\mathcal{R}(\theta_j \tau_j^4 \cdot \overline{\Delta} \cdot h)\overline{\Delta}d\lambda$$

so

$$P^*(h) = \sum_{j=0}^{\infty} \frac{\tau_j^4}{\overline{\Delta}}\mathcal{R}(\theta_j \tau_j^4 \overline{\Delta}h).$$

To check that P^* is of weak type (1-1) let us take $\beta \in L_\infty(\mathbb{T}, \lambda)$ with $\beta \geq 0$ and $\|\beta\|_\infty = 1$. Then we have

$$\int |P^*h|^{\frac{1}{2}}\beta\overline{\Delta}d\lambda \leq C \int \sum_{j=0}^{\infty} |\tau_j|^2 |\mathcal{R}(\theta_j \tau_j^4 \overline{\Delta}h)|^{\frac{1}{2}}\beta\sqrt{\overline{\Delta}}d\lambda$$

$$\leq C \sum_{j=0}^{\infty} \sqrt{c_j} \int |\tau_j|^{3/2}\beta|\mathcal{R}(\theta_j \tau_j^4 \overline{\Delta}h)|^{\frac{1}{2}}d\lambda \qquad \text{by (22)}$$

$$\leq C \sum_{j=0}^{\infty} \sqrt{c_j} \left(\int |\theta_j \tau_j^4 \overline{\Delta}h|d\lambda\right)^{\frac{1}{2}} \left(\int |\tau_j|^{3/2}\beta d\lambda\right)^{\frac{1}{2}} \text{by (25)}$$

$$\leq C \sum_{j=0}^{\infty} \left(\int |\tau_j^4 h|\overline{\Delta}d\lambda\right)^{\frac{1}{2}} \left(\int |\tau_j|^{3/2}c_j\beta d\lambda\right)^{\frac{1}{2}}$$

$$\leq C \left(\int \left(\sum_{j=0}^{\infty} |\tau_j|^4\right)|h|\overline{\Delta}d\lambda\right)^{\frac{1}{2}} \left(\int \left(\sum_{j=0}^{\infty} |\tau_j|^{3/2}c_j\right)\beta d\lambda\right)^{\frac{1}{2}}$$

$$\leq C \left(\int |h|\overline{\Delta}d\lambda\right)^{\frac{1}{2}} \left(\int \beta\overline{\Delta}d\lambda\right)^{\frac{1}{2}} \qquad \text{by (19)}.$$

From Lemma 11 we infer that P^* is of weak type (1-1) with respect to the measure $\overline{\Delta}d\lambda$. ∎

12.   Now let us see how we can apply our considerations to estimate integral norms of absolutely summing operators on the disc algebra.

Theorem.   Let $T: A \to X$ be a p-absolutely summing operator, $1 < p < \infty$. Then

(a) T is p-integral and $i_p(T) \leq C \max(p, \frac{1}{p-1}) \cdot \pi_p(T)$,

(b) for every q with $\infty > q > p$, and every θ, $0 < \theta < \phi$ where $\phi = 1 - \frac{p}{q}$
we have

$$i_q(T) \leq \frac{c(p)}{\phi - \theta} \|T\|^\theta \pi_p(T)^{1-\theta}. \tag{26}$$

Proof: Let $\mu = \Delta d\lambda + \mu_s$, be a Pietsch measure for T (see Theorem III.F.8). Applying Theorem 10 to the weight Δ we get a measure $\tilde{\mu} = \overline{\Delta} d\lambda + \mu_s$ which is also a dominating measure for T with $\|\tilde{\mu}\| \leq C$. Proposition 6 and Theorem 10 show that there is a projection P from $L_p(\tilde{\mu})$ onto $H_p(\tilde{\mu})$ of norm $\leq C \max(p, \frac{1}{(p-1)})$ and also of weak type (1-1).

Let us consider the commutative diagram

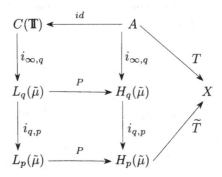

where id and i with various subscripts denote the formal identity acting between spaces indicated, and $\tilde{T}$ is the natural extension of T which exists because T is p-absolutely summing. This diagram clearly shows that T is p-integral with

$$i_p(T) \leq \|P\|_p \cdot \|\tilde{T}\| \leq C \max(p, \frac{1}{p-1}) \pi_p(T)$$

so we have (a).

In order to show (b) let us note that

$$i_q(T) \leq \|\tilde{T} \circ i_{q,p} \circ P \circ i_{\infty,q} \circ id\| \leq \|\tilde{T} \circ i_{q,p} \circ P\| = \|P^* \circ i^*_{q,p} \circ \tilde{T}^*\|.$$

Given $x^* \in X^*$ with $\|x^*\| = 1$ we define $\varphi = P^* i^*_{q,p} \tilde{T}^*(x^*) \in L_{q'}(\tilde{\mu})$, so

$$i_q(T) \leq \sup\{\|\varphi\|_{q'} : \|x\| = 1\}. \tag{27}$$

Applying the Hölder inequality we get

$$\|\varphi\|_{q'} = \left(\int |\varphi|^{q'} d\tilde{\mu} \right)^{\frac{1}{q'}} \tag{28}$$

$$\leq \left(\int |\varphi|^{q'\theta u} d\tilde{\mu} \right)^{\frac{1}{uq'}} \left(\int |\varphi|^{q'(1-\theta)v} d\tilde{\mu} \right)^{\frac{1}{q'v}} = \|\varphi\|_{\alpha}^{\theta} \cdot \|\varphi\|_{p'}^{1-\theta}$$

where $\frac{1}{u} + \frac{1}{v} = 1$, $q'\theta u = \alpha$ and $q'(1-\theta)v = p'$. Note that $\alpha < 1$. Since clearly (see the diagram) $\varphi = P^*\widetilde{T}^*(x^*)$ we get

$$\|\varphi\|_{p'} \leq \|P^*\| \cdot \|\widetilde{T}^*\| \leq C_p \pi_p(T). \tag{29}$$

Every $a^* \in A^*$ can be extended to a measure ν on $\mathbb{T}$ and if $\nu = g\tilde{\mu} + \nu_s$ we define an operator $Q \colon A^* \to L_\alpha(\tilde{\mu})$ by $Q(a^*) = P^*(g)\tilde{\mu}$. Since any other extension of a^* to a measure on $\mathbb{T}$ is of the form $g_1\tilde{\mu} + \nu_s$ with $P^*(g) = P^*(g_1)$ we get that Q is a continuous, linear operator from A^* into $L_\alpha(\tilde{\mu})$ with $\|Q\| \leq \frac{C}{(1-\alpha)^{\frac{1}{\alpha}}}$ (use weak type (1-1)). One also sees that as an element of $L_\alpha(\tilde{\mu})$ the function φ equals $QT^*(x^*)$. This gives

$$\|\varphi\|_\alpha \leq \frac{C}{(1-\alpha)^{\frac{1}{\alpha}}} \|T^*(x^*)\| \leq \frac{C}{(1-\alpha)^{\frac{1}{\alpha}}} \|T\|. \tag{30}$$

Putting together (27), (28), (29) and (30), we get

$$i_q(T) \leq C_p (1-\alpha)^{-\frac{\theta}{\alpha}} \|T\|^\theta \pi_p(T)^{1-\theta}.$$

Since $\alpha = p(p - 1 + \frac{\phi}{\theta})^{-1}$ (this requires a small calculation) a routine but a bit tedious calculation gives (26). ■

13 Corollary. *Let $p \geq 2$ and let Y be a space such that $L(C(\mathbb{T}), Y) = \Pi_p(C(\mathbb{T}), Y)$. Then also $L(A, Y) = \Pi_p(A, Y)$.*

Proof: Let $T \colon A \to Y$ be a finite rank operator and let $\gamma_\infty(T)$ denote the infimum of norms of extensions of T to an operator $\widetilde{T} \colon C(\mathbb{T}) \to Y$ (see III.B.3). Clearly for every $q > p$, we have $\gamma_\infty(T) \leq i_q(T)$. Using (26) we get $\gamma_\infty(T) \leq (c(p)/(\phi - \theta)) \|T\|^\theta \pi_p(T)^{1-\theta}$. Note that for fixed p and $q \to \infty$ we have $\phi \to 1$ so passing to the limit we have

$$\gamma_\infty(T) \leq \frac{c(p)}{1-\theta} \|T\|^\theta \pi_p(T)^{1-\theta}. \tag{31}$$

Since $L(C(\mathbb{T}), Y) = \Pi_p(C(\mathbb{T}), Y)$ there is a constant C such that $\pi_p(T) \leq C\gamma_\infty(T)$ for every finite rank operator $T \colon A \to Y$. Thus (31) yields $\pi_p(T) \leq C\|T\|$. Since A has the bounded approximation property (see II.E.5(b)) the same inequality holds for an arbitrary operator $T: A \rightarrow Y$.   ∎

14 Corollary.   *The space* L_1/H_1 *(and so also* A^**) has cotype 2 and*

$$\Pi_1(L_1/H_1, \ell_2) = L(L_1/H_1, \ell_2).$$

Proof: Corollary 13 and Theorem III.F.29 give that $\Pi_2(A, \ell_1) = L(A, \ell_1)$ so the standard localization (using the fact that A has b.a.p. (see II.E.5(b)) and Proposition III.F.28 give $\Pi_1(A^*, \ell_2) = L(A^*, \ell_2)$. This fact implies that A^* has the Orlicz property. Take any unconditionally convergent series $\sum_{n=1}^{\infty} x_n$ in A^* and let $\varphi_n \in A^{**}$ be such that $\|\varphi_n\| = 1$ and $\varphi_n(x_n) = \|x_n\|$ for $n = 1, 2, \ldots$. Take any $\alpha = (\alpha_n)_{n=1}^{\infty} \in \ell_2$ and define $T_\alpha: A^* \rightarrow \ell_2$ by $T_\alpha(x^*) = (\alpha_n \varphi_n(x^*))_{n=1}^{\infty}$. Clearly $\|T_\alpha\| \leq \|\alpha\|_2$ so $\pi_1(T_\alpha) \leq C\|\alpha\|_2$. Thus $\sum_{n=1}^{\infty} |\alpha_n|\|x_n\| \leq \sum_{n=1}^{\infty} \|T_\alpha(x_n)\| \leq C$ so $\sum_{n=1}^{\infty} \|x_n\|^2 < \infty$. Remark III.E.13 and Proposition III.A.24 show that A^* has cotype 2.   ∎

15 Corollary.   *Every rank n operator* $T: A \rightarrow Y$ *has an extension* $\tilde{T}: C(\mathbb{T}) \rightarrow Y$ *with* $\|\tilde{T}\| \leq C\|T\| \cdot \log n$.

Proof: It follows easily from the Auerbach lemma II.E.11 (or use III.G Exercise 14(a)) that $\pi_2(T) \leq n(T) \leq \|T\| \cdot n$. From (31) we get

$$\gamma_\infty(T) \leq \frac{c(2)}{1-\theta}\|T\|^\theta\|T\|^{1-\theta}n^{1-\theta} = C\|T\|\frac{n^{1-\theta}}{1-\theta}.$$

Taking $\theta = 1 - \frac{1}{\log n}$ we get the claim.   ∎

Note that this is an optimal result. It follows from Theorem III.B.22 that every extension to $C(\mathbb{T})$ of the projections defined in III.E.15 has norm greater that $C \log n$.

Before we proceed further we will present some modest applications of the results obtained so far.

16 Proposition.   *Every linear operator* $T: A \rightarrow L_p(\Omega, \mu)$, $0 < p \leq 2$, *factorizes strongly through* $L_2(\Omega, \mu)$.

Proof: This is a direct consequence of Corollary 14 and Proposition III.H.15.   ∎

17 Corollary.   *The sequence* $\Lambda = (\lambda_n)_{n=0}^{\infty}$ *is a coefficient multiplier from A into* $\ell_p, 0 < p \leq 2$ *if and only if* $(\lambda_n)_{n=1}^{\infty} \in \ell_{\frac{2p}{(2-p)}}$.

Proof: From Proposition 16 we obtain that Λ is a coefficient multiplier from A into ℓ_p, $0 < p \leq 2$, if and only if Λ is a multiplier from ℓ_2 into ℓ_p. The rest is a clear application of the Hölder inequality.   ∎

18 Corollary.   *Let* $(\varphi_n)_{n=1}^{\infty}$ *be a complete orthonormal system in* $L_2(\mathbb{T})$. *Then there exists* $f \in A$ *such that* $\sum_{n=1}^{\infty} |\langle \varphi_n, f \rangle|^p = \infty$ *for all* $p < 2$.

Proof: Repeat the proof of Theorem III.A.25 using the fact that every operator from A into ℓ_p, $p < 2$, is compact. This follows from Proposition 16.

19.   Let us now consider the space $A \widehat{\otimes} A$. (The definition of $X \widehat{\otimes} Y$ is given in III.B.25). There is a natural 'identity' map $A \widehat{\otimes} A \rightarrow C(\mathbb{T}) \widehat{\otimes} C(\mathbb{T})$. We want to show that this is an isomorphic embedding. Let us start with $id: A \widehat{\otimes} A \rightarrow A \widehat{\otimes} C(\mathbb{T})$. By Corollary III.B.27 this map is an isomorphic embedding if and only if every operator from A into A^* extends to an operator from $C(\mathbb{T})$ into A^*. To show this it is enough to show $L(A, A^*) = \Pi_2(A, A^*)$ (cf. III.F. 9(c)). Corollary 13 gives that it is enough to check that $L(C(\mathbb{T}), A^*) = \Pi_2(C(\mathbb{T}), A^*)$ so by the duality Theorem III.F.27 and Corollary III.F.25 it is enough to check that every operator of the form

$$\ell_{\infty}^N \xrightarrow{T} A^* \xrightarrow{S} \ell_{\infty}^N$$

with $\pi_2(S) \leq 1$ is nuclear with $n_1(ST) \leq C\|T\|$ (C independent of N and T). But S admits a factorization through a Hilbert space (see III.F.8) so from Corollary 14 we get $\pi_1(S) \leq C$. Thus by the remark after III.F.22 and III.F.24 we get

$$n_1(ST) = i_1(ST) = \pi_1(ST) \leq \|T\| \cdot \pi_1(S) \leq C\|T\|.$$

So $A \widehat{\otimes} A$ is closed in $A \widehat{\otimes} C(\mathbb{T})$. Repeating basically the same argument we get $A \widehat{\otimes} C(\mathbb{T})$ closed in $C(\mathbb{T}) \widehat{\otimes} C(\mathbb{T})$ so putting things together we get $A \widehat{\otimes} A$ closed in $C(\mathbb{T}) \widehat{\otimes} C(\mathbb{T})$. We do not, however, have, equality of the norms. Since for any tensor $t \in A \widehat{\otimes} A$ we have $\|t\|_{A \widehat{\otimes} A} \geq \|t\|_{A \widehat{\otimes} C(\mathbb{T})} \geq \|t\|_{C(\mathbb{T}) \widehat{\otimes} C(\mathbb{T})}$ it is enough to check that $id: A \widehat{\otimes} A \rightarrow A \widehat{\otimes} C(\mathbb{T})$ is not an isometry. By III.B.27 we have to exihibit an operator $T: A \rightarrow A^*$ such that every extension $\tilde{T}: C(\mathbb{T}) \rightarrow A^*$ will have $\|\tilde{T}\| \geq c\|T\|$ for some

$c > 1$. Since ℓ_1^2 is 1-complemented in A^* it suffices to consider the operator $T: A \to \ell_1^2$ given by $T(f) = (\hat{f}(0), \frac{1}{2}\hat{f}(1))$. We have

$$\|T\|_{A \to \ell_1^2} = \sup\{|\hat{f}(0)| + \frac{1}{2}|\hat{f}(1)| : \|f\|_\infty \le 1\}$$

$$= \sup\left\{\frac{1}{2\pi}\int f(\theta)\left(\alpha + \frac{1}{2}\beta e^{-i\theta}\right)d\theta\right.$$

$$\left. : |\alpha| \le 1, |\beta| \le 1, \|f\|_\infty = 1\right\}$$

$$= \left\|1 + \frac{1}{2}e^{i\theta}\right\|_{L_1/H_1} \le \left\|\left(\frac{1}{2}e^{-i\theta} + 1 + \frac{1}{2}e^{i\theta}\right)\right\|_{L_1} = 1.$$

On the other hand if $\tilde{T}: C(\mathbb{T}) \to \ell_1^2$ is an extension of T then, by a standard averaging argument we get $\|\tilde{T}\| \ge \|T_1\|$ where $T_1(f) = \left(\hat{f}(0), \frac{1}{2}\hat{f}(1)\right)$ for $f \in C(\mathbb{T})$. But

$$\|T_1\| = \left\|1 + \frac{1}{2}e^{i\theta}\right\|_{L_1} = \frac{1}{2\pi}\int_0^{2\pi}\sqrt{1.25 + \cos\theta}\, d\theta > 1.$$

20.   We would now like to discuss the reflexive subspaces of L_1/H_1. We start with the following proposition which is analogous to Corollary III.C.18.

Proposition.   Let $X \subset L_1/H_1$ be a closed subspace which does not have type p for any $p > 1$. Then X contains ℓ_1.

Proof: This follows directly from Theorem III.C.16 and III.D.31. ∎

21 Theorem.   Let $X \subset L_1/H_1$ be a reflexive subspace and let q denote the natural quotient map from L_1 onto L_1/H_1. There exists a reflexive subspace $\overline{X} \subset L_1$ such that $q|\overline{X}$ is an isomorphism from $\overline{X}$ onto X.

Proof: It follows immediately from Proposition 20 that X has type p for some $p > 1$. We see from Proposition III.H.14 and Corollary 13 that $L(A, X^*) = \Pi_q(A, X^*)$ for some $q < \infty$. In particular the operator $\varphi: A \to X^*$ defined by $\varphi(f)(x) = x(f)$ for $x \in X$ is q-absolutely summing, so by Theorem 12 q-integral. Thus φ extends to an operator $\tilde{\varphi}: C(\mathbb{T}) \to X^*$. Let us take $\tilde{\varphi}^*: X \to M(\mathbb{T})$. For $f \in A$ and $x \in X$ we have $\tilde{\varphi}^*(x)(f) = x(\tilde{\varphi}(f)) = x(\varphi(f)) = x(f)$ so $\tilde{\varphi}^*(X) \subset L_1(\mathbb{T})$ and $q \circ \tilde{\varphi}^* = id_X$. This shows that $\tilde{\varphi}^*$ is an isomorphic embedding and we can put $\overline{X} = \tilde{\varphi}^*(X)$. ∎

22.   Our next major theorem will be Theorem 25 which says that every operator defined on a reflexive subspace $X \subset L_1/H_1$ going into H_∞ extends to an operator from L_1/H_1 into H_∞. We will prove this using III.B.27 so our analysis will deal with functions of two variables.

Suppose (Ω_1, μ_1) and (Ω_2, μ_2) are two probability measure spaces. Assume that $R^i \colon L_p(\Omega_i, \mu_i) \to L_p(\Omega_i, \mu_i)$, $i = 1, 2$, are continuous linear operators. By $R^1 \otimes R^2$ we mean the operator from $L_p(\Omega_1 \times \Omega_2, \mu_1 \times \mu_2)$ into itself defined as 'an operator R^1 acting in first variable and R^2 acting in the second'. More precisely on a function of the form $f(\omega_1, \omega_2) = \sum_j g_j^1(\omega_1) \cdot g_j^2(\omega_2)$ with the series convergent in $L_2(\Omega_1 \times \Omega_2, \mu_1 \times \mu_2)$ we write $R^1 \otimes R^2(f)(\omega_1, \omega_2) = \sum_j R^1(g_j^1)(\omega_1) \cdot R^2(g_j^2)(\omega_2)$. One easily checks that $R^1 \otimes R^2$ is bounded so, since functions of the above form are dense in $L_p(\Omega_1 \times \Omega_2, \mu_1 \times \mu_2)$, it extends to the whole space. This also shows that given $f(\omega_1, \omega_2)$ we can compute $R^1 \otimes R^2(f)(\omega_1, \omega_2)$ applying first R^2 to each function $f(\omega_1, \cdot)$, treat the result as a function of two variables, $F(\omega_1, \omega_2)$, and apply R^1 to each function $F(\cdot, \omega_2)$. The result will be the same if we reverse the order in this procedure, i.e. first apply R^1 and next R^2.

We have the following

Proposition.   Let $\Delta_i \in L_1(\mathbb{T})$ be such that $\Delta_i \geq 0$ and $\int \Delta_i = 1$ for $i = 1, 2$. Let $\overline{\Delta}_i$ and P_i be given by Theorem 10 and let us define $Q_i = I - P_i$ and $Q = Q_1 \otimes Q_2$. If $1 < p < q < \infty$ and $\frac{1}{p} = \theta + \frac{(1-\theta)}{q}$ then

$$\|Q(\varphi)\|_p \tag{32}$$
$$\leq C\|Q(\varphi)\|_q^\theta \cdot \inf\{\|\varphi - f\|_1 \colon \operatorname{supp} \hat{f}(n, m) \subset (\mathbb{Z} \times \mathbb{Z}) \backslash (\mathbb{Z}_- \times \mathbb{Z}_-)\}^{1-\theta}$$

where all the norms are with respect to the measure $\overline{\Delta}_1(x)dx \times \overline{\Delta}_2(y)dy$ on $\mathbb{T}^2$.

Let us recall that $\mathbb{Z}$ denotes the set of integers and $\mathbb{Z}_-$ the set of negative integers.

The proof of this proposition follows from the explicit form of projections P_i and the analogous fact for Riesz projections. So we start our analysis of the Riesz projection with

23 Lemma.   If $\mathcal{R}$ denotes the Riesz projection and $\mathcal{R}_- = I - \mathcal{R}$ and $1 < p < q < \infty$ and $\frac{1}{p} = \theta + \frac{(1-\theta)}{q}$ then

$$\|\mathcal{R}_-(\varphi)\|_p \leq C\|\varphi\|_{1,\infty}^\theta \|\mathcal{R}_-(\varphi)\|_q^{1-\theta}. \tag{33}$$

♦

Proof: Fix a number $\lambda > 0$ to be specified later and consider the outer function τ with $|\tau| = \min(1, \frac{\lambda}{|\varphi|})$. Since $\varphi = \tau\varphi + (1-\tau)\varphi$ we have

$$\mathcal{R}_-(\varphi) = \mathcal{R}_-(\tau\varphi) + \mathcal{R}_-((1-\tau)\varphi) = \mathcal{R}_-(\tau\varphi) + \mathcal{R}_-((1-\tau)\mathcal{R}_-(\varphi)) \quad (34)$$

so

$$\|\mathcal{R}_-(\varphi)\|_p \le C_p \|\tau\varphi\|_p + C_p \|(1-\tau)\mathcal{R}_-(\varphi)\|_p \quad (35)$$

because $\mathcal{R}_-$ is continuous in $L_p(\mathbb{T})$.

From the definition of $|\tau|$ we see that

$$\|\tau\varphi\|_p^p \le \int_{\{|\varphi| \le \lambda\}} |\varphi(\theta)|^p d\theta + \lambda^p |\{|\varphi| > \lambda\}| \le C\lambda^{p-1} \|\varphi\|_{1,\infty} \quad (36)$$

because

$$\int_{\{|\varphi| \le \lambda\}} |\varphi|^p = \int_0^\lambda x^p dm_\varphi(x) \le \left| x^p m_\varphi(x) \Big|_0^\lambda \right| + \left| p \int_0^\lambda x^{p-1} m_\varphi(x) dx \right|$$

$$\le \lambda^{p-1} \|\varphi\|_{1,\infty} + p \|x m_\varphi(x)\|_\infty \int_0^\lambda x^{p-2} dx.$$

From the Hölder inequality we get

$$\|(1-\tau)\mathcal{R}_-(\varphi)\|_p^p \le \left(\int |1-\tau|^{\frac{pq}{(q-p)}} \right)^{1-\frac{p}{q}} \left(\int |\mathcal{R}_-(\varphi)|^q \right)^{\frac{p}{q}}. \quad (37)$$

Using the explicit form of τ (see I.B.23) and the boundedness of the trigonometric conjugation (see I.B.22) we have

$$\int |1-\tau|^{\frac{pq}{(q-p)}} \le C \int |1-\tau|^p \le C\left(|\{|\varphi| > \lambda\}| + \int_{\{|\varphi| \le \lambda\}} |1-\tau|^p \right)$$

$$\le C|\{|\varphi| > \lambda\}| + C \int_{\{|\varphi| \le \lambda\}} |\widetilde{\log |\tau|}|^p$$

$$\le C_p\left(|\{|\varphi| > \lambda\}| + \int |\log|\tau||^p \right) = \quad (38)$$

$$\le C_p\left(|\{|\varphi| > \lambda\}| + \int_{\{|\varphi| > \lambda\}} \left(\log\frac{|\varphi|}{\lambda} \right)^p \right) \le C_p\lambda^{-1}\|\varphi\|_{1,\infty}$$

because

$$\int_{\{|\varphi| > \lambda\}} \left(\log\frac{|\varphi|}{\lambda} \right)^p = \int_\lambda^\infty \left(\log\frac{x}{\lambda} \right)^p dm_\varphi(x)$$

$$= -\int_\lambda^\infty m_\varphi(x) p \left(\log\frac{x}{\lambda} \right)^{p-1} \frac{dx}{x}$$

$$\le C_p \sup_{x \ge \lambda} |m_\varphi(x) \cdot x|\lambda^{-1} \int_1^\infty \frac{\log^{p-1} x}{x^2} dx$$

$$\le C_p\lambda^{-1}\|\varphi\|_{1,\infty}.$$

Putting together (35), (36), (37) and (38) we get

$$\|\mathcal{R}_-(\varphi)\|_p^p \le C_p\left(\lambda^{p-1}\|\varphi\|_{1,\infty} + \left(\lambda^{-1}\|\varphi\|_{1,\infty}\right)^{1-\frac{p}{q}}\|\mathcal{R}_-(\varphi)\|_q^p\right).$$

Choosing $\lambda = \|\varphi\|_{1,\infty}^{-q'/q}\|\mathcal{R}_-(\varphi)\|_q^{q'}$ we get the desired estimate.   ∎

24 Lemma. Define $\mathcal{K} = \mathcal{R}_-^{(1)} \otimes \mathcal{R}_-^{(2)}$ where $\mathcal{R}_-^{(i)}$ is the operator $\mathcal{R}_-$ acting in the i-th variable, $i = 1, 2$. Then for $1 < p < q < \infty$ and $\frac{1}{p} = \theta + \frac{(1-\theta)}{q}$ we have

$$\|\mathcal{K}(\varphi)\|_p \le C\|\varphi\|_q^{1-\theta} \cdot \inf\{\|\varphi - f\|_1^\theta : \operatorname{supp}\hat{f}(n,m) \subset (\mathbb{Z} \times \mathbb{Z})\setminus(\mathbb{Z}_- \times \mathbb{Z}_-)\} \tag{39}$$

for $\varphi \in L_p(\mathbb{T} \times \mathbb{T})$.

Proof: Since for arbitrary φ and f as in the infimum in (39) we have $\mathcal{K}(\varphi) = \mathcal{K}(\varphi - f)$ it is enough to show

$$\|\mathcal{K}(\varphi)\|_p \le C\|\varphi\|_q^{1-\theta}\|\varphi\|_1^\theta. \tag{40}$$

If $\varphi = \varphi(x_1, x_2)$ then applying $\mathcal{R}_-^{(2)}$ in the second variable we infer from Lemma 23 that for each x_1 we have

$$\|\mathcal{R}_-^{(2)}\varphi(x_1, \cdot)\|_p^p \le C\|\varphi(x_1, \cdot)\|_{1,\infty}^{p\theta}\|\mathcal{R}_-^{(2)}\varphi(x_1, \cdot)\|_q^{(1-\theta)p} \tag{41}$$

where all the norms are computed with respect to the second variable. Integrating (41) with respect to x_1 and applying the Hölder inequality (note that $\frac{1}{p\theta} > 1$) we get

$$\|\mathcal{R}_-^{(2)}(\varphi)\|_p^p$$
$$\le C \int \|\varphi(x_1, \cdot)\|_{1,\infty}^{p\theta} \cdot \|\mathcal{R}_-^{(2)}\varphi(x_1, \cdot)\|_q^{p(1-\theta)} dx_1$$
$$\le C\left(\int \|\varphi(x_1, \cdot)\|_{p,\infty} dx_1\right)^{p\theta}$$
$$\quad \cdot \left(\int \left(\int |\mathcal{R}_-^{(2)}\varphi(x_1, x_2)|^q dx_2\right)^{\frac{1}{q}(1-\theta)p\frac{1}{1-p\theta}} dx_1\right)^{1-p\theta} \tag{42}$$
$$\le C\left(\int \|\varphi(x_1, \cdot)\|_{1,\infty} dx_1\right)^{p\theta} \cdot \left(\int\int |\mathcal{R}_-^{(2)}\varphi(x_1, x_2)|^q dx_1 dx_2\right)^{\frac{(1-\theta)p}{q}}$$
$$\le C\|\varphi(\cdot, \cdot)\|_{1,\infty}^{p\theta} \cdot \|\mathcal{R}_-^{(2)}\varphi(\cdot, \cdot)\|_q^{(1-\theta)p}.$$

Applying (42) for $\mathcal{R}_-^{(1)}(\varphi)$ we get

$$\|\mathcal{K}\varphi\|_p^p \le C\|\mathcal{R}_-^{(1)}\varphi\|_{1,\infty}^{\theta p}\|\mathcal{K}\varphi\|_q^{(1-\theta)p}. \tag{43}$$

But the weak type (1-1) inequality for $\mathcal{R}_-$ gives that for every x_2

$$|\{x_1: |\mathcal{R}_-^{(1)}(\varphi)(x_1,x_2)| > \lambda\}| \le C\frac{\|\varphi(\cdot,x_2)\|_1}{\lambda} \tag{44}$$

so integrating (44) with respect to x_2 we get

$$|\{(x_1,x_2): |\mathcal{R}_-^{(1)}(\varphi)(x_1,x_2)| > \lambda\}| \le C\frac{\|\varphi\|_1}{\lambda}$$

so

$$\|\mathcal{R}_-^{(1)}\varphi\|_{1,\infty} \le C\|\varphi\|_1. \tag{45}$$

Since $\mathcal{K}$ is continuous in $L_q(\mathbb{T}\times\mathbb{T})$ we see that (43) and (45) give (40). ∎

Now we are ready for the

Proof of Proposition 22.   Using the explicit definition of the projection P (see the proof of Theorem 10) we get

$$Q_1 \otimes Q_2\varphi(x,y)$$
$$= \sum_s \sum_i \theta_s^1(x)\theta_i^2(y)\tau_{s,1}^4(x)\tau_{i,2}^4(y)\mathcal{R}_-^{(1)}\otimes\mathcal{R}_-^{(2)}(\tau_{s,1}^4(x)\tau_{i,2}^4(y)\varphi(x,y)).$$

Thus applying (18) and (22) and then Lemma 24 we get

$$I = \int\int |Q_1\otimes Q_2(\varphi)(x,y)|^p\overline{\Delta}_1(x)\overline{\Delta}_2(y)dxdy$$
$$\le C\sum_s\sum_i c_s^{(1)}c_i^{(2)}\int|\mathcal{R}_-^{(1)}\otimes\mathcal{R}_-^{(2)}(\tau_{s,1}^4(x)\tau_{i,2}^4(y)\varphi(x,y))|^p dxdy$$
$$\le C\sum_s\sum_i c_s^{(1)}c_i^{(2)}\inf_f\{\|\tau_{s,1}^4(x)\tau_{i,2}^4(y)(\varphi-f)\|_1^{\theta p}\}$$
$$\|\tau_{s,1}^4(x)\tau_{i,2}^4(y)\varphi(x,y)\|_q^{(1-\theta)p}.$$

Using the Hölder inequality and the definitions of $\overline{\Delta}_1$ and $\overline{\Delta}_2$ we further get

$$I \le C\left(\sum_s\sum_i c_s^{(1)}c_i^{(2)}\inf_f\int\int|\tau_{s,1}^4(x)\tau_{i,2}^4(y)(\varphi-f)|dxdy\right)^{\theta p}$$

$$\cdot \left(\sum_s \sum_i c_s^{(1)} c_i^{(2)} \int\int |\tau_{s,1}^4(x)\tau_{i,2}^4(y)\varphi(x,y)|^q dxdy \right)^{p(1-\theta)/q}$$

$$\leq C \inf_f \left(\int\int |\varphi - f|\overline{\Delta}_1(x)\overline{\Delta}_2(y)dxdy \right)^{\theta p}$$

$$\left(\int\int |\varphi(x,y)|^q \overline{\Delta}_1(x)\overline{\Delta}_2(y)dxdy \right)^{p(1-\theta)/q}. \qquad \blacksquare$$

25 Theorem.   If $Y \subset L_1/H_1$ is a reflexive subspace then every bounded linear operator $T: Y \to H_\infty$ extends to an operator $\widetilde{T}: L_1/H_1 \to H_\infty$.

The reflexivity assumption is needed as is indicated in Exercise 6.

Proof: Since $(L_1/H_1)^* = H_\infty$ we infer from Corollary III.B.27 that it is enough to show that $i: Y\widehat{\otimes}(L_1/H_1) \to (L_1/H_1)\widehat{\otimes}(L_1/H_1)$ is an isomorphism, i.e. for every tensor $\sum_{j=1}^n y_j \otimes x_j$ we have

$$C \left\| \sum y_j \otimes x_j \right\|_{Y\widehat{\otimes}(L_1/H_1)} \leq \left\| \sum_j y_j \otimes x_j \right\|_{(L_1/H_1)\widehat{\otimes}(L_1/H_1)}. \qquad (46)$$

But clearly if $x_j = q(f_j)$, $j = 1,\ldots,n$ (note that q is the natural quotient map from L_1 onto L_1/H_1) then

$$\left\| \sum_j y_j \otimes x_j \right\|_{Y\widehat{\otimes}(L_1/H_1)} \leq \left\| \sum_j y_j \otimes f_j \right\|_{Y\widehat{\otimes}L_1}. \qquad (47)$$

Let $\overline{Y} \subset L_1(\mathbb{T})$ be the subspace given by Theorem 21. We see (see III.B.28) that

$$\left\| \sum_j y_j \otimes f_j \right\|_{Y\widehat{\otimes}L_1(\mathbb{T})} \leq C \left\| \sum_j \bar{y}_j \otimes f_j \right\|_{\overline{Y}\widehat{\otimes}L_1(\mathbb{T})} =$$

$$\leq C \int\int |\bar{y}_j(x) \cdot f_j(y)| dxdy. \qquad (48)$$

So comparing (46), (47) and (48) we see that to prove the theorem it is enough to show that

$$I \stackrel{df}{=} \inf \left\{ \int\int |\bar{y}_j(x)f_j(y)|dxdy : q(f_j) = x_j \right\} \leq C \sum_j \|y_j\| \cdot \|x_j\|. \qquad (49)$$

From Rosenthal's theorem (Corollary III.H.13) we infer that there exists $\Delta_1(x) > 0$ with $\int \Delta_1 = 1$ and $r > 1$, such that

$$\left(\int \left(\frac{|\bar{y}|}{\Delta_1} \right)^r \Delta_1 \right)^{\frac{1}{r}} \leq C \|\bar{y}\|_1 \tag{50}$$

for all $\bar{y} \in \overline{Y}$.

Let us fix the sequence $(f_j)_{j=1}^n \subset L_1(\mathbb{T})$ such that $q(f_j) = x_j$, $j = 1, 2, \ldots, n$ and

$$\int \int |\bar{y}_j(x) \cdot f_j(y)| dx dy \leq 2I. \tag{51}$$

We define $\Delta_2(y) = I^{-1} \int |\sum_j \bar{y}_j(x) f_j(y)| dx$.

Applying Theorem 10 to the weights $\Delta_1(x)$ and $\Delta_2(y)$ we obtain weights $\overline{\Delta}_1(x)$ and $\overline{\Delta}_2(y)$ and projections P_1 and P_2. We put $Q_i = I - P_i$, $i = 1, 2$. We also define outer functions ϕ_i, $i = 1, 2$ such that on $\mathbb{T}$ we have $|\phi_i| = \overline{\Delta}_i$. We put

$$f_j' = \phi_1 Q_1(\phi_1^{-1} f_j), \qquad j = 1, 2, \ldots, n \tag{52}$$

and

$$y_j' = \phi_2 Q_2(\phi_2^{-1} \bar{y}_j), \qquad j = 1, 2, \ldots, n. \tag{53}$$

Since $f_j - f_j' = \phi_1(\phi_1^{-1} f_j - Q_1(\phi_1^{-1} f_j)) = \phi_1 P_1(\phi_1^{-1} f_j)$ we get $q(f_j') = q(f_j)$, $j = 1, 2, \ldots, n$, and analogously we see that $q(y_j') = q(\bar{y}_j) = y_j$, $j = 1, 2, \ldots, n$.

In particular for every sequence of scalars $(\alpha_j)_{j=1}^n$ we get

$$\left\| \sum_j \alpha_j \bar{y}_j \right\|_1 \leq C \left\| \sum_j \alpha_j y_j \right\|_{L_1/H_1}$$
$$= C \left\| q \left(\sum_j \alpha_j y_j' \right) \right\|_{L_1/H_1}$$
$$\leq C \left\| \sum_j \alpha_j y_j' \right\|_1. \tag{54}$$

Thus we have (see (49) and (54))

$$I \leq \int \int |\sum_j \bar{y}_j(x) f_j'(y)| dx dy \leq C \int \int |\sum_j y_j'(x) f_j'(y)| dx dy. \tag{55}$$

If we write $F(x,y) = \sum_j \bar{y}_j(x) f_j(y)$ and $Q = Q_1 \otimes Q_2$ then from (52), (53) and (55) we get

$$I \leq C \int \int |Q(\phi_1^{-1}(x)\phi_2^{-1}(y)F(x,y))|\overline{\Delta}_1(x)\overline{\Delta}_2(y) dx dy.$$

Now let us fix q, $1 < q < r$ and θ such that $\frac{1}{q} = \theta + \frac{(1-\theta)}{r}$. Proposition 22 and the continuity of Q in $L_r(\overline{\Delta}_1 dx \otimes \overline{\Delta}_2 dy)$ give

$$I \leq C \left(\int \int |\phi_1^{-1}(x)\phi_2^{-1}(y)F(x,y)|^r \overline{\Delta}_1(x)\overline{\Delta}_2(y) dx dy \right)^{\frac{\theta}{r}} \quad (56)$$

$$\cdot \inf \left\{ \int \int |\phi_1^{-1}(x)\phi_2^{-1}(y)F(x,y) \right.$$

$$\left. - f(x,y)|\overline{\Delta}_1(x)\overline{\Delta}_2(y) dx dy \colon \operatorname{supp} \hat{f} \subset \mathbb{A} \right\}^{1-\theta}$$

where $\mathbb{A}$ denotes the set $(\mathbb{Z} \times \mathbb{Z}) \backslash (\mathbb{Z}_- \times \mathbb{Z}_-)$.
Using (50) and the definition of Δ_2 the first factor in (56) is majorized by

$$\left(\int \left(\int \left| \sum_j f_j(y)\bar{y}_j(x) \right|^r \overline{\Delta}_1^{1-r}(x) dx \right) \overline{\Delta}_2^{1-r}(y) dy \right)^{\frac{\theta}{r}}$$

$$\leq C \left(\int \left(\int \left| \sum_j f_j(y)\bar{y}_j(x) \right| dx \right)^r \overline{\Delta}_2^{1-r}(y) dy \right)^{\frac{\theta}{r}} \quad (57)$$

$$= C \left(\int I^r \overline{\Delta}_2^r(y)\overline{\Delta}_2^{1-r}(y) dy \right)^{\frac{\theta}{r}} \leq C I^\theta.$$

Since $\phi_1^{-1}(x) \cdot \phi_2^{-1}(y)$ is analytic the second factor in (56) is majorized by

$$\inf \left\{ \int \int |F(x,y) - f(x,y)| dx dy \colon \operatorname{supp} \hat{f} \subset \mathbb{A} \right\}^{1-\theta}$$

$$\leq \inf \left\{ \int \int \left| \sum_j (\bar{y}_j(x) - h_j^1(x))(f_j(y) \right. \right.$$

$$\left. \left. - h_j^2(y)) \right| dx dy \colon h_j^1, h_j^2 \text{ are analytic} \right\}^{1-\theta}$$

$$\leq \left(\sum_j \|y_j\| \cdot \|f_j\| \right)^{1-\theta}. \quad (58)$$

Putting together (56), (57) and (58) we get (49).   ∎

26.   One of the reasons the above theorem may be relevant to the more classical analysis is the following.

Proposition.   *Given $\phi \in H_\infty(\mathbb{D} \times \mathbb{D})$ we define $T_\phi: L_1/H_1 \to H_\infty$ by $T_\phi(u)(z) = \langle u, \phi(z, \cdot) \rangle$. The correspondence $\phi \leftrightarrow T_\phi$ establishes an isometry between $L(L_1/H_1, H_\infty)$ and $H_\infty(\mathbb{D} \times \mathbb{D})$.*

Proof: Clearly T_ϕ is a well defined linear operator into H_∞. Since $(L_1/H_1)^* = H_\infty$ we have

$$\|T_\phi\| = \sup_{z \in \mathbb{D}} \sup_{\|u\| \leq 1} |\langle u, \phi(z, \cdot) \rangle| = \sup_{z \in \mathbb{D}} \|\phi(z, \cdot)\|_\infty = \|\phi\|_\infty,$$

so the map $\phi \mapsto T_\phi$ is an isometry. If we are given $T: L_1/H_1 \to H_\infty$ then we define $\phi(z_1, z_2) = T(\delta_{z_1})(z_2)$ where for $z \in \mathbb{D}$ the symbol δ_z denotes the functional on A 'value at z'. The Poisson formula shows that $\delta_z \in L_1/H_1$. Obviously $\|\phi\|_\infty \leq \|T\|$. Since $\delta_{z_1} \in L_1/H_1$ for $z_1 \in \mathbb{D}$ we get that $\phi(z_1, z_2)$ is analytic in z_2. But $T(\delta_{z_1})(z_2) = T^*(\delta_{z_2})(z_1)$ where $T^*: H_\infty^* \to H_\infty$ so ϕ is also analytic in z_1. This shows that $\phi \in H_\infty(\mathbb{D} \times \mathbb{D})$.   ∎

27.   Let us recall that a set of integers Λ is called a Λ_p set, $1 < p < \infty$, if there exists a constant C such that for all sequences of scalars $(\alpha_j)_{j \in \Lambda}$ we have

$$\left\| \sum_{j \in \Lambda} \alpha_j e^{ij\theta} \right\|_p \leq C \left\| \sum_{j \in \Lambda} \alpha_j e^{ij\theta} \right\|_1.$$

Corollary.   *If Λ is a Λ_2 subset of positive integers and $(\varphi_k(z))_{k \in \Lambda}$ is a sequence of H_∞ functions such that*

$$\sup_{z \in \mathbb{D}} \sum_{k \in \Lambda} |\varphi_k(z)|^2 < \infty \tag{59}$$

then there exists a function $\phi(z, w) \in H_\infty(\mathbb{D} \times \mathbb{D})$ such that

$$\int \phi(z, e^{i\theta}) e^{-ik\theta} d\theta = \varphi_k(z) \quad \text{for} \quad k \in \Lambda.$$

Proof: One easily checks that $Y = \text{span}\{e^{-ik\theta}\}_{k \in \Lambda} \subset L_1/H_1$ is isomorphic to ℓ_2. The operator $T: Y \longrightarrow H_\infty$ given by $T(\sum_{k \in \Lambda} \alpha_k e^{-ik\theta}) =$

$\sum_{k\in\Lambda}\alpha_k\varphi_k(z)$ is well defined by (59) and by Theorem 25 admits an extension $\tilde{T}\colon L_1/H_1 \to H_\infty$. Using Proposition 26 this extension yields the desired function ϕ. ∎

Notes and Remarks.
Clearly this chapter contains the most recent and the most advanced results of this book. It shows the very intricate interplay between the general theory and concrete analytical problems. The main results presented in this chapter are due to J. Bourgain. It is my feeling that they are not fully understood yet and their power remains to be explored.

Proposition 2 and *Corollary 3* are due to Kislyakov [P]. This is the solution of the problem posed by Glicksberg [1964], whether a closed, proper, point-separated subalgebra of $C(S)$ can be complemented. The earlier work on the Glicksberg problem is presented in Pełczyński [1977], Ch. 5. Building on ideas of Kislyakov [P], Garling [P] has generalized *Corollary 3* and has shown that a proper uniform algebra is not a quotient of any C^*-algebra. *Proposition 4* is an observation of Milne [1972]. *Theorem 10* and its proof (i.e. *Lemma 8* and *Theorem 9* and *Lemma 11*) are due to Bourgain [1986] but build on his earlier work, Bourgain [1984]. This paper contains also our *Theorem 12(b)*, *Corollaries 13, 14, 15*, and *Theorem 21*. *Theorem 25*, its proof and consequences are taken from Bourgain [1986].

Actually later J. Bourgain developed still another approach to these results. He proved the following theorem (see Dechamp-Gondim [1985] or Bourgain-Davis [1986]).

Theorem A.   Let $1 < p < \infty$ and $0 < \alpha < 1$ and let (Ω, Σ, μ) be a *probability measure space and let $\mathcal{R}$ denote the Riesz projection. The operator $Id \otimes \mathcal{R}$ acts as a continuous operator from $L_p(\mathbb{T}, L_1(\Omega))$ into $L_p(\mathbb{T}, L_\alpha(\Omega))$.*

This Theorem implies *Corollary 14* (see Dechamp-Gondim [1985] or Bourgain-Davis [1986]) and *Theorem 25* (see Kislyakov [1987]). It is perfectly possible and even shorter to present the main results of this chapter avoiding *Theorem 10* and *Proposition 22* and using *Theorem A* instead. We have chosen to present longer arguments for the following reasons.

(a)   We feel that the approach to p-summing operators on the disc algebra A described in **7** is very natural.

(b)   We feel that *Theorem 10* is very interesting in its own right (but this is clearly also true about *Theorem A*).

(c) There are some interesting applications of *Theorem 9*. To be more precise, Bourgain [1984a] uses *Theorems 9* and *10* to prove the following

Theorem B. *Assume n is a positive integer and $(\phi_k, x_k)_{k=1}^n$ form a biorthogonal sequence in $A \times A^*$ such that*

$$\|\phi_k\|_\infty \leq M \quad for \quad k = 1, 2, \ldots, n, \tag{1}$$

$$\left\| \sum_{k=1}^n a_k x_k \right\| \leq M \cdot \left(\sum_{k=1}^n |a_k|^2 \right)^{\frac{1}{2}} \quad for\ all\ sequences\ of\ scalars. \tag{2}$$

Then for some $z \in \mathbb{C}, |z| < 1$ we have

$$\frac{1}{n} \sum_{m=1}^n \left\| \sum_{k=1}^m \phi_k(z) x_k \right\| \geq C \log n$$

This is the disc algebra version (and generalization) of the results of Olevskii (see Olevskii [1975]).

The papers Bourgain [1984] and [1986] contain answers to problems asked in Pełczyński [1977] and by N. Varopoulos. Varopoulos, motivated by his important theory of tensor algebras (see e.g. Graham-McGehee [1979] Chapter 11) asked the question whether $A \widehat{\otimes} A$ is closed in $C(\mathbb{T}) \widehat{\otimes} C(\mathbb{T})$. We saw in **19** that the positive answer routinely follows from *Corollary 13*. The observation that $id: A \widehat{\otimes} A \to C(\mathbb{T}) \widehat{\otimes} C(\mathbb{T})$ is not an isometry is due to Kaijser [1978].

Naturally the work of Bourgain subsumes many earlier particular results. We will not discuss these here in any detail, but let us point out that the whole effort in *Theorem 10* is directed at obtaining the weak type estimate. Projections bounded in $L_p, 1 < p < \infty$, are much easier to construct but they lead at best to the conclusion that L_1/H_1 has cotype q for any $q > 2$ (see Kislyakov [1981b]).

The concrete applications of our main results which we present are mostly routine. Some more are to be found in the Exercises. *Corollary 27* is a two-dimensional analogue of III.E.9. Let us also recall that *Theorem 10* has applications to operators on Hilbert space which we discussed in Notes and remarks to Chapter III.F. It was also instrumental in constructing counterexamples to some old conjectures of Grothendieck (see Pisier [1983]).

The very important technical aspect of our work in this chapter is the extrapolation argument based on interpolation inequalities. The first time this appeared in this book was in the proof of *Theorem III.F.27*.

In this chapter we have two interpolation inequalities, the abstract one namely *Theorem 12(b)* and a very concrete one namely *Lemma 23*. It was recognized by Kislyakov [1987a] that inequalities like in *Lemma 23* hold for other natural operators and can sometimes serve as a substitute for the weak type (1-1) inequality. This has some nice consequences.

It is also interesting to note that our work is essentially restricted to spaces of analytic functions of one variable. Bourgain [1985] and [1986] constructed for every $p > 2$ an operator on $A(\mathbb{B}_d), d > 1$, and on $A(\mathbb{D}^n), n > 1$, which is p-absolutely summing but not q-integral for any q. This shows that most of the results presented in this chapter about the disc algebra are false both for $A(\mathbb{B}_d)$ and for $A(\mathbb{D}^n)$ with $d > 1$ and $n > 1$. It seems to be a very interesting problem to figure out what goes on in several variables.

Added in proof (May 31, 1990) After this book was submitted I received a very interesting work of Kislyakov [P_1]. It contains a new, simple proof of Theorem 12 b). The main truncation lemma (refered to in the title) asserts the following:

If $\Delta \geq \delta > 0$ is such that $\widetilde{\Delta^{1/2}} \leq c\Delta^{1/2}$ for some c then every $f \in H_1(\Delta d\lambda)$ can be represented as $f = f_1 + f_2$ where
(i) $f_1 \in H_\infty$ and $\|f_1\|_\infty \leq 24c^2$
(ii) for every $s, 1 \leq s < \infty$ if $f \in H_s(\Delta d\lambda)$ then $f_2 \in H_s(\Delta d\lambda)$ and $(\int |f_2|^s \Delta d\lambda)^{1/s} \leq 106c^4 \left(\int_{|f| \geq 1} |f|^s \Delta d\lambda \right)^{1/s}$.

Some further extensions are contained in [P_2].

Exercises

1.
Show (without appeal to Theorem 10) that every multiplier $T: \ell_2 \to A$ (i.e. any operator of the form $T(\xi_n) = \sum_{n=0}^\infty \xi_n t_n z^n$) has 1-absolutely summing adjoint.

2.
Let $T: A \to H_2$ be a multiplier (i.e. $T(\sum_{n=0}^\infty a_n z^n) = \sum_{n=0}^\infty a_n t_n z^n$). Show that the following conditions are equivalent:

(a) T is p-absolutely summing for some $p < 1$.;

(b) T is 1-nuclear;

(c) T factors through an $L_1(\mu)$ space;

(d) $(t_n)_{n=0}^\infty \in \ell_2$.

Note that it follows from this exercise and Paley's inequality I.B.24 (or Hardy's inequality I.B.25) that the map $id: A(\mathbb{T}) \to H_1(\mathbb{T})$ does not factor through any L_1-space. A second such example is in Exercise III.G.10(d).

3. Let X be a finite dimensional Banach space and let us fix numbers K and p, $1 < p < \infty$. Show that the following conditions are equivalent:

   (a) $i_p(T) \le K\pi_p(T)$ for every operator $T: X \to Z$, where Z is an arbitrary Banach space;

   (b) for every subspace $Y \subset L_{p'}(\mu)$, $\frac{1}{p} + \frac{1}{p'} = 1$, and for every linear operator $T: Y \to X$ there exists an extension $\widetilde{T}: L_{p'}(\mu) \to X$ with $\|\widetilde{T}\| \le K\|T\|$.

4. Let $X \subset C(S)$ be a closed subspace. Show that the following conditions are equivalent:

   (a) $X \widehat{\otimes} C(S)$ is closed in $C(S) \widehat{\otimes} C(S)$;

   (b) every operator $T: X \to L_1(\mu)$ extends to an operator $\widetilde{T}: C(S) \to L_1(\mu)$;

   (c) $\Pi_2(X, \ell_1) = L(X, \ell_1)$ and X^* has cotype 2.

5. Show that every reflexive subspace of $C^1(\mathbb{T}^2)^*$ embeds into some L_p space for some $p > 1$.

6. (a) Define an operator $T: H_1(\mathbb{T}) \to H_\infty(\mathbb{T})$ by $T(\sum_{n=0}^{\infty} a_n z^n) = \sum_{n=0}^{\infty} \frac{a_n}{n+1} z^n$. Show that T is continuous and does not have an extension to $\widetilde{T}: L_1(\mathbb{T}) \to H_\infty(\mathbb{T})$.

   (b) Show that, if $X \subset L_1(\mathbb{T})$ is a non-reflexive subspace, then there exists an operator $T: X \to H_\infty(\mathbb{T})$ which does not have an extension to $\widetilde{T}: L_1(\mathbb{T}) \to H_\infty(\mathbb{T})$.

   (c) Show that, if $X \subset L_1/H_1$ is a non-reflexive subspace, then there exists an operator $T: X \to H_\infty(\mathbb{T})$ which does not have an extension to $\widetilde{T}: L_1/H_1 \to H_\infty(\mathbb{T})$.

7. Show that Theorem 25 holds with the space L_1/H_1 replaced by the space $L_1(\Omega, \mu, L_1/H_1)$ of Bochner integrable (L_1/H_1)-valued functions (see III.B.28 for definitions).

8. Let R be a reflexive subspace of $L_1(\Omega, \mu)$ and let $T: R \to H_\infty(\mathbb{T})$. Show that T extends to an operator $\widetilde{T}: L_1(\Omega, \mu) \to H_\infty(\mathbb{T})$.

9. (a) Suppose that $(f_j)_{j=0}^\infty \subset L_1(\mathbb{T})$ and define an operator T from $H_\infty(\mathbb{T}^2)$ into ℓ_1 by

$$T(g) = (\langle g_j, f_j \rangle)_{j=0}^\infty \quad \text{where} \quad g(z,w) = \sum_{j=0}^\infty g_j(z)w^j.$$

Show that T is bounded if and only if

$$\inf \left\{ \int_{\mathbb{T}} \left(\sum_{j=0}^\infty |f_j + h_j|^2 \right)^{\frac{1}{2}} : h_j \in H_1^0 \text{ for } j = 0, 1, 2, \ldots \right\} < \infty.$$

(b) Show that the matrix $(a_{n,m})$ $n, m \geq 0$ is a coefficient multiplier from $A(\mathbb{D}^2)$ into $\ell_1(\mathbf{N} \times \mathbf{N})$ if and only if $\sum_{n,m \geq 0} |a_{n,m}|^2 < \infty$.

(c) Suppose that $M = (m_{k_1,\ldots k_N})_{k_1,\ldots,k_N \geq 0}$ is a coefficient multiplier from $A(\mathbb{D}^N)$ into $\ell_1(\mathbf{N}^N)$. Show that there exists a constant C such that for every $K \in \mathbf{N}$

$$\left(\sum_{k_1,\ldots,k_N \leq K} |a_{k_1,\ldots,k_N}|^2 \right)^{\frac{1}{2}} \leq C\|M\|(N \log Nk)^{\frac{1}{2}}.$$

10. Let $\Lambda \subset \mathbf{Z}$ be $\{-2^k\}_{k=1}^\infty$. Let $X \subset C(\mathbb{T})$ be the span$\{e^{in\theta}\}_{n \in \Lambda \cup \mathbf{N}}$. Define an operator R on X by $R(f) = (\hat{f}(-2^k))_{k=1}^\infty$.

(a) Show that R maps X onto $\ell_2(\Lambda)$.

(b) Show that R is p-absolutely summing for $p < 1$.

(c) Show that there is no projection from X onto A.

(d) Note that this gives an alternative proof of Corollary III.F.35.

11. If R is a reflexive subspace of $L_1(\mathbb{T})$, then R is isomorphic to a subspace of L_1/H_1 and to a subspace of $H_1(\mathbb{T})$.

12. An n dimensional projection $P\colon A(\mathbb{T}) \to A(\mathbb{T})$ is called interpolating if there exist distinct points $(t_j)_{j=1}^n \subset \mathbb{T}$ such that $Pf(t_j) = f(t_j)$ for $j = 1, 2, \ldots, n$ and every $f \in A(\mathbb{T})$. Show that there does not exist any sequence of interpolating projections $P_n\colon A(\mathbb{T}) \to A(\mathbb{T})$ such that $P_n(f) \to f$ for every $f \in A(\mathbb{T})$.

13. Show that $A(\mathbb{D})$ is not isomorphic to $A(\mathbb{D}^2)$.

14. Show that the ball algebra $A(\mathbb{B}_d)$ for $d > 1$ is not isomorphic to the disc algebra $A(\mathbb{B}_1)$.

Hints For The Exercises

II.A. 1. If there is a metric, the balls have to be unbounded in norm. This would give a weakly null sequence which is not norm-bounded. 2. Each basic neighbourhood $\mathcal{U}(0; \varepsilon, x_1^*, \ldots, x_n^*)$ restricts only countably many coordinates from Γ. This implies that if $\mathcal{U}_1 \supset \mathcal{U}_2 \supset \cdots$ are weakly open sets in $B_{\ell_2(\Gamma)}$ and $0 \in \mathcal{U}_j$ for $j = 1, 2, \ldots$ then $\bigcap_{j=1}^{\infty} \mathcal{U}_j$ has continuum cardinality. 3. Use the Riesz representation theorem I.B.11 and the dominated convergence theorem. 4. Take $X = c_0$ and $(x_n^*)_{n=1}^{\infty}$ the unit vectors in $\ell_1 = c_0^*$. $X = \ell_1$ and $x_n^* \in \ell_\infty, x_n^* = \sum_{j=n}^{\infty} e_j$ also works. 5. Both topologies are metrizable so it is enough to check the convergence of sequences. Work with Taylor coefficients. 6. Use Exercise 5. Note also that T is 1-1 so $T^{-1}(B_{A(\mathbf{D})})$ does not contain a line, so it is not $\sigma(H_\infty, L_1/H_1)$-open. 7. Show that the products are positive and have integral 1. Use the Fourier coefficients to show that the cluster point is unique. 8. This is basically the same as Exercise 7. 9. First show that for $\gamma \in \Gamma$ we have $T\gamma = a_\gamma \gamma$. Take as μ the w^*-limit of $T(g_n)$ where $g_n = \|\chi_{A_n}\|_1^{-1} \chi_{A_n}$ where A_n are neighbourhoods of the neutral element in G and $|A_n| \to 0$. 10. Put $f_r(e^{i\theta}) = f(re^{i\theta})$ and note that functions f_r are uniformly bounded in $L_1(\mathbf{T})$. Take the $\sigma(M(\mathbf{T}), C(\mathbf{T}))$-limit.

II.B. 1. See Proposition 3. 2. (a) Think what it means in terms of the unit balls. The proof is in Pełczyński [1960]. (b) Note that if $\|x + y\|_2 = \|x\|_2 + \|y\|_2$ then $x = \lambda y$ for $\lambda \geq 0$. (c) Show that $||| \cdot |||$ of (b) is strictly convex. (d) Consider the subspace $\ell^0 = \{x \in \ell_\infty(\Gamma): \text{card}\{\gamma: x(\gamma) \neq 0\} \leq \aleph_0\}$. For $x, y \in \ell^0$ such that $\|x\|_\infty = \|y\|_\infty = 1$ write $x < y$ if $y(\gamma) = x(\gamma)$ for all $\gamma \in \Gamma$ such that $x(\gamma) \neq 0$. Put $F_x = \{y \in \ell^0: x < y\}$. Let $||| \cdot |||$ be any equivalent norm on ℓ^0. Put $m_x = \inf\{|||y|||: x < y\}$ and $M_x = \sup\{|||y|||: x < y\}$. Use transfinite induction to get $z \in \ell^0, \|z\|_\infty = 1$ such that for all $y > z$ we have $m_y = m_z$ and $M_y = M_z$. Then $m_z = M_z$. This is only a glimpse of renorming theory. For detailed exposition the reader can consult Diestel [1975]. 3. Look at the partial sum projections. Consider $f_n = \sum_{j=1}^{n} e_j$ in c_0. 4. Examine how in the proof of Corollary 18 we used the assumption that (x_n) is weakly null. 5. If the sum $X + Y$ is not closed the map $(x + y) \mapsto x$ is unbounded so there are $x \in X$ and $y \in Y$ with $\|x\| = \|y\| = 1$ and $\|x - y\|$ arbitrarily small. 6. (a) Approximate the Faber-Schauder system. (b) Approximate the Haar functions, the n-th function in L_{p_n} where $p_n \nearrow \infty$. Examine what happens if $\delta \geq 1$ in Proposition 15. **7.** (a) The Haar system is basic in $L_\infty[0,1]$. (b) Interpret Haar functions as functions on Δ. **8.** Find the coefficient functionals explicitly. This shows $|a_n| < cn^{-\alpha}$ for $f \in Lip_\alpha[0,1]$. Conversely write $f = \sum_{n=1}^\infty a_n\varphi_n = \sum_{k=1}^\infty f_k$ where $f_k = \sum_{n=2^k+1}^{2^{k+1}} a_n\varphi_n$ and for t,s such that $|t-s| \sim 2^{-N}$ estimate separately $\sum_{k=1}^N f_k$ and $\sum_N^\infty f_k$. Compare with III.D.27. **9.** (a) Permute the trigonometric system and use the Riesz projection. This was shown by Boas [1955]. (b) The derivative maps $Lip_1[0,1]$ onto $L_\infty[0,1]$ and is almost an isomorphism. (c), (d) Look at (b). **10.** (a) Show that S contains a closed subset homeomorphic to Δ. This is done in Kuratowski [1968] III§36.V and Lacey [1974]. (b) Compare the dual spaces. **11.** Use the decompositoin method. For (a) represent $(\Sigma C[0,1])_0$ as a subspace of $C[0,1]$. **12.** Note that for $1 \leq p < \infty, p \neq 2$ two functions f, g such that $\|f + \alpha g\|_p = (\|f\|_p^p + \|g\|_p^p)^{\frac{1}{p}}$ for all scalars α with $|\alpha| = 1$ have to be disjointly supported. **13.** Use Theorem 4 and Exercise II.A.3. **14.** Start first with finite Σ_1. For existence in the general case use the Radon-Nikodym theorem. Such projections are called conditional expectations, and are of fundamental importance in probability theory. They are studied in almost every introductory book on probability. **15.** (a) Take any countable dense set in B_X and map the unit vectors onto this set. (b) $\ell_1(\Gamma)$ is a subspace of $C[0,1]^*$. Use (a). (c) On ℓ_∞ there exists a sequence of functionals $(x_n^*)_{n=1}^\infty$ such that if $x_n^*(x) = 0$ for $n = 1, 2, \ldots$ then $x = 0$. **16.** Try to repeat the proof of Proposition 6.

II.C.   **1.** Define $P(x^{***}) = x^{***}|i(X) \in X^*$. **2.** Use the dominated convergence theorem. Construct Rademacher-like functions. **3.** Find a sequence $(x_n)_{n=1}^\infty \subset X$ with $\|x_n\| = 1$ for $n = 1, 2, \ldots$ such that $(Tx_n)_{n=1}^\infty$ is a basic sequence in Y and there is an $y^* \in Y^*$ such that $y^*(Tx_n) \geq \delta > 0$ for $n = 1, 2, \ldots$. This was proved in Lindenstrauss-Pełczyński [1968]. **4.** (c)$\Rightarrow$(b) follows from the Fejér theorem I.B.16 and for (a)$\Rightarrow$(c) consider $f_\varepsilon = (2\varepsilon)^{-1}\chi_{(-\varepsilon,\varepsilon)}$ and show that ω^*-$\lim_{\varepsilon\to 0} T_\mu(f_\varepsilon) \in L_1(\mathbb{T})$. **5.** Note that $T_K: L_2[0,1] \to L_\infty[0,1]$. To see that T_{K_0} is not weakly compact look at the images of the Haar functions. **6.** Suppose $|\hat\mu(n_j)| > \varepsilon$ for $n_j \to \infty$. Let μ_∞ be the ω^*-cluster point of $\{e^{in_j\theta}\mu\}_{j=1}^\infty$ and let ν be the ω^*-cluster point of $\{e^{-in_j\theta}\mathcal{D}_{n_j} * \mu\}_{j=1}^\infty$. The F.-M. Riesz theorem yields ν absolutely continuous. Also $\hat\mu_\infty(n) = \hat\nu(n)$ for $n \leq 0$, so μ_∞ is also absolutely continuous. On the other hand writing $\mu = f dt + \mu_s$ one checks that μ_∞ is singular. This is a result of Helson [1954]. Compare with Exercise 7. **7.** If not, take $n(p)$ and $m(p)$ in $\mathbb{N}$ tending to ∞ with p so that $\hat\mu(n(p) - j) = \hat\mu(m(p) - j)$ for $j = 1, 2, \ldots, p$ but $\hat\mu(n(p)) \neq \hat\mu(m(p))$.

Write $\mu = f dt + \mu_s$ and look at $\{(e^{-in(p)\theta} - e^{-im(p)\theta})\mu_s\}_{p=1}^{\infty} \subset L_1(|\mu_s|)$. The weak cluster point μ_∞ exists, belongs to $L_1(|\mu_s|)$ and is not zero. On the other hand F.-M. Riesz Theorem gives that μ_∞ is absolutely continuous. This is from Helson [1955]. Compare with Exercise 6. **8.** (a) The very definition of a shrinking basis gives that $(x_n^*)_{n=1}^{\infty}$ is a basis in X^*. It is boundedly complete by the Alaoglu theorem. (b) Boundedly completeness give the *weak compactness of the unit ball. (c) Put together (a) and (b). (d) By (a) $(x_n^*)_{n=1}^{\infty}$ is a basis in X^* so every $x^{**} \in X^{**}$ can be identified with a sequence of scalars. This is old and well known to specialists (see Lindenstrauss-Tzafriri [1977]). Some parts are already in Karlin [1948]. **9.** (a) Cauchy sequences in $\| \cdot \|_J$ are coordinatewise Cauchy. (b) Consider vectors $(1, \ldots, 1, 0, 0, \ldots)$. (c) Show that if $n_1 < m_1 < n_2 < m_2 < \cdots$ and $x_k = \sum_{j=n_k}^{m_k} \alpha_j e_j$ then $\|\sum_{k=1}^{N} x_k\|_J \leq (\sum_{k=1}^{N} \|x_k\|_J^2)^{\frac{1}{2}}$. (d) Use Exercise 8 (d). (e) Look at the isomorphism between c_0 and c given in II.B.2(a). (f) The number $\dim(J^{**}/J)$ is an isomorphic invariant. All this except (f) can be found in James [1950]. (f) is due to Bessaga-Pełczyński [1960a]. **10.** This and much more can be found in Davis, Figiel, Johnson, Pełczyński [1974].

II.D. 1. Use the form of the partial sum projection as given in the proof of II.B.10. **2.** Look at $\sum_{k=0}^{\infty} \sum_{2^k+1}^{2^{k+1}} a_n \varphi_n$. **3.** Note that the series is weakly unconditionally convergent and use Proposition 5. **4.** Use Theorem 13 and Theorem 6. **5.** If $T: C(K) \to \ell_1$ is not compact then $\sum_{n=1}^{\infty} T^*(e_n)$ is a weakly unconditionally convergent but not unconditionally convergent series in $M(K)$. Proposition 5 and Theorem 6 lead to a contradiction. **6.** Consider the Orlicz property. **7.** (a) It is enough to consider Hilbert space. It is possible to prove it by induction on n. (b) Show that if $x \in X$ is a limit of some subsequence of partial sums of the series $\sum_{n=1}^{\infty} x_n$ then $x \in \mathcal{U}(x_n)$. Both (a) and (b) are due to Steinitz [1913]. (c) Take functions $\{\pm 2^{-\frac{n}{2}}|h_n|\}_{n=0}^{\infty}$. Show that they all can be ordered into a series whose sum is 0 and into another series whose sum is 1. Since each function takes only values 0 and 1 every sum will take integer values. (d) The example is in Kadec-Woźniakowski [P]. It is a bit too complicated to repeat it here. **8.** (a) This is almost obvious. (b) Show that for every N there is a measure preserving transformation of [0,1] which transforms $\{g_{n,k}\}$, $n = 1, \ldots, N, k = 1, \ldots, 2^n$ onto the first Haar functions. (c) Like in the proof of Theorem 10 produce a block-basic sequence as in (b). (d) Find blocks of $(\varphi_n)_{n=1}^{\infty}$ behaving like those considered $(g_{n,k})$ in (b). This basically reduces the problem to the Haar system. This exercise shows the fundamental role played by the Haar system in the study of $L_p[0,1]$-spaces and in theory of orthonormal series. (d) is a result of Olevskii (see Olevskii [1975] p. 75). A Banach space theoretical version of these phenomena is presented in Lindenstrauss-Pełczyński [1971]. **9.** (a) Look at the formulas and compute carefully. (b) Apply to the d_k's the procedure applied to the Haar functions in order to get the $g_{n,k}$'s of Exercise 8(b). Note that the conclusions of (a) hold. Represent it as blocks of the Haar system. (c) This follows directly from Exercise 8(c). All this is due to Burkholder [1982] and [1984].

II.E. **1.** (a) This is just reformulation of the definition. (b) Take $x^{**} \in X^{**}, \|x^{**}\| = 1$ and a net $(x_\gamma)_{\gamma \in \Gamma} \subset B_X$ tending to x^{**} in the $\sigma(X^{**}, X^*)$-topology. Use (a) to show that this net converges in norm. This is a classical result of D.P. Milman. This proof is due to Ringrose [1959]. (c) One has to show the Clarkson [1936] inequalities $\left\|\frac{(u+v)}{2}\right\|_p^p + \left\|\frac{(u-v)}{2}\right\|_p^p \leq \frac{1}{2}(\|u\|_p^p + \|v\|_p^p)$ for $2 \leq p < \infty$ and $\left\|\frac{(u+v)}{2}\right\|_p^{p'} + \left\|\frac{(u-v)}{2}\right\|_p^{p'} \leq \left(\frac{1}{2}\|u\|_p^p + \frac{1}{2}\|v\|_p^p\right)^{p'-1}$ for $1 < p \leq 2$. The first one is the integration of the corresponding numerical inequality while the second follows from the appropriate numerical inequality and the inequality $\|\,|u| + |v|\,\|_q \geq \|u\|_q + \|v\|_q$ valid for $q \leq 1$. We apply it to $q = p - 1$. (d) Observe that if $\|z + v\| \leq 1$ and $\|z - v\| \leq 1$ then $\|z\| \leq 1 - \varphi(\|v\|)$. Apply this observation inductively to the finite sums. This is due to Kadec [1956]. (e) This is quite obvious. Use (b) to show the existence of the best approximation. **2.** (a) Replace max by the average in the definition and estimate from below $\int_0^{2\pi} |1 + be^{i\theta}| d\theta$. (b) Use the ideas of the proof of Exercise 1.d. The notion of complex uniform convexity was first studied by Globevnik [1975]. **3.** Take $(p_n)_{n=1}^\infty$ and $(q_n)_{n=1}^\infty$ two disjoint sequences, dense in $[1, 1.5]$ such that $p_1 = 1$. Take $X = \left(\sum_{n=1}^\infty \ell_{p_n}^5\right)_2$ and $Y = \left(\sum_{n=1}^\infty \ell_{q_n}^5\right)_2$. Show that Y does not contain ℓ_1^5 isometrically. **4.** (a) Adapt the proof of Theorem 9. This is a correct estimate (see III.B.22). (b) Consider everything on $[-\pi, \pi]$. Look at the translation of the square of the Dirichlet kernel. If $\mathcal{D}_r(t) = \sin(r+2)t/2\sin\frac{1}{2}t$ then $\mathcal{D}_r(0) = r + \frac{1}{2}$ and $\mathcal{D}_r(\theta_s) = 0$ for $\theta_s = \frac{2s\pi}{(2r+1)}, s = \pm 1, \ldots, \pm r$. Put $f_{r,s}(t) = [\mathcal{D}_r(t - \theta_s)]^2, s = 0, \pm 1, \ldots, \pm r$. Then $\sum_{s=-r}^r f_{r,s} = (r + \frac{1}{2})^2$ and this helps to show that $\{(r + \frac{1}{2})^{-2} f_{r,s}\}_{s=-r}^r$ is isometrically equivalent to the unit vector basis in ℓ_∞^{2r+1}. This is a classical interpolation problem (see Natanson [1949]). **5.** Apply Theorem 9 twice. **6.** Every open ball contains infinitely many disjoint balls of equal radii. **7.** This is a compactness argument. For each $\alpha \in \Gamma$ and $x^* \in X^*$ define a function on Y by the formula

$$\varphi_\alpha(x^*)(y) = \begin{cases} 0 & \text{if } y \notin Y_\alpha, \\ x^*(S_\alpha(y)) & \text{if } y \in Y_\alpha. \end{cases}$$

Taking a pointwise cluster point we find $\varphi\colon X^* \to Y^*$ which is bounded and linear and $T^*\varphi$ is a projection onto $T^*(Y^*)$. **8.** (a) Note that $\left(\sum_{n=1}^\infty \ell_2^n\right)_\infty = \left(\sum_{n=1}^\infty \ell_2^n\right)_1^*$ and use Exercise 7. (b) If $d(X,Y)$ is small one can represent X and Y as norms on $\mathbb{R}^n$ (or $\mathbb{C}^n$) such that the unit balls are close, so the norms are close as functions on $\mathbb{R}^n$ (or $\mathbb{C}^n$). Now we see that the limit exists, so we have completeness. For total boundedness use the Auerbach Lemma. (c) Use (b) and Exercise 7. This and Exercise 7 can be found in Johnson [1972]. Exercise 7 is an improvement of an earlier result of C. Stegall. **9.** (a) On each finite dimensional subspace of X we have a uniformly convex norm, uniformly close to the original. Use a compactness argument. (b) Similar to (a). **10.** Show that if $E \subset L_p[0,1]$ is finite dimensional, $1 \le p \le \infty$, then there exists $F \subset L_p[0,1], F \cong \ell_p^n$ such that E is close to a subspace of F, where both n and 'closeness' are controlled. The case $p = \infty$ is relatively easy (use Lemma 11 or Proposition 10). For the case $p < \infty$ show that $f(t) = \sup\{|x(t)|\colon x \in E, \|x\| \le 1\}$ is in $L_p[0,1]$ and consider $g(t) = \max(f(t),1)$. Take the isometry $I\colon L_p[0,1] \to L_p([0,1], g^p dt)$ defined by $Ih = h \cdot g^{-1}$. Note that $IB_E \subset B_{L_\infty([0,1],g^p dt)}$. Now we can follow the case $p = \infty$. This is taken from Pełczyński-Rosenthal [1975]. **11.** Take the quotient map from ℓ_1 onto ℓ_2^n (see Exercise II.B.15 (a) and dualize. Or use II.B.4. Show that a finite dimensional subspace of c_0 is also a subspace of ℓ_∞^N (for some N), so has a finite number of extreme points.

III.A. **1.** Note that if $f \le g$ and $f \ne g$ in $L_p(\mu)$ then $\int |f-g|^p d\mu > 0$. Since everything is below g we can reach the max in a countable number of steps. **2.** Estimate $\int |x|^q$ using the Hölder inequality. **3.** Assume $L_p(\mu) = L_p[0,1]$ and find in Y a block basis $(y_n)_{n=1}^\infty$ of the Haar system equivalent to the unit vector basis in ℓ_p. Take $(y_n^*)_{n=1}^\infty$, the sequence of biorthogonal functionals such that $y_n^*(y_n) \ge c_p\|y_n^*\|\|y_n\|$ and such that y_n^* are in the same block of the Haar system as y_n for $n = 1, 2, \dots$. The projection $P(f) = \sum_{n=1}^\infty y_n^*(f)y_n$ works. **4.** Consider sets $M_\varepsilon = \{f \in L_p[0,1]\colon \|f\|_p \le \varepsilon\|f\|_2\}$ for $\varepsilon > 0$. If $X \subset M_\varepsilon$ for some $\varepsilon > 0$ then $X \sim \ell_2$ and is complemented. If $f \in M_\varepsilon$ then there exists a set $A \subset [0,1]$ with $|A| < \varepsilon$ and $\|f \cdot \chi_A\|_p \ge (1-\varepsilon^p)^{\frac{1}{p}}\|f\|_p$. From this, if X is not in any M_ε we can find a sequence in X close to the sequence of disjointly supported

328   Hints For The Exercises

functions. This is a result of Kadec-Pełczyński [1962]. **5.** Each norm-1
projection in $L_p(\mu)$ is a conditional expectation projection (for definition
see Exercise II.B.14). To see this is a rather tedious process. We check
that if $f \in ImP$ and supp $g \subset$ supp f then supp $Pg \subset$ supp f. We
also check that $P(h \text{ sgn } f) = |P(h \text{ sgn } f)| \text{ sgn } f$ for $f \in ImP$ and
$h \geq 0$. The details and references are in Lacey [1974]. **6.** Use the finite
dimensional version of Proposition 7 and Theorem 6. **7.** Use Exercise 6.
8. If $P_1, \ldots, P_n \in \mathcal{P}$, $Q_1, \ldots, Q_n \in \mathcal{Q}$ with $P_k P_j = 0$ and $Q_k Q_j = 0$ for
$k \neq j$ then the norm of $\sum_{j,k} P_k Q_j$ can be estimated by twice applying
the Khintchine inequality. Note that $\sum_{j=1}^{n} \pm P_j$ is uniformly bounded.
This is due to McCarthy [1967]. **9.** From the Khintchine inequality
we get $\| \sum_{n=1}^{\infty} a_n f_n \|_p \sim (\int_0^1 (\sum_{n=1}^{\infty} |a_n|^2 r^{2^{k+1}})^{\frac{p}{2}} r dr)^{\frac{1}{p}}$. To estimate it
from above we use the $\frac{p}{2}$ convexity of the $L_{p/2}$ norm (for $p \leq 2$) or
the Hölder inequality (for $p \geq 2$). To estimate from below we replace
$r^{2^{n+1}}$ by $r^{2^{n+1}} \cdot \chi_{E_n}$ where $E_n = \{t : 1 - 2^{-n} \leq t \leq 1 - 2^{-n-1}\}$. **10.**
Follow the proof of Theorem 8 $(P = P_0)$. For (a) note that P is
a selfadjoint (and so orthogonal) projection. This is classical, due to
Bergman. A similar exposition on $\mathbb{B}_d$ can be found in Rudin [1980].
11. The operator $T_g : B_p(\mathbb{D}) \to B_p(\mathbb{D})$ is compact (see Exercise 16).
Consider the spectrum of T_g. This is due to Axler [1985]. **12.** Start
with $n = 2$ and write P explicitly, then use the multiplier theorem I.B.32.
13. Apply Proposition 9. **14.** Apply Proposition 9c with $y(x) = x^\alpha$
for right α. **15.** Use Proposition II.B.17 (or see Exercise II.B.4) or its
modifications for $p < 1$ to show that the existence of a non-compact
operator $T : \ell_p \to \ell_q$ implies that $id : \ell_p \to \ell_q$ is bounded. **16.** All except
(e) are variants of Theorem 25 and can be found in Wojtaszczyk [1988].
For (a), (b), (d) repeat the proof. For (c) apply (a) inductively. For
(e) take a system such that $\sum_{n=1}^{\infty} \int |\varphi_n| < \infty$ (e.g. a subsequence of
the Haar system. **17.** Apply definitions. **18.** Use Remark 20. **19.**
(a) is an example of Schreier [1930]. The original construction requires
some familiarity with ordinal numbers. The other way is to invent any
Banach space with the sequence violating (a) and use Theorem II.B.4.
One such example is to define $\|(x_j)_{j=1}^{\infty}\| = \sup\{\sum_{k=1}^{n} |x_{j_k}| : n = j_1 <
j_2 < \cdots < j_n\}$. For (b) use (a) and apply Theorem II.C.5 to the
operator $T : \ell_1 \to C[0,1]$ given by $T(e_n) = f_n$, $n = 1, 2, \ldots$.

III.B. **1.** Show that $L_\infty[0,1]$ embeds into ℓ_∞ (see Exercise II.B.15(a),
use Theorem 2 and the decomposition method. This is due to Pełczyński
[1958]. **2.** span$\{z^k \omega^{n-k} : k = 0, 1, \ldots, n\}$ is such a subspace. **3.**
Identify t^k with $\cos^k \theta \in T_n^\infty$, $k = 0, 1, \ldots, n$. This is a classical device; see Natanson [1949]. **4.** Compute the relative projec-

tion constants of subspaces of polynomials of degree at most n. **5.**
(a) Use estimates for $d(\ell_p^n, \ell_2^n)$ and $d(\ell_p^n, \ell_\infty^n)$. (b) Dualize. **6.** Let
$\Delta_n = \{-1,1\}^n$. Note that ℓ_1^n is isometric to the span$\{r_j\}_{j=1}^n \subset C(\Delta_n)$,
where $r_j(\varepsilon_1, \ldots, \varepsilon_n) = \varepsilon_j$. Observe that this can be identified with
the span of the first n Rademacher functions in $C(\Delta)$. Apply Theo-
rem 13. Projection constants of ℓ_p^n spaces can be found in Theorem
VII.1.9 of Tomczak-Jaegermann [1989]. **7.** Identify Y with c_0 and put
$(x_n^*)_{n=1}^\infty$ the norm-preserving extension of coordinate functionals. Find
$z_n^* \in X^* \cap Y^\perp$ such that $x_n^* - z_n^* \to 0$ in $\sigma(X^*, X)$-topology. Put
$P(x) = (x_n^*(x) - z_n^*(x))_{n=1}^\infty$. This fact is due to Sobczyk [1941] and
the proof indicated here to Veech [1971]. **8.** (a) Use the following set-
theoretical result due to W. Sierpiński. If N is a countable set then there
exists a family $\{A_\gamma\}_{\gamma \in [0,1]}$ of infinite subsets of N such that $A_{\gamma_1} \cap A_{\gamma_2}$
is finite for all $\gamma_1 \neq \gamma_2$. For the proof of this identify N with the set
of rationals in [0,1] and put A_γ any sequence of rationals tending to γ.
(b) Use Exercise II.B.15c. This is a classical result of Phillips [1940].
The argument indicated here is taken from Whitley [1966]. (c) Suppose
P is a projection onto X and $i: c_0 \xrightarrow{\text{onto}} X$ is an isomorphism. Extend i
to $j: \ell_\infty \to \ell_\infty$ and show that $i^{-1}Pj$ is a projection onto c_0. **9.** Start
the induction in the proof of Theorem 21 with the polynomial p. **10.**
Show Proposition 19 for φ being a lower semi-continuous function on
$\overline{\mathbb{B}}_d$ and with the inequality in (a) holding on $\overline{\mathbb{B}}_d$. This can be found
in Rudin [1986]. **11.** Take $f \in H_1(\mathbb{B}_d)$, $f = \sum_{n=0}^\infty f_n$ where f_n is a
homogeneous polynomial of degree n. First note that $Rf = \sum_{n=0}^\infty n f_n$
and next show that $\sum_{n=1}^\infty n^{-d} \|f_n\|_\infty \leq \|f\|_1$. For this use the Hardy
inequality on one-dimensional complex subspaces of $\mathbb{C}^d$ and estimate the
ratio between $\|f_n\|_1$ and $\|f_n\|_\infty$ like in the proof of Proposition 18. This
is taken from Ahern-Bruna [1988]. **12.** Dualise and use the weak type
(1-1) of the Cauchy projection. For details see Wojtaszczyk [1982]. **13.**
(a) This is a direct calculation. (b) Use functions from (a). This can
be verified by the direct calculation or by appeal to Corollary III.H.16.
14. Use the ideas from Exercise 13 and the polynomials constructed in
Proposition 18. Better results can be found in Ullrich [1988a].

III.C.   **1.** Reduce to the case $T: c_0 \to X, \|Te_n\| \leq n^{-2}$. This is
not a semi-embedding because $\sum_{n=1}^\infty Te_n \in X$. This can be found in
Bourgain-Rosenthal [1983]. **2.** For each t consider $(2\varepsilon)^{-1}\chi_{[t-\varepsilon, t+\varepsilon]}$ and
let $x_t^* \in X^*$ be a $\sigma(X^*, X)$-cluster point. Show that there are uncount-
ably many t's so that x_t^* are far apart. This is a classical result of
Gelfand. Modern generalizations can be found in Diestel-Uhl [1977]. **3.**
Take two sequences convergent to different limits. The desired sequence

consists of long stretches of one sequence separated by long stretches of the other. **4.** If it is not so, build in H the unit vector basis in ℓ_1. **5.** Observe that $\operatorname{span} T(L_1(\Omega,\mu)) = \operatorname{span} T(L_2(\Omega,\mu))$. **6.** Simply a uniformly integrable sequence convergent in measure converges in norm. **7.** One possible candidate is $\{f \in L_1[0,1] : \sum_n \left(\int_{\frac{1}{(n+1)}}^{\frac{1}{n}} |f(t)|^{1+\frac{1}{n}} dt\right)^{\frac{2n}{(n+1)}} \leq 1\}$.
8. Factor $f_n = B_n F_n$ and show that a subsequence of B_n and F_n converge weakly. Consider $\sqrt{F_n}$ in $H_2(\mathbb{T})$. This is due to Newman [1963].
9. Modify the proof of Lemma 15. This is a result of James [1964]. **10.** If ℓ_1 were finitely representable in X^* then the ℓ_∞^n's would be uniform quotients of X. But X has type $p, p > 1$ so by Exercise III.A.17 ℓ_∞^n would also have type p. **11.** Instead of characteristic functions use their smooth approximations. **12.** Find closed, disjoint sets $F_n \subset [0,1]$ and functions $h_n \in H$ such that $\inf_n \int_{F_n} |h_n(t)| dt > 0$. Find $f_n \in C[0,1]$ such that $\int_{F_n} |h_n(t)| dt \sim \int_{F_n} h_n(t) f_n(t) dt, \|f_n\|_\infty \leq 1$ and $f_n \mid F_k = 0$ for $k < n$. Put $\varphi_n = \prod_{k=1}^{n-1}(1 - |f_k|) f_n$. This is due to Pełczyński [1962]. More general results are in III.D. **13.** Use Lemma 10 and Proposition III.A.5.

III.D.   **1.** Write $T \in L(\ell_p)$ as $T(x) = \sum_{i=1}^{\infty} f_i^T(x) e_i$. For $F \in L(\ell_p)^*$ define $G \in L(\ell_p)^*$ by $G(T) = \sum_{i=1}^{\infty} F(T_i)$ where $T_i(x) = f_i^T(x) e_i$. Show $\|G\| = \|F\|$ and on compact operators G agrees with F. Show that $\|F\| = \|G\| + \|F - G\|$. This is due to Hennefeld [1973]. For $p = 2$ see Alfsen-Effros [1972]. **2.** For $h^* \in H^*$ define $Eh^* = \mu|S$ where μ is any measure on T which extends h^* to $C(T)$. Use the definitions to check that it makes sense. **3.** For $K = [0,1]$ the Faber-Schauder system shows this. For the general case use the same ideas. **4.** Show that $dist(f, I_\varphi(C[0,1])) = \frac{1}{2}\sup\{|f(s') - f(s'')| : \varphi(s') = \varphi(s'')\}$ so $C(\Delta)/I_\varphi(C[0,1]) \cong c_0$, with unit vectors corresponding to points $s' \neq s'' : \varphi(s') = \varphi(s'')$. If there is a projection we can lift these unit vectors to $f_n \in C(\Delta)$. On the other hand since those points are dense in Δ we can find a subsequence such that $\left\|\sum_{j=1}^{n} f_{n_j}\right\| \geq cn$. This was proved by M.I. Kadec. The proof is in Pełczyński [1968] §9. **5.** One example is: K_1 is a disjoint union of the interval [0,1] and the interval [0,1] with circle attached at each end. K_2 is the disjoint union of two intervals each with one circle attached at one of the ends. Check that it works. The details are in Cohen [1975]. **6.** For (a) and (b) reduce to the case of selfadjoint operators and write $\langle Ax, x \rangle$ explicitly. (c) follows from (b). More details can be found in Kwapień-Pełczyński [1970] and Bennett [1977]. **7.** If n_k is very lacunary then you can analyse sgn $e^{in_k\theta}$ and conclude that $(e^{in_k\theta})_{k=1}^{\infty}$ is in sup-norm equivalent to the unit vector basis. As a model think about Rademacher functions. A more efficient way is to use Riesz products (see Exercise II.A.7). **8.** Assume $\|T_n\| = 1$ and take $p(x)$ such that $p(x_0) = \|p\|$. Then $T_n^*(\delta_{x_0})(p) \to 1$ so $T_n^*(\delta_{x_0}) \to \delta_{x_0}$ in ω^*-topology. Note also that the mass of $T_n^*(\delta_{x_0})$ has to concentrate around x_0. From this get the convergence of $T_n(f)$ for smooth f's. This is an improvement of the original Korovkin theorem (see Korovkin [1959] or Wulbert [1968]). **9.** (a) Note that $\sum_{k=0}^n \binom{k}{n}(1-x)^{n-k}x^k = 1$. (b) Use the Korovkin theorem (Exercise 8) and compute that $B_n(1) = 1$, $B_n(x) = x$, $B_n(x^2) = x^2 + \frac{(x-x^2)}{n}$. This is a modern version of S.N. Bernstein's proof of the Weierstrass approximation theorem. **10.** Use the remark after III.A.12 to show that $X_s \sim \ell_\infty$ and that $(X_s^0)^{**} = X_s$. The fact that $X_s^0 \sim c_0$ is more involved. Analogously as in the proof of Theorem III.A.11 show that X_s^0 is isomorphic to a complemented subspace of c_0. **11.** If not then $R: A(\mathbb{B}_d) \to B_1(\mathbb{D}_d)$. Like in III.A.11 we show that $B_1(\mathbb{B}_d)$ is isomorphic to a subspace of ℓ_1. Thus (use DP and Pełczyński property) R is compact. But for the polynomials $p_n(z)$ constructed in III.B.18 we have $\|Rp_n\| \geq c > 0$ for all n. This contradicts the compactness. **12.** (a) The desired embedding of T_n^∞ into ℓ_∞^{4n+1} is given by $p \mapsto (p(\exp(2\pi k\theta/(4n+1)))_{k=0}^{4n}$ (see II.E.9). This is due to Marcinkiewicz [1937a] (see also Zygmund [1968] chapter X §7). (b) Take small δ and a maximal $\frac{\delta}{\sqrt{n}}$ separated subset of the unit sphere $\mathbf{S}_d \subset \mathbb{C}^d$ considered with the quasi-metric $\rho(\zeta, \eta) = 1 - |\langle \zeta, \eta \rangle|$. The embedding into ℓ_∞ is given via the point evaluations (see Wojtaszczyk [1986]). (c) Identify ℓ_∞^N with $L_\infty(\{1, \ldots, N\}, \mu)$ where $\mu(\{k\}) = \frac{1}{N}$ for $k = 1, 2, \ldots, N$. Take a maximal set $(x_j)_{j=1}^k \subset E$ such that

$$\|x_j\|_\infty = 1, \quad \|x_j\|_2 = \sqrt{\frac{2}{n}} \quad \text{and} \quad \left\| \sum_{j=1}^k |x_j| \right\|_\infty \leq 1 + \delta.$$

Then $\text{span}(x_j)_{j=1}^k = G$. To estimate k show that if $\left| \{ i: \sum_{j=1}^k |x_j(i)| > \delta \} \right| \leq \frac{1}{2}n$ then III.B.9 implies that $(x_j)_{j=1}^k$ is not maximal. This is Corollary 6.2 of Figiel-Johnson [1980]. **13.** Instead of Λ_j's consider the basic splines b_j, i.e. functions such that b_j is continuous and $b_j|(s_k, s_{k+1})$ is a quadratic polynomial for all k and $b_j(s_j) = 1$ and b_j is non-zero only in three of the intervals (s_k, s_{k+1}). (The numbers s_k are those defined in **20**.) Show that such b_j's exist and check their properties. Follow the proof of Proposition 21. **14.** (a) Note that $B \subset L_1[0,1]$. (b) Show that $\|f_n\|_B \leq C(n+1)^{-\frac{1}{2}}$. To do this expand f_n into the Haar series. The antiderivative of $\sum_{j=1}^n \langle f_n, h_j \rangle h_j$ is a piecewise linear function. Write it in terms of Λ_j's as defined in **20** and differentiate back. Conversely it is enough to estimate the Franklin coefficients of a special atom. Estimate separately small coefficients ($n \leq -\log_2 |I|$) and big ones. For details see Wojtaszczyk [1986a]. **15.** The general strategy is similar to the proof of Theorem 27. The details and generalizations can be found in Ciesielski [1975]. **16.** As a simple model show directly that the Haar system is not an unconditional basic sequence in $L_\infty[0,1]$. Next use the same idea to show that derivatives of the Franklin system are not an unconditional basic sequence in $L_\infty[0,1]$. **17.** Use Proposition 21b) to show that

$$f \mapsto \sup_N \Big| \sum_{n=0}^{N} \langle f, f_n \rangle f_n \Big|$$

is a weak type 1-1 map. This can be found in Ciesielski [1966]. **18.** (a) We can follow the proof for the Franklin function or perform a direct and rather explicit calculation. (b) The orthonormality follows easily from the definitions. The completeness follows from the fact that continuous functions are dense in $L_2(\mathbb{R})$. This and much more can be found in Strömberg [1983]. **19.** Suppose $x_n \xrightarrow{\omega} 0$ and y_n are bounded and such that $\|x_n\| = x_n(y_n)$. Find a weakly Cauchy subsequence y_{n_k} and consider $T: X \to c$ defined as $x \mapsto x(y_{n_k})$. **20.** Use the Dunford-Pettis property. **21.** One example is $\big(\sum_{n=1}^{\infty} \ell_2^n\big)_1$; use Exercise II.E.8. **22.** Modify implications (e)⇒(d) and (d)⇒(b) of Theorem 31. This requires the use of nets. The argument is in Bourgain [1984b]. **23.** Use the Pełczyński property of ℓ_∞ and Exercise III.B.7 to show that the existence of an operator that is not weakly compact would imply that ℓ_∞ has a complemented subspace Y isomorphic to c_0. This is impossible; see Exercise III.B.8(c). **24.** Use the Ascoli theorem. **25.** Identify $C^1(\mathbb{T}^2)$ with a subspace X of $C(\mathbb{T}^2, \ell_2^3)$ by $f \mapsto (f, \partial_1 f, \partial_2 f)$. Consider the annihilator $X^\perp$ of X, $X^\perp \subset M(\mathbb{T}^2, \ell_2^3)$, where $M(\mathbb{T}^2, \ell_2^3)$ is the space of measures with values in ℓ_2^3. Consider the space $G \subset M(\mathbb{T}^2, \ell_2^3)$ of all measures μ such that $\lim_{n \to \infty} \int f_n d\mu = 0$ for all sequences $f_n \in X$ such that if $f_n = (g_n, \partial_1 g_n, \partial_2 g_n)$ then $\partial_1 g_n$ and $\partial_2 g_n$ tend to zero pointwise on $\mathbb{T}^2$. Show that G is complemented in $M(\mathbb{T}^2, \ell_2^3)$ (use that it is a $C(\mathbb{T}^2)$ module). Show that $G/X^\perp$ separable. Also show that the kernel of a projection onto G is isomorphic to $M(\mathbb{T})$. The details can be found in Pełczyński [1989]. **26.** The closedness follows from Lemma 6 like in the proof of Corollary 7. To show that it is an algebra, use III.B.20 to show that for a Lipschitz function $\varphi \in C(\mathbf{S})$ and $h \in H_\infty(\mathbf{S})$ the function $\varphi \cdot h \in H_\infty + C$. This is from Rudin [1975]. **27.** The proof of closedness is similar to the case $n = 1$ (see Corollary 7). To show that $H_\infty(\mathbb{T}^n) + C(\mathbb{T}^n)$ is not an algebra for $n > 1$ take $f \in H_\infty(\mathbb{T}) \setminus A(\mathbb{T})$

and show that $\bar{z}_n f(z_1) \in L_\infty(\mathbb{T}^n)$ but is not in $H_\infty(\mathbb{T}^n) + C(\mathbb{T}^n)$. This is from Rudin [1975]. that one can assume they are both in $L_1(\mathbb{T})$. This implies that there is $\varphi \in H_\infty(\mathbb{T})$ such that both $\varphi\mu_1 \geq 0$ and $\varphi\mu_2 \geq 0$. Since for every $\psi \in H_\infty(\mathbb{T})$, $\int \psi\varphi d\mu_1 = \int \psi\varphi d\mu_2$ we infer $\mu_1 = \mu_2$. **3.** If you have a set of extensions that is not relatively weakly compact (it is enough to assume these extensions are in $L_1(\mathbb{T})$) then use Theorem III.C.12 and Lemma III.C.20 (or Exercise III.C.11) to produce a c_0-sequence showing that the original set of functionals was not relatively weakly compact. Use the methods of Theorem III.D.31 or Exercise III.C.12. **4.** (a) Use the Hardy inequality. (b) Easily follows from the fact that diagonal multiplication by $\frac{1}{\sqrt{n}}$ in ℓ_2 is not 2-absolutely summing (see III.G.12). To find an elementary example look at lacunary series. **5.** The isometry in both cases is given as $[f] \mapsto (\alpha_{ij})_{i,j\geq 0}$ with $\alpha_{ij} = \hat{f}(-(i+j))$. For $f \in L_\infty(\mathbb{T})$ consider the operator $H_f(g) = P(f \cdot g)$ where $g \in H_2(\mathbb{T})$ and P is an orthogonal projection from $L_2(\mathbb{T})$ onto $\overline{H_2(\mathbb{T})}$. To evaluate $\|H_f\|$ use the canonical factorization I.B.23. These are classical results of Nehari and Hartman (see Nikolskiĭ[1980]). **6.** (a) Use condition (b) of Theorem 4. These examples are due to Hayman and Newman and can be found in Hoffmann [1962]. For (b) take $(\lambda_n)_{n=1}^\infty$ such that $\overline{\{\lambda_n\}} \subset \overline{\mathbb{D}}$ is such that $\overline{\{\lambda_n\}} \cap \mathbb{T}$ has positive measure but is not $\mathbb{T}$. **7.** (a) For $F(z) = \frac{1}{2}(z + \frac{1}{z})$ we have $F(\mathbb{P}_r) = E_r$ is an ellipse, so $A(E) \cong A(\mathbb{D})$. The function F induces an isometric embedding of $A(E)$ into $A(\mathbb{P}_r)$ and the image is 1-complemented. (b) Consider the map $z \mapsto e^z$ from an appropriate strip onto $\mathbb{P}_r$. Consider $A(\mathbb{P}_r)$ as a space of functions on this strip. Use the ideas of the proof of Theorem 12. This still requires some effort. For details of (a) and (b) see Wolniewicz [1980]. (c) The exact computation of this norm is in Voskanjan [1973]. Consider a very thin annulus. **8.** (a) Use the canonical factorization, Theorem I.B.23. (b) Show that $\int_{-\pi}^{\pi} \left(\overline{\lim_{r\to 1}} \int_{-\pi}^{\pi} \log |\omega_\varphi(re^{it})| dt\right) d\varphi = 0$ and use (a). (c) Use (b) to approximate inner functions by Blaschke products. Use Exercise III.B.9 to approximate an arbitrary function by the inner functions. (d) From (c) follows that functions $f(z) = B(rz)$ with B a finite Blaschke product and $r < 1$ are dense in B_A. Represent explicitly $B_\alpha(rz) = \frac{(rz-\alpha)}{(1-\bar{\alpha}rz)}$ as a convex combination of Blaschke products. (e) Show that for a Möbius transformation $p(z) = (z - \lambda)(1 - \bar{\lambda}z)^{-1}$ we have $\|p(T)\| \leq 1$. Use (d) to show that it extends to any $f \in A$. (b) is a classical result of Frostman (see Koosis [1980] IV.9 or Garnett [1981]). (c) is even older, it goes back to Nevanlinna. (d) is a result of Fisher [1968]. (e) is a classical and important result of von Neumann [1951]. The proof we indicate here is from Drury [1983] and is close to the original. For more about this inequality see III.F.15. A different

proof is indicated in Exercise III.H.19. **9.** (a) Put $\text{Max}(f) = \{t \in$
$\mathbb{T}: |f| = \|f\|\}$. Show that $\{f \in ImP: |\text{Max}(f)| = 0\}$ is dense in E.
Find a sequence $(e_1, \ldots e_n) \subset E, (e_1^*, \ldots, e_n^*) \subset E^*, n = \dim E$ such
that $\text{Max}(e_j) = 0, j = 1, \ldots, n, \|e_j\| = \|e_j^*\| = \|e_j^*(e_j)\| = 1$ and the
matrix $(e_i^*(e_j))_{i,j=1}^n$ is non-singular. Then $f_i^* = e_i^* \circ P, \ i = 1, \ldots, n$,
span ImP^*. (b) If the disc algebra is a π_1-space then (a) and the F.-
M. Riesz theorem show that A locally looks like ℓ_∞^n but this is not the
case. This is taken from Wojtaszczyk [1979a] (see also Exercise III.I.12.).
10. (a) Use the $\sigma(L_\infty, L_1)$-compactness of the closed ball in H_∞. (b)
Regularize using the Poisson kernel. (c) Use duality to find $F \in H_1^0(\mathbb{T})$
such that $\|F\| = 1$ and $(2\pi)^{-1} \int_{\mathbb{T}} Ff = dist(f, H_\infty)$. This gives that for
any best approximation $g \in H_\infty$ to f we have $f - g = \frac{\overline{F}}{|F|}$. (d) Take $f =$
$\sum_{n=1}^\infty a_n \varphi_n$ where $\|\varphi_n\|_\infty = 1$ and supp $\varphi_n \subset (\frac{1}{(n+1)}, \frac{1}{n}) \subset (-\pi, \pi] = \mathbb{T}$
and $a_n \to 0$ slowly enough. All this is quite old. A nice presentation,
references and much more can be found in Garnett [1981].

III.F. **1.** Work with the definitions and the Hölder inequality. **2.**
Think of ℓ_1 as a span of Rademacher functions in $L_\infty[0,1]$ and ℓ_2 as
a span of Rademacher functions in $L_p[0,1]$, $1 \le p < \infty$. **3.** Consider
the Haar system. This gives (a) and can be used to get the lower estimate in (b). To get the upper estimate in (b) you can follow the
proof of III.H.24. **4.** Use the fact that $(\frac{1}{\sqrt{n+1}})_{n=1}^\infty \notin \ell_2$. **5.** Pietsch's
theorem shows that T must map some $L_1([0,1], \mu)$ into $C[0,1]$. What
can be said about μ? **6.** Use the Pietsch theorem and the fact that
$L_p[0,1]$ is not equal to any $L_q[0,1]$. **7.** (a) Use the factorization.
(b) First note that every $T \in I_p(X, Y)$ is compact (use the Dunford-
Pettis property). Next use the Pietsch theorem and arguments like in
Lemma III.A.12 to show that T is a sum of absolutely convergent series in $N_p(X, Y)$. This is due to Persson [1969]. **8.** Use Corollary 9.
9. (a) Since $T_\mu(L_1(m)) \subset L_1(m)$, we see that T_μ is 1-integral. Since
$\hat{\mu}(\gamma) \to 0, T_\mu$ is compact (look at $L_2(G)$). If T_μ is 1-nuclear, then the
definition yields (look also at T_μ^*) that $T_\mu(f)(x) = \int_G K(x,y)f(y)dm(y)$
for some $K \in L_1(G \times G, m \times m)$, but this is impossible for singular μ.
(b) Use Proposition 12. Show also that translation invariant, nuclear
operator on $C(\mathbb{T})$ is a limit in the nuclear norm of operators of convolution with a polynomial. Compare with III.G.18. **10.** Take $(x_j)_{j=1}^n \subset \ell_1^n$
such that $\sum_{j=1}^n |x^*(x_j)| \le C\|x^*\|$ and apply Theorem 14 to the matrix
$(x_j(i))_{i,j=1}^n$. **11.** Estimate $\int_{\mathbb{T}} \left(\sum_{k=1}^N e^{-i2^k\theta}\right) f(\theta) d\theta$ from above using
the Hölder inequality. The Khintchine inequality will yield the esti-
mate from below for $\|f\|_\infty$. The details are in Bourgain [1987]. **12.**

(a) Note that if $x = \sum_{n=1}^{\infty} a_n x_n$ then $\sum_{n=1}^{\infty} |a_n| \|T x_n\| \le C\|x\|$. (b) $id: C[0,1] \to L_1[0,1]$ does not factor through ℓ_1. (c) Let X be either L_p^F or C_F. Fix $x \in X, \xi \in X^*$ and $(\varepsilon_n)_{n \in F}, \varepsilon_n = \pm 1$. Define $A: L_2^F \to X$ by $A\big(\sum_{n \in F} a_n e^{in\theta}\big) = \sum_{n \in F} a_n \hat{x}(n) \varepsilon_n e^{in\theta}$ and $B: X \to L_2^F$ by $B(x) = \sum_{n \in F} \xi(e^{in\theta}) \hat{x}(n) e^{in\theta}$. Show that $\pi_1(A^*) \le C\|x\|$ and $\pi_1(B) \le C\|\xi\|$ and next $|\mathrm{tr}\, BA| \le \pi_1(A^*) \pi_1(B)$. This is a weak form of a result of Pisier [1978]. **13.** Show that if id_F is p-integral then the orthogonal projection from $L_p(\mathbb{T})$ onto L_p^F is bounded. Then use II.D.9.

III.G. **1.** Use Proposition 4 in one direction. For the other use the Schmidt decomposition. **2.** For selfadjoint $A_1, A_2, \ldots, A_n \in \sigma_1(\ell_2)$ such that $\sum_{j=1}^{n} \sigma_1(A_j)^2 = 1$ take selfadjoint $B_1, \ldots, B_n \in L(\ell_2)$ such that $\sum_{j=1}^{n} \|B_j\|^2 = 1 = \sum_{j=1}^{n} \mathrm{tr}(A_j B_j)$. For $k = 2^{k_1} + 2^{k_2} + \cdots + 2^{k_j}$ with $0 \le k_1 < k_2 < \cdots < k_j \le n$ we define $\Phi(t) = \sum_k B_{k_1} \ldots B_{k_j} r_{k_1}(t) \ldots r_{k_j}(t)$. Estimate $\|\Phi(t)\|$ and evaluate $\int_0^1 \mathrm{tr}\big(\Phi(t) \sum_{j=1}^{n} r_j(t) A_j\big) dt$. This is taken from Tomczak-Jaegermann [1974]. **3.** (a) is obvious. For (b) compare the σ_p-norm of an $n \times n$ unitary matrix all of whose entries have absolute value $\frac{1}{\sqrt{n}}$ with the σ_p-norm of the matrix all of whose entries are $\frac{1}{\sqrt{n}}$. (c) follows from Exercise III.A.8 and (b). **4.** Use 3(a) and gliding hump arguments. Note that if $u_j \in \sigma_\infty(\ell_2)$ are such that $P_{A_j} u_j P_{A_j} = u_j \ne 0$ for some disjoint sets $A_j \subset \mathbb{N}$, $j = 1, 2, \ldots$ then $\mathrm{span}(u_j)_{j=1}^{\infty} \sim c_0$ and is complemented. On the other hand if $P_B u_j P_{A_j} = u_j \ne 0$ for some finite set $B \subset \mathbb{N}$ and disjoint $A_j \subset \mathbb{N}$ then $\mathrm{span}(u_j)_{j=1}^{\infty} \sim \ell_2$ and is complemented. This is due to Holub [1973]. More results of this type are in Arazy-Lindenstrauss [1975]. **5.** Use the projections P_A of Exercise 3 and the decomposition method II.B.23. **6.** Consider the diagonal operators. If $A^1(\ell_\infty, \ell_1)$ had an equivalent norm, then every diagonal operator would be in $A^1(\ell_\infty, \ell_1)$. To see that this is not so, show that for $id_n: \ell_\infty^n \to \ell_1^n$ we have $a_k(id_n) \ge (n - k + 1)^{-1}$ for $k = 1, 2, \ldots, n$. **7.** Write $\lambda_n = \alpha_n \cdot \beta_n$ with $(\alpha_n) \in \ell_2$ and $(\beta_n) \in c_0$. Define $T(x, y) = (\alpha(y), \beta(x))$ where $\alpha(y) = (\alpha_n^2 y_n)_{n=1}^{\infty}$ and $\beta(x) = (\beta_n^2 x_n)_{n=1}^{\infty}$. This example is due to Kaiser-Retherford [1984]. **8.** If $K(x, y)$ is square integrable this follows from Proposition 13 or Theorem 19. The general operator differs from this case by one dimension. **9.** Show that T_K admits a factorization $L_{q'} \xrightarrow{\alpha} L_\infty \xrightarrow{\beta} L_{q'}$ where β is an operator of multiplication by a function. Use Theorem 19. **10.** (a) For the lower estimate take $e_j \otimes e_i$ and see that $\sum_{i,j} |x^*(e_j \otimes e_i)| = \sum_{i,j} |x^*(i,j)|$ where x^* is really an operator on ℓ_2^n with $n_1(x^*) \le 1$. Use Exercise 12. For the upper estimate note that $\sigma_2(u) \le cn \int_0^1 |\langle u(u_t), v_t \rangle| dt$ where $u_t = n^{\frac{1}{2}}(r_1(t), \ldots, r_n(t))$ and $v_t = n^{\frac{1}{2}}(r_{n+1}(t), \ldots, r_{2n}(t))$ where

$(r_j(t))_{j=1}^{2n}$ are Rademacher functions. (b) Let $\mathcal{U}$ be an $n \times n$ unitary matrix with $|u(i,j)| = \frac{1}{\sqrt{n}}$ and let I_t have diagonal $r_1(t), r_2(t), \ldots, r_n(t)$, zeros otherwise and let J_t have diagonal $r_{n+1}(t), \ldots, r_{2n}(t)$, zeros otherwise. Show that $\sigma_2(u) \leq 3\sqrt{n} \int_0^1 |tr(uI_t\mathcal{U}J_t)| dt$. c) Passing to the adjoint note that $\gamma_1(J_n) = n_\infty(id : \sigma_2(\ell_2^n) \to \sigma_\infty(\ell_2^n))$. Use the trace duality III.F.24. (d) Glue together the finite dimensional operators from (c). (e) This is almost the same as Exercise III.F.12a. (f) Use (e) and (c). All this is taken from Gordon-Lewis [1974]. **11.** Use III.G (13). **12.** Consider one-dimensional operators. **13.** (a) Factorize T as $C(K) \xrightarrow{id} L_2(\mu) \xrightarrow{id} L_1(\mu) \xrightarrow{\alpha} \ell_2$ and write α $id : L_2(\mu) \to \ell_2$ as $\sum_{n=1}^{\infty} \alpha_n$ with α_n of finite rank and $\sigma_2(\alpha_n) \leq 2^{-n}$. Show that $n_1(\alpha_n id) = i_1(\alpha_n id) \leq 2^{-n}$. (b) Consider the factorization $C(S) \xrightarrow{onto} C^1(\mathbb{T}^2) \xrightarrow{i} W_1^1(\mathbb{T}^2) \xrightarrow{j} L_2(\mathbb{T}^2)$ where i is the identity and j is also the identity (see I.B.31). Show that it is not nuclear and use (a) or show that it fails (c). (c) Given $V_g \subset L_1(\mu)$ and $T : L_1(\mu) \to L_2(\nu)$ consider $M_g : L_\infty(\mu) \to L_1(\mu), M_g(f) = g \cdot f$ and use the fact that TM_g is nuclear by (a). **14.** (a) First do it for ℓ_2^n. For general E factor $id : E \to E$ as $E \xrightarrow{\alpha} \ell_2^n \xrightarrow{\beta} E$ with $\pi_2(\alpha) = 1$ and $\|\beta\| = \pi_2(id)$. But then $\beta\alpha = id_{\ell_2^n}$. This is a result of Garling-Gordon [1971] but the proof indicated is due to Kwapień. It can be found in Pisier [1986]. (b) Use (a), (c) and (d). Use also III.F.8. **15.** Assume that one norm is given by the usual scalar product on $\mathbb{R}^n$ (or $\mathbb{C}^n$) and show that the other can be chosen to be $\langle x, y \rangle_2 = \sum_{j=1}^n \alpha_j x_j \bar{y}_j$ with $\alpha_j > 0$. Take X_1 spanned by an appropriate block basis. **16.** (a) Take x to be an extreme point in B_E. (b) Identify ℓ_∞^N with $L_2(N, \mu)$ where $N = \{1, 2, \ldots, N\}$ and μ is a probability counting measure. On E we have two Hilbertian norms, from $L_2(N, \mu)$ and the one given by the distance. Use (a) and Exercise 15. **17.** (a), (b) Use Exercise 16a.

III.H. 1. Use the well known fact (see Katznelson [1968], Zygmund [1968] etc.) that Mf is of weak type 1-1 but not continuous on $L_1[0,1]$. **2.** It is enough to work with finite dimensional spaces. Apply III.F.33 and dualize. This works fine for $q > 1$. The case $q = 1$ requires more care. This shows that Proposition 15 is actually an equivalence. Like Proposition 15 this is from Maurey [1974]. **3.** If i^* is p' summing use Pietsch theorem and dualize. This shows that i factors through L_p. Use Proposition 10 to show that this is impossible. This was observed by Kwapień [1970]. **4.** Use III.F.29 and Corollary 11. **5.** and **6.** Use Proposition 5. These can be found in Maurey [1974]. **7.** For $p < 1$ there exist positive stable random variables (where stable is understood in a more general sense than in III.A.14) for which III.A.16 holds. The construction can be found in Feller [1971] or Lukacs [1970] or in other books on probability theory. **8.** Modify the proof of Proposition 10. The details are in Pisier [1986a]. **9.** Use Proposition 16 and Exercise III.A.15. **10.** Show that if $(a_i)_{i=1}^\infty$ are numbers such that $c_n = \sum_{i=1}^n a_i$ converges then $\sum_{i=1}^n \left(\frac{i}{n}\right) a_i \to 0$ as $n \to \infty$. Apply this and the Menchoff-Rademacher theorem to $a_n = n^{-1} f_n(\omega)$. This is due to Banach [1919]. The argument indicated here was shown to me by Mr Wojciechowski. **11.** If $\sum_{n=1}^\infty |f_n| < \infty$ a.e. then the map $(\xi_n)_{n=1}^\infty \mapsto \sum_{n=1}^\infty \xi_n f_n$ is a continuous operator from ℓ_∞ into $L_0[0,1]$. Use Corollary 16. **12.** If every $f' \in N$ then the map $f \mapsto f'(e^{i\theta})$ would be well defined into $L_0(\mathbb{T})$, so would admit a factorization through $H_p(\mathbb{T}), p < 1$ (use Corollary 7). This is impossible. **13.** Note that a system $(f_n)_{n \geq 1}$ is a system of convergence in measure if and only if there exists $T: \ell_2 \to L_0[0,1]$, a continuous, linear operator such that $f_n = T(e_n)$. Use Proposition 5, Corollary 11 and the dilation theorem 19. This is due to Nikishin [1970]. **14.** Use Proposition 5. **15.** If so then the double Riesz projection $\sum_{n,m=-\infty}^{+\infty} a_{nm} e^{in\theta} e^{in\varphi} \mapsto \sum_{n,m \geq 0} a_{nm} e^{in\theta} e^{im\varphi}$ would be of weak type (1-1) (use Corollary 7). The Marcinkiewicz theorem (see I.B.7) gives a bound on the norm in $L_p(\mathbb{T}^2), p > 1$, which is false. **16.** Use Corollary 11 and show using the structure of the system (more precisely the ergodicity of measure preserving maps $t \mapsto nt$) that the multiplication operator equals the identity. Compute the Fourier coefficients of sgn sin x. This yields the coefficients of sgn sin nx. Use it to estimate the L_2 norm of $\sum_{n=1}^N$ sgn sin nx. **17.** Use Corollary 16. **18.** Use Corollary 16. **19.** Check the von Neumann inequality for unitary maps. **20.** (a) Use Theorem III.C.16 and Lemma III.C.15. (b) Put $V = (Y \oplus Z)_1/H$ where $H = \{(x, -Tx): x \in X\}$. (c) If $T: X \to \ell_2$ use (b) to extend T to $T_1: C(S) \to V$ and use (a) to show that V^* has some type > 1. Use Proposition 14 and Lemma III.F.37. (d) Use similar arguments to III.A.25. An analogue of Lemma III.A.26 follows from (c) and the Pietsch factorization theorem. **21.** This is like Theorem 30. For (b) observe that $B_p(\mathbb{D})^* = B_r(\mathbb{D})$ where $\frac{1}{p} + \frac{1}{r} = 1$, $2 \leq p < \infty$ (this follows from the boundedness of the Bergman projection in $L_p(\mathbb{D})$, $1 < p < \infty$; see Exercise III.A.10). We use Proposition 29 twice.

III.I. 1. By III.F.36 it is enough to show $\pi_2(T^*) < \infty$, so by duality (see III.F.27 and III.F.25) we have to show that αT^* is nuclear for $\alpha: \ell_2 \to L_1/H_1$, a 2-absolutely summing multiplier. Factorize α and use the fact that $\left(\sum_{n=0}^\infty |t_n|^2\right)^{\frac{1}{2}} \leq \|T\|$. This is due to Kwapień-Pełczyński [1978]. **2.** (d)$\Rightarrow$(c) and (b)$\Rightarrow$(c) are obvious while (c)$\Rightarrow$(a) follows from

III.F.35. For (a)$\Rightarrow$(d) take $\alpha \mapsto f_\alpha \in A$ with $f_\alpha(t) = \sum_{j=0}^{n} e^{ijt} e^{ij\alpha}$, apply III.F.33 and use the Kolmogorov theorem I.B.20. For (d)$\Rightarrow$(b) note that T extends to $L_1(\mathbb{T})$, so use Exercise III.G.13(a). This is due to Kwapień-Pełczyński [1978]. **3.** The proof is based on duality theory and clever diagram chasing. This is due to Maurey [1972] (see also Pełczyński [1977]). **4.** (c)$\Rightarrow$(b) follows from elementary properties of 2-absolutely summing maps. (b)$\Leftrightarrow$(a) by III.B.27 and (b)$\Rightarrow$(c) follow from the Grothendieck theorem III.F.29. **5.** Use Exercise III.A.12 and Theorem III.D.31 to show that such a subspace embeds into some L_p for $p < 1$. **6.** (a) Show by averaging that if there is an extension, then the invariant extension is bounded. This is false. (b) By III.C.18 and III.C.16 the space X contains ℓ_1^n almost isometrically, so complemented, even in $L_1(\mathbb{T})$. Use the finite dimensional version of (a). (c) Is almost the same as (b) but one has to use different theorems. This is noted in Bourgain [1986] to show that the reflexivity assumption in III.I.25 and III.I.Ex.8 is needed. **7.** Use Remark III.E.13 and observe that all arguments in the proof of Theorem 25 are local. **8.** Follow the proof of Theorem 25. One does not need Proposition 22. This is due to Bourgain [1986]. **9.** (a) the 'if' part is easy. For the 'only if' part use Exercise 7 and follow the ideas of sections 26 and 27 with L_1/H_1 replaced by $L_1(\mathbb{T}, L_1/H_1)$. For (b) use (a). The details are in Bourgain [1986]. (c) Like in Exercise III.D.12 show that span$\{z_1^{k_1} \dots z_N^{k_N}\}_{k_1 \leq K}$ is uniformly a subspace of ℓ_∞^s with $s = (100KN)^N$. Next show that if $X \subset \ell_\infty^s$ and $T: X \to \ell_1$ then $\pi_2(T) \leq C\sqrt{\log s}\|T\|$. This uses III.F.37 and the estimate $c_p \leq c\sqrt{p}$ for the constant c_p appearing in III.F.37. This estimate follows from the estimate given without proof in Remark III.A.20. This is due to Kislyakov [1981]. **10.** (a) Dualize and use the F.-M. Riesz theorem and Paley's projection. (b) Take X_p, the closure of X in $L_p(\mathbb{T}), p < 1$, and show that $X_p = $ span$\{e^{in\theta}\}_{n \in \Lambda} \oplus H_p(\mathbb{T})$. (c) Average and note that $X/A \sim \ell_2$ but span$\{e^{in\theta}\}_{n \in \Lambda} \subset C(\mathbb{T})$ is isomorphic to ℓ_1. This can be found in Kislyakov [1981a]. **11.** Use III.H.13 and Exercise III.A.2 to show that R is isomorphic to a subspace Y of $H_p(\mathbb{T})$, $p > 1$, such that for $y \in Y$ we have $\|y\|_p \leq c\|y\|_1$. **12.** If such a sequence exists then $\sup d(Im\ P_n, \ell_\infty^n) < \infty$. This would imply that 1-absolutely summing operators on $A(\mathbb{D})$ behave like 1-absolutely summing operators on $C(\mathbb{T})$. (see also Exercise III.E.9b). **13.** Show that Theorem 12(a) fails for $id: A(\mathbb{D}^2) \to H_1(\mathbb{D}^2)$. There are many other ways. The result was first proved in Henkin [1967a]. **14.** The idea is to observe that Proposition 6 implies that if $T: A(\mathbb{D}) \to \ell_2$ is a 1-absolutely summing operator onto then $T^*(\ell_2)$ is 'essentially contained' in L_1/H_1. On the other hand in $\mathbb{C}^2$ we have a continuum of complex lines L_α such that $\mathbf{S}_2 \cap L_\alpha$ are disjoint, so we have a continuum of 1-absolutely summing operators $A(\mathbb{B}_2) \to \ell_2$ defined by $T_\alpha(f) = P(f|S_2 \cap L_\alpha)$ where P is a Paley operator. Also $S_2 \cap L_\alpha$ are disjoint peak sets so $T_\alpha^*(\ell_2)$ tend to behave like an ℓ_1 sum. This is too much to fit into L_1/H_1. The details (quite complicated) are in Mitiagin-Pełczyński [1975].

List of symbols

General symbols

δ_x	the Dirac measure at the point x
$\delta_{n,m}$	the Kronecker symbol; 1 if $n = m$, 0 otherwise
$|A|$	the absolute value of a number; the cardinality of a finite set; otherwise the Lebesgue measure
$\hat{f}(n)$	the n-th Fourier coefficient of a function f
$[x]$	integer part of a real number x
χ_A	indicator function of a set A
σ	the normalized Lebesgue measure on $\mathbf{S}_d$
ν	when it denotes a measure on $\mathbb{B}_d$ it is the normalized Lebesgue measure
$r_n(t)$	Rademacher functions, I.B.8.

Sets

$\mathbb{B}_d$	open unit (euclidean) ball in $\mathbb{C}^d$
$\mathbb{C}$	complex plane
$\mathbb{D}$	unit disc in $\mathbb{C}$
$\mathbb{N}$	natural numbers
$\mathbb{R}$	real numbers
$\mathbb{R}_+$	nonnegative real numbers
$\mathbf{S}_d$	unit (euclidean) sphere in $\mathbb{C}^d$
$\mathbb{T}$	unit circle in $\mathbb{C}$
$\mathbb{Z}$	integers
$\mathbb{Z}_+$	nonnegative integers
Δ	Cantor set, if it is a set; sometimes Δ means something else

Functional analytical symbols

$X \oplus Y$	direct sum of Banach spaces, II.B.19
$X \widehat{\otimes} Y$	projective tensor product of Banach spaces, III.B.25
$X \sim Y$	isomorphic Banach spaces, II.B.1

$X \cong Y$	isometric Banach spaces, II.B.1
$\langle \cdot, \cdot \rangle$	inner product in a Hilbert space and sometimes the duality between linear spaces
$X^{\perp}$	annihilator of a subspace $X \subset Y$, i.e. $\{y^* \in Y^* : y^*|X = 0\}$
span	the linear closed span of a set of vectors, I.A.1
conv	the convex hull of a set of vectors, I.A.21
B_X	the closed unit ball in the space X I.A.2
$\sigma(T)$	the spectrum of an operator T

Spaces

$A(\cdot)$	the disc or ball or polydisc algebra, I.B.26 and I.B.28
$A_0(\cdot)$	subspace of $A(\cdot)$ of functions vanishing at 0, I.B.26 or I.B.28
$C(\cdot)$	continuous functions, I.B.9
$C_0(\cdot)$	continuous functions vanishing at ∞, I.B.9
$C^k(\mathbb{T}^s)$	functions on $\mathbb{T}^s$ all of whose derivatives up to order k are continuous, I.B.30
c	convergent sequences, I.B.9
c_0	null sequences, I.B.9
$H_p(\cdot)$	Hardy space, I.B.19 and I.B.21
$L_p(\cdot)$	p-integrable functions, I.B.2
$L_0(\cdot)$	all measurable, a.e. finite functions, I.B.2
$L_\infty(\cdot)$	essentially bounded functions, I.B.2
ℓ_p	p-summable, infinite sequences, I.B.5
ℓ_p^n	$\mathbb{R}^n$ or $\mathbb{C}^n$ with the ℓ_p-norm, I.B.5
ℓ_∞	all bounded sequences, I.B.5
ℓ_∞^n	$\mathbb{R}^n$ or $\mathbb{C}^n$ with the sup-norm, I.B.9
$Lip_\alpha(\cdot)$	functions satisfying the Hölder condition of order α, I.B.29
$M(K)$	regular, Borel measures on K, I.B.11.
$W_p^k(\mathbb{T}^s)$	Sobolev space of functions on $\mathbb{T}^s$ all of whose derivatives up to order k are in L_p, I.B.30

Norms

$\|\cdot\|$	norm in general
$\|\cdot\|_X$	norm in the space X

$\|\cdot\|_p$	norm in some L_p
$\|\cdot\|_\infty$	the supremum norm
$\|\cdot\|_{p,\infty}$	norm in $L_{p,\infty}$, I.B.7
$a_p(T)$	p-approximable norm of an operator T, III.G.3
$i_p(T)$	p-integral norm of an operator T, III.F.21
$n_p(T)$	p-nuclear norm of an operator T, III.F.19
$\pi_p(T)$	p-absolutely summing norm of an operator T, III.F.2
$\sigma_p(T)$	norm in the Schatten-von Neuman class σ_p, III.G.6

Sets of operators

$A_p(X,Y)$	p-approximable operators from X into Y, III.G.3
$I_p(X,Y)$	p-integral operators from X into Y, III.F.21
$K(X,Y)$	compact, linear operators from X into Y
$L(X,Y)$	continuous, linear operators from X into Y
$N_p(X,Y)$	p-nuclear operators from X into Y, III.F.19
$\Pi_p(X,Y)$	p-absolutely summing operators from X into Y, III.F.2
σ_p	the Schatten-von Neuman class, III.G.6

Operators

$\mathcal{C}$	the Cauchy projection, I.B.21
$\mathcal{D}_n$	the Dirichlet kernel or corresponding projection, I.B.15
$\mathcal{F}_n$	the Fejér kernel or corresponding operator, I.B.16
$\mathcal{P}_r$	the Poisson kernel or corresponding operator, I.B.18
$\mathcal{R}$	the Riesz projection, I.B.20

Constants

$bc(\cdot)$	basis constant, II.B.6
$ubc(\cdot)$	unconditional basis constant, Exercise II.D.8
$d(X,Y)$	Banach-Mazur distance, II.E.6
$\gamma_\infty(\cdot)$	constant of factorization through L_∞, III.B.3
$\lambda(X)$	projection constant, III.B.3
$e(X)$	extension constant, III.B.3
$T_p(X)$	type p constant, III.A.17
$C_p(X)$	cotype p constant, III.A.17

References

Adams R.A.
[1975] *Sobolev spaces*, Academic Press, New York

Ahern P. and Bruna J.
[1988] On holomorphic functions in the ball algebra with absolutely
continuous boundary values, *Duke Math. J.* 56 N^o1
pp.129-142

Akilov G. P.
[1947] On extension of linear operations, *Doklady AN SSSR* 57 N^o3
pp.643-646 (in Russian)

Alaoglu L.
[1940] Weak topologies of normed linear spaces, *Annals of Math.* 41
pp.252-267

Aldous D. J.
[1977] Limit theorems for subsequences of arbitrarily-dependent
sequences of random variables, *Zeitschrift für
Wahrscheinlichkeitstheorie verw. Gebiete* 40 pp.59-82

Aleksandrov A.B.
[1982] The existence of inner functions in the ball, *Mat. Sbornik*
118(160) N^o2 pp.147-163 (in Russian)
[1984] Inner functions on compact spaces, *Funktsional. Anal. i
Prilozen.* 18 N^o2 pp.1-13 (in Russian)
[1987] Inner functions: results, methods, problems, in *Proc. Intern.
Congress of Math. Berkeley* 1986 pp.699-707, Amer. Math.
Soc. Providence R.I.

Alexander H.
[1982] On zero sets for the ball algebra, *Proc. Amer. Math. Soc.* 86
N^o1 pp.71-74

Alfsen E.M. and Effros E.G.
[1972] Structure in real Banach spaces I, II, *Annals of Math.* 96 N^o1
pp.98-173

Allahverdiev D.E.

[1957] On the rate of approximation of completely continuous
operators by finite dimensional operators, *Azerbaidzhan Gos.
Univ. Učon. Zap. Seria Fiz-Mat i Chim Nauk* 2 pp.27-35 (in
Russian)

Amir D.

[1965] On isomorphisms of continuous function spaces, *Israel J.
Math.* 3 pp.205-210

Ando T.

[1978] On the predual of H_∞ , *Commentationes Math. Tomus
specialis in honorem Ladislai Orlicz* pp.33-40

Arazy J. and Lindenstrauss J.

[1975] Some linear-topological properties of the spaces C_p, of
operators on Hilbert spaces, *Compositio Math.* 30 pp.81-111

Axler S.

[1985] Zero multipliers of Bergman spaces, *Canadian Math. Bull.*
28(2) pp.237-242

[1988] Bergman spaces and their operators, in *Survey of some recent
results in operator theory*, ed. B. Conway and B. Morrel,
Pitman Research Notes in Math.

Axler S., Berg I. D., Jewell N. and Shields A.

[1979] Approximation by compact operators and the space $H_\infty + C$,
Annals of Math. 109 pp.601-612

Banach S.

[1919] Sur la valeur moyenne des fonctions orthogonales, *Bull.
International de l'Acad. Pol. des Sciences et Lettres, Classe
des Sci. math. et natur. Série A.* pp.66-72

[1931] Teorja operacyj, Warszawa, Wydawnictwo Kasy im.
Mianowskiego (in Polish)

[1932] *Théorie des opérations linéaires*, Warszawa

Banach S. and Mazur S.

[1933] Zur theorie der lineare Dimension, *Studia Math.* 4
pp.100-112

Banach S. and Saks S.
[1930] Sur la convergence forte dans le champ L_p, *Studia Math.* 2 pp.51-57

Beauzamy B.
[1985] *Introduction to Banach spaces and their geometry*, North Holland, Math. Studies 68, (second edition)

Behrends E.
[1979] *M-structure and the Banach-Stone theorem*, Lecture Notes in Mathematics 736, Springer Verlag

Bennett C. and Sharpley R.
[1988] *Interpolation of operators*, Pure and Appl. Math. vol. 129, Academic Press

Bennett G.
[1976] Unconditional convergence and almost everywhere convergence, *Zeitschrift für Wahrscheinlichkeitstheorie verw. Gebiete* 34 pp.135-155
[1977] Schur multipliers, *Duke Math. J.* 44 $N°3$ pp.603-640

Berkes I.
[1986] An extension of the Komlos Subsequence Theorem, *Longhorn Notes, The University of Texas at Austin Functional Analysis Seminar* 1985-86, pp.75-82

Bernstein S.
[1914] Sur la convergence absolue des séries trigonométriques, *Comptes Rendues Acad. Sci. Paris* 158 pp.1661-1663

Bessaga C. and Pełczyński A.
[1958] On bases and unconditional convergence of series in Banach spaces , *Studia Math.* 17 pp.151-164
[1960] Spaces of continuous functions IV (On isomorphic classification of spaces C(S), *Studia Math.* 19 pp.53-62
[1960a] Banach spaces non-isomorphic to their Cartesian squares I, *Bull. Acad. Pol. Sci. Série Math. Astr.et Phys.* 9 $N°2$ pp.77-80

Bichteler K.
[1981] Stochastic integration and L^p-theory for semi-martingales, *Annals of Prob.* 9 N^o1 pp.49-89

Boas R. P.
[1955] Isomorphism between H_p and L_p, *Amer. J. of Math.* 77 pp.655-656

Bochner S.
[1937] Stable laws and completely monotone functions, *Duke Math. J.* 3 pp.726-728

Bočkarev S. V.
[1974] Existence of a basis in the space of functions analytic in the disc and some properties of the Franklin system, *Mat. Sbornik* 95(137) N^o1 pp.3-18 (in Russian)
[1978] A method of averaging in the theory of orthogonal series and some problems in the theory of bases, *Trudy Steklov Institut* 146 (in Russian)
[1985] Construction of dyadic interpolating basis in the space of continuous functions using the Féjer kernels, *Trudy Steklov Institut* 172 pp.29-59 (in Russian)

Borsuk K.
[1933] Über Isomorphie der Funktionalraumen, *Bull. Acad. Pol. Sci. et Letters* pp.1-10

Bourbaki N.
[1938] Sur les espaces de Banach, *Comptes Rendues Acad. Sci. Paris* 206, pp.1701-1704

Bourgain J.
[1983] On weak completeness of the dual of spaces of analytic and smooth functions, *Bull. Soc. Math. Belg. Serie B* 35 N^o1 pp.111-118
[1984] New Banach space properties of the disc algebra and H_∞, *Acta math.* 152 pp.1-48
[1984a] On bases in the disc algebra, *Trans. Amer. Math. Soc.* 285 N^o1 pp.133-139
[1984b] The Dunford-Pettis property for the ball-algebra, the polydisc-algebras and Sobolev spaces, *Studia Math.* 77 N^o3 pp.245-253

[1985] Applications of the spaces of homogenous polynomials to some problems on the ball algebra, *Proc. Amer. Math. Soc.* 93 $N°2$ pp.277-283

[1986] Extension of H_∞-valued operators and bounded bianalytic functions, *Trans. Amer. Math. Soc.* 286 $N°1$ pp.313-337

[1986a] On the similarity problem for polynomially bounded operators on Hilbert space, *Israel J. Math.* 54 $N°2$ pp.227-241

[1987] A remark on entropy of Abelian groups and the invariant uniform approximation property, *Studia Math.* 86 pp.79-84

Bourgain J. and Davis W. J.

[1986] Martingale transforms and complex uniform convexity, *Trans. Amer. Math. Soc.* 294 pp.501-515

Bourgain J. and Milman V.

[1985] Dichotomie du cotype pour les espaces invariants, *Comptes Rendues Acad. Sci. Paris* 300 $N°9$ pp.263-266

Bourgain J. and Rosenthal H. P.

[1983] Applications of the theory of semi-embeddings to Banach space theory, *J. Funct. Anal.* 52 pp.149-188

Bourgain J., Rosenthal H. P., Schechtman G.

[1981] An ordinal L^p-index for Banach spaces, with applications to complemented subspaces of L^p, *Annals of Math.* 114 pp.193-228

Bożejko M.

[1987] Littlewood functions, Hankel multipliers and power bounded operators on a Hilbert space, *Colloquium Math.* 51 pp.35-42

Burkholder D. L.

[1982] A nonlinear partial differential equation and the unconditional constant of the Haar system in L^p, *Bull. Amer. Math. Soc.* 7 $N°3$ pp.591-595

[1984] Boundary value problems and sharp inequalities for martingale transforms, *Annals of Probability* 12 $N°3$ pp.647-702

[1986] *Martingales and Fourier analysis in Banach spaces*, Lecture Notes in Mathematics 1206, pp.61-108, Springer Verlag

[1988] A proof of Pełczyński conjecture for the Haar system, *Studia Math.* 91 pp.79-83

Cambern M.
[1967] On isomorphisms with small bounds, *Proc. Amer. Math. Soc.* 18 pp.1062-1066

Carleman T.
[1918] Über die Fourierkoeffizienten einer stetigen Funktion, *Acta Math.* 41 pp.377-384

Carleson L.
[1957] Representations of continuous functions, *Math. Zeit.* 66 pp.447-451
[1966] On convergence and growth of partial sums of Fourier series, *Acta Math.* 116 pp.135-157

Casazza P. G., Pengra R. W., Sundberg C.
[1980] Complemented ideals in the disc algebra, *Israel J. Math.* 37 pp.76-83

Chevet S.
[1969] Sur certains produits tensoriels topologiques d'espaces de Banach, *Zeitschrift für Wahrscheinlichkeitstheorie verw. Gebiete* 11 pp.120-138

Choi Man Duen and Effros E.G.
[1977] Lifting problems and the cohomology of C^*-algebras, *Canadian Math. J.* 29 pp.1092-1111

Ciesielski Z.
[1960] On isomorphisms of the spaces H_α and m, *Bull. Acad. Pol. Sci. Serie Math. Astr. et Phys.* 8 $N°4$ pp.217-222
[1963] Properties of the orthonormal Franklin system, *Studia Math.* 23 pp.141-157
[1966] Properties of the orthonormal Franklin system II, *Studia Math.* 27 pp.289-323
[1975] Constructive function theory and spline systems, *Studia Math.* 53 pp.277-302

Ciesielski Z. and Domsta J.
[1972] Construction of an orthonormal basis in $C^m(I^d)$ and $W_p^m(I^d)$, *Studia Math.* 41 pp.211-224

Clarkson J. A.
[1936] Uniformly convex spaces, *Trans. Amer. Math. Soc.* 40 pp.396-414

Cohen H. B.
[1975] A bound-two isomorphism between C(X) Banach spaces, *Proc. Amer. Math. Soc.* 50 pp.215-217

Davie A. M.
[1976] Classification of essentially normal operators, Lecture Notes in Mathematics 512, Springer Verlag, pp.31-55

Davis W. J., Figiel T., Johnson W. B., Pełczyński A.
[1974] Factoring weakly compact operators, *J. Funct. Anal.* 17 $N°3$ pp.311-327

Day M.M.
[1958] *Normed linear spaces*, Springer Verlag

Dechamps-Gondim M.
[1985] Analyse harmonique, analyse complexe et géometrié des espaces de Banach (d'après Jean Bourgain), *Séminaire Bourbaki,* Astérisque 121-122 pp.171-195

Delbaen F.
[1977] Weakly compact operators on the disc algebra, *J. of Algebra* 45 pp.284-294

Demko S.
[1977] Inverses of band matrices and local convergence of spline projections, *SIAM J. Numer. Anal.* 14 $N°4$ pp.616-619

Demko S., Moss W. F., Smith Ph. W.
[1984] Decay rates for inverses of band matrices, *Mathematics of Computation* 43 $N°168$ pp.491-499

Diestel J.

[1975] *Geometry of Banach spaces - selected topics*, Lecture Notes in Mathematics 485, Springer Verlag

[1980] A survey of results related to Dunford-Pettis property, in *Contemporary Math. 2*, Amer. Math. Soc. Providence, Proc. of the Conf. on Integration, Topology and Geometry in Linear Spaces pp.15-60

Diestel J and Uhl J.J.

[1977] *Vector measures*, Mathematical Surveys 15, Amer. Math. Soc. Providence

Drury S. W.

[1983] Remarks on von Neumann's inequality, in *Banach spaces, harmonic analysis and probability theory*, Lecture Notes in Mathematics 995, Springer Verlag pp.14-32

Dubinsky E., Pełczyński A., Rosenthal H. P.

[1972] On Banach spaces X for which $\Pi_2(L_\infty, X) = B(L_\infty, X)$, *Studia Math.* 44 pp.617-648

Dugundji J.

[1951] An extension of Tietze's theorem, *Pacific J. Math.* 1 pp.353-367

Dunford N.

[1939] A mean ergodic theorem, *Duke Math. J.* 5 pp.635-646

Dunford N. and Pettis B. J.

[1940] Linear operators on summable functions, *Trans. Amer. Math. Soc.* 47 pp.323-392

Dunford N. and Schwartz J. T.

[1958] *Linear operators. Part I, General theory*, Interscience Publishers Inc. N.Y.

Duren P. L.

[1970] *Theory of H_p-spaces*, Academic Press, New York and London

Eberlein W. F.

[1947] Weak compactness in Banach spaces I, *Proc. Nat. Acad. USA* 33 pp.51-53

Edwards R. E.
[1965] *Functional analysis, theory and applications*, Holt, Reinhart and Winston, New York

Enflo P.
[1973] A counterexample to the approximation problem in Banach spaces, *Acta Math.* 130 pp.309-317

Faber G.
[1910] Über die Orthogonalfunktionen des Herrn Haar, *Jahresber. Deutsch. Math. Verein.* 19 pp.104-112
[1914] Über die interpolatorische Darstellung stetiger Funktionen, *Jahresber. Deutsch. Math. Verein.* 23 pp.192-210

Feller W.
[1971] *An introduction to probability theory and its applications*, Wiley, New York

Figiel T. and Johnson W. B.
[1980] Large subspaces of l_∞^n and estimates of the Gordon-Lewis constant, *Israel J. Math.* 37 pp.92-111

Figiel T. and Pisier G.
[1974] Séries aléatoires dans les espaces uniformément convexes ou uniformément lisses, *Comptes Rendues Acad. Sci. Paris Série A* 279 pp.611-614

Fisher S.
[1968] The convex hull of the finite Blaschke products, *Bull. Amer. Math. Soc.* 74 pp.1128-1129

Foguel S. R.
[1964] A counter example to a problem of Sz-Nagy, *Proc. Amer. Math. Soc.* 15 pp.788-790

Folland J. B. and Stein E. M.
[1982] *Hardy spaces on homogenous groups*, Mathematical Notes 28, Princeton University Press

Forelli F. and Rudin W.
[1976] Projections on spaces of holomorphic functions in balls, *Indiana Univ. Math. J.* 24 N^o6 pp.593-602

Fournier J. J. F.
[1974] An interpolation problem for coefficients of H_∞ functions,
    Proc. Amer. Math. Soc. 42 N^o2 pp.402-407

Franklin Ph.
[1928] A set of continuous orthogonal functions, *Math. Annalen* 100
    pp.522-529

Fréchet M.
[1935] Sur la définition axiomatique d'une classe d'espaces vectoriels
    distanciés applicables vectoriellement sur l'espaces de Hilbert,
    Annals of Math. 36 pp.705-18

Fredholm I.
[1903] Sur une classe d'équations fonctionelles, *Acta Math.* 27
    pp.365-390

Gamelin T. W.
[1969] *Uniform algebras*, Prentice Hall Inc., Englewood Cliffs N.J.

Gamelin T. W., Marshal D. E., Younis R., Zame W. R.
[1985] Function theory and M-ideals, *Arkiv för Matematik* 23 N^o2
    pp.261-279

Gantmacher V.
[1940] Über schwache totalstetige Operatoren, *Mat. Sbornik* 7
    pp.301-307

Garcia-Cuerva J. and Rubio de Francia J. L.
[1985] *Weighted norm inequalities and related topics*, North Holland
    Mathematical Studies 116

Garling D. J. H.
[1970] Absolutely p-summing operators in Hilbert space, *Studia
    Math.* 38 pp.319-331
  [P] On the dual of the proper uniform algebra, *Bull. London.
    Math. Soc.* 21 N^o3 (1989) pp.279-284

Garling D. J. H. and Gordon Y.
[1971] Relations between some constants associated with finite
    dimensional Banach spaces, *Israel J. Math.* 9 pp.346-361

Garnett J. B.
[1977] Two remarks on interpolation by bounded functions, Lecture Notes in Mathematics 604 pp.32-40, Springer Verlag
[1981] *Bounded analytic functions*, Academic Press

Glicksberg I.
[1964] Some uncomplemented function algebras, *Trans. Amer. Math. Soc.* 111 pp.121-137

Globevnik J.
[1975] On complex strict and uniform convexity, *Proc. Amer. Math. Soc.* 47 pp.175-178

Gohberg I. C. and Krein M. G.
[1965] *Introduction to the theory of linear non-selfadjoint operators*, Nauka, Moscow (in Russian)

Goldstine H. H.
[1938] Weakly complete Banach spaces, *Duke Math. J.* 4 pp.125-131

Goodner D. B.
[1950] Projections in normed linear spaces, *Trans. Amer. Math. Soc.* 69 pp.89-108

Gordon Y. and Lewis D. R.
[1974] Absolutely summing operators and local unconditional structures, *Acta Math.* 133 pp.27-48

Gorin E. A., Hruščov S. V., Vinogradov S. A.
[1981] Interpolation in H_∞ along P. Jones' lines, *Zapiski Nauč. Sem. LOMI 113*, Issl. po lin. op. i teorii funkciĭ XI pp.212-213 (in Russian)

Graham C. and McGehee O. C.
[1979] *Essays in commutative harmonic analysis*, Springer Verlag

Grothendieck A.
[1953] Sur les applications linéaires faiblement compactes d'espaces du type C(K), *Canadian Math. J.* 5 pp.129-173

[1955] Produits tensoriels topologiques et espaces nucléaires, *Memoirs Amer. Math. Soc.* 16

[1956] Résumé de la théorie métrique des produits tensoriels topologiques, *Bol. Soc. Mat. São Paulo* 8, pp.1-79

[1956a] Erratum au mémoire: Produits tensoriels topologiques et espaces nucléaires, *Ann. Inst. Fourier* 6 pp.117-120

Grünbaum B.
[1960] Projection constants, *Trans. Amer. Math. Soc.* 95 pp.451-465

de Guzman M.
[1981] *Real variable methods in Fourier analysis*, North Holland Mathematical Studies 46

Gurariĭ V. I., Kadec M. I., Macaev V. I.
[1965] On Banach-Mazur distance between certain Minkowski spaces, *Bull. Acad. Pol. Sci.* 13 pp.719-722

Haagerup U.
[1987] A new upper bound for the complex Grothendieck constant, *Israel Math. J.* 60 pp.199-225

Haar A.
[1910] Zur theorie der orthogonalen Funktionensysteme, *Math. Annalen* 69 pp.331-371

Halmos P. R.
[1950] *Measure theory*, van Nostrand Comp. Toronto

Havin V. P.
[1973] Weak completeness of the space L^1/H_o^1, *Vestnik Leningrad. Univ.* 13 pp.77-81 (in Russian)

Helson H.
[1954] Proof of a conjecture of Steinhaus, *Proc. Nat. Acad. USA* 40 pp.205-206

[1955] On a theorem of Szego, *Proc. Amer. Math. Soc.* 6 pp.235-242

Henkin G. M.
[1967] A proof of non-isomorphism of the spaces of smooth functions on the interval and on the square, *Doklady AN SSSR* 172 pp.48-51 (in Russian)

[1967a] Non-isomorphism of certain spaces of functions of different
        number of variables, *Funct. Anal. Prilož.* 1 N^o4 pp.57-68 (in
        Russian)

Hennefeld J.
  [1973] A decomposition for $B(X)^*$ and unique Hahn-Banach
         extensions, *Pacific Math. J.* 46 pp.197-199

Hoffman K.
  [1962] *Banach spaces of analytic functions*, Prentice Hall, Englewood
         Cliffs N.J.

Holub J. R.
  [1973] On subspaces of norm ideals, *Bull. Amer. Math. Soc.* 79
         pp.446-448

Hruščov S. V. and Peller V. V.
  [1982] Hankel operators, best approximations and stationary
         Gaussian processes, *Uspekhi Mat. Nauk* 37 N^o1 pp.53-124 (in
         Russian)

Hunt R. A.
  [1968] On the convergence of Fourier series, *Proc. Conf. Orthogonal
         Expansions and Continuous Analogues,* Southern Ill. Univ.
         Press pp.235-255

Hutton C. V.
  [1974] On the approximation numbers of an operator and its adjoint,
         Math. Annalen 210 pp.277-280

James R. C.
  [1950] Bases and reflexivity of Banach spaces, *Annals of Math.* 52
         pp.518-527
  [1964] Uniformly non-square Banach spaces, *Annals of Math.* 80
         pp.542-550
  [1981] Structure of Banach spaces; Radon-Nikodym and other
         properties, in *General topology and modern analysis,* ed. L.F.
         McAuley and M.M. Rao, Academic Press, pp.347-364

Jameson G. J. O.
  [1987] *Summing and nuclear norms in Banach space theory*, London
         Math. Soc. Student Text 8, Cambridge University Press

Jarosz K.
[1985] *Perturbations of Banach algebras*, Lecture Notes in Math.
1120, Springer Verlag

John F.
[1948] Extremum problems with inequalities as subsidiary
conditions, in *R. Courant anniversary volume*, Interscience
N.Y. pp.187-204

Johnson W. B.
[1972] A complementary universal conjugate Banach space and its
relation to the approximation problem, *Israel J. Math.* 13
$N°$3-4 pp.301-310

Johnson W. B., König H., Maurey B., Retherford J. R.
[1979] Eigenvalues of p-summing and l_p-type operators in Banach
spaces, *J. Funct. Anal.* 32 pp.353-380

Jordan P. and von Neumann J.
[1935] On inner products in linear metric spaces, *Annals of Math.* 36
pp.719-723

Kadec M. I.
[1956] Unconditional convergence of series in uniformly convex
Banach spaces, *Uspekhi Mat. Nauk.* 11 $N°$5 pp.185-190 (in
Russian)
[1958] On linear dimension of the spaces L_p, *Uspekhi Mat. Nauk.* 13
pp.95-98 (in Russian)
[1974] On relative projection constant and a theorem of Z. A.
Čanturia, *Bull. Acad. Sci. Georgian SSR* 74 $N°$3 pp.541-543
(in Russian)

Kadec M. I. and Pełczyński A.
[1962] Bases, lacunary sequences and complemented subspaces in
the spaces L_p, *Studia Math.* 21 pp.161-176

Kadec M. I. and Snobar M. G.
[1971] Certain functionals on the Minkowski compactum, *Mat.
Zametki* 10 pp.453-458 (in Russian)

Kadec M.I. and Woźniakowski K.
[P] On series whose permutations have only two sums, to appear in *Bull Acad. Pol. Sci* 37 N^o1-6 (1989) pp.15-21

Kahane J.-P.
[1985] *Some random series of functions,* second edition, Cambridge Studies in Advanced Math. 5, Cambridge University Press

Kahane J.-P., Katznelson Y., de Leeuw K.
[1977] Sur les coefficients de Fourier des fonctions continues, *Comptes Rendues Acad. Sci. Paris* 285 N^o16 pp.A1001-1003

Kahane J.-P. and Salem R.
[1963] *Ensembles parfaits et séries trigonométriques,* Herman, Paris

Kaijser S.
[1978] Some results in the metric theory of tensor products, *Studia Math.* 63 pp.157-170

Kaiser R. J. and Retherford J. R.
[1984] Nuclear cyclic diagonal mappings, *Math. Nachr.* 119 pp.129-135

Kakutani S.
[1939] Weak topology and regularity of Banach spaces , *Proc. Imp. Acad. Tokyo* 15 pp.169-173

Kalton N. J. and Trautman D. A.
[1982] Remarks on subspaces of H_p when $0 \le p \le 1$, *Michigan Math. J.* 29 pp.163-170

Kantorovič L. V.
[1935] On semi-ordered linear spaces and their applications to the theory of linear operations, *Doklady AN SSSR* 4 pp.11-14 (in Russian)

Karlin S.
[1948] Bases in Banach spaces, *Duke Math. J.* 15 pp.971-985

Kashin B. S.
[1974] On unconditional convergence in the space L_1, *Mat. Sbornik* 94(136) pp.540-550 (in Russian)

Kashin B. S. and Saakian A. A.
[1984] *Orthogonal series*, Nauka, Moscow (in Russian)

Katznelson Y.
[1968] *An introduction to harmonic analysis*, Wiley, New York

Kechris A. S. and Louveau A.
[1987] *Descriptive set theory and the structure of sets of uniqueness*, London Math. Soc. Lecture Notes 128, Cambridge University Press

Kelley J. L.
[1952] Banach spaces with the extension property, *Trans. Amer. Math. Soc.* 72 pp.323-326

Kislyakov S. V.
[1975] On the conditions of Dunford-Pettis, Pełczyński and Grothendieck, *Doklady AN SSSR* 225 pp.1252-1255 (in Russian)
[1976] Sobolev embedding operators and non-isomorphism of certain Banach spaces, *Funkts. Anal. Priloz.* 9 N^o4 pp.22-27 (in Russian)
[1977] p-absolutely summing operators, in *Geometry of linear spaces and operator theory* pp.114-174, Yaroslavl Gos. Univ. Yaroslavl (in Russian)
[1979] On reflexive subspaces of C_A^*,*Funkts. Anal. Priloz.* 13 pp.21-30 (in Russian)
[1981] Fourier coefficients of boundary values of functions analytic in the disc and bidisc, *Trudy Steklov Institut* 155, pp.77-94 (in Russian)
[1981a] What is needed for a 0-absolutely summing operator to be nuclear, in *Complex analysis and spectral theory*, Lecture Notes in Math. 864 pp.336-365, Springer Verlag
[1981b] Two remarks on the equality $\Pi_p(X,.) = I_p(X,.)$, *Zapiski Nauč. Sem. LOMI 113*, Issled. po lin. op. i teorii funktsiǐ XI, pp.135-148 (in Russian)

[1987] Simplified proof of a theorem of J. Bourgain, *Zapiski Nauč. Sem. LOMI 157*, Issled. po lin. op. i teorii funktsiĭ XVI, pp.146-150 (in Russian)

[1987a] Interpolatory inequalities for Fourier multipliers and their applications, *Doklady AN SSSR* 292 N^o1 pp.29-32 (in Russian)

[P] The proper uniform algebras are uncomplemented, *Doklady AN SSSR* 309 N^o4 (1989) pp.795-798 (in Russian)

[P_1] Truncating functions in weighted H^p and two theorems of J. Bourgain, *Uppsala Univ. Dept. of Math. Report 1989:10*

[P_2] $(q.p)$-summing operators on the disc algebra and a weighted estimates for certain outer functions, *preprint LOMI E-11-89*

Kolmogorov A. N.

[1923] Une série de Fourier-Lebesgue divergente presque partout, *Fund. Math.* 4 pp.324-328

[1925] Sur les fonctions harmoniques conjuguées et les séries de Fourier, *Fund. Math.* 7 pp.23-29

[1926] Une série de Fourier-Lebesgue divergente partout, *Comptes Rendues Acad. Sci. Paris* 183 pp.1327-1328

Komlós J.

[1967] A generalisation of a problem of Steinhaus, *Acta. Math. Acad. Sci. Hung.* 18 pp.217-229

König H

[1985] Spaces with large projection constant, *Israel Math. J.* 50 pp.181-188

[1986] *Eigenvalue distribution of compact operators*, Birkhauser Verlag, Basel-Boston-Stuttgart, Operator theory: Advances and Applications 16

König H. and Lewis D. R.

[P] A strict inequality for projection constant *J. Funct. Anal.* 74 N^o2 (1987) pp.328-332

König H. and Tomczak-Jaegermann N.

[P] Bounds for projection constants and 1-summing norms *Trans. Amer. Math. Soc.* 320 (1990) pp.799-832

Koosis P.
[1980] *Introduction to H_p-spaces*, London Math. Soc. Lecture Notes
40, Cambridge University Press
[1981] A theorem of Khrushchev and Peller on restrictions of
analytic functions having finite Dirichlet integral to closed
subsets of the unit circumference, in *Conference on harmonic
analysis in honor of Antoni Zygmund*, ed. W.Beckner,
A.P.Calderon, R.Fefferman, P.W.Jones, Wadsworth Int.
Belmont pp.740-748

Kopp P.E.
[1984] *Martingales and stochastic integrals*, Cambridge University
Press

Korovkin P. P.
[1959] *Linear operators and approximation theory*, Fizmatgiz,
Moscow (in Russian)

Krein M. and Šmulian V.
[1940] On regularly convex sets in the space conjugate to a Banach
space, *Annals of Math.* 41 pp.556-583

Krein S. G., Milman D. P., Rutman L.A.
[1940] On the property of a basis in Banach space, *Zapiski
Nauĭ-Jssl. Just. Mat. Kharkov. Mat. O6šč. serie 4, vol. XVI*
4 pp.106-110 (in Russian and in English)

Krein S. G., Petunin Ju. I., Semenow E. M.
[1978] *Interpolation of linear operators*, Nauka, Moscow (in Russian)

Kuratowski K.
[1968] *Topology*, Academic Press and Państwowe Wydawnictwo
Naukowe

Kwapień S.
[1970] On a theorem of L. Schwartz and its applications to
absolutely summing operators, *Studia Math.* 38 pp.193-201

Kwapień S. and Pełczyński A.
[1970] The main triangle projection in matrix spaces and its
applications, *Studia Math.* 34 pp.43-68

[1978] Remarks on absolutely summing, translation invariant operators from the disc algebra and its dual into a Hilbert space, *Michigan Math. J.* 25 pp.173-181

Lacey H. E.
[1974] *The isometric theory of classical Banach spaces*, Springer Verlag, Die Grundlehren der mathematischen Wissenschaften band 208

Lebow A.
[1968] A power bounded operator that is not polynomially bounded, *Michigan Math. J.* 15 pp.397-399

Levy P.
[1925] *Calcul des probabilités*, Gauthier-Villars, Paris

Lewis D. R.
[1978] Finite dimensional subspaces of L_p, *Studia Math.* 63 pp.207-212

Lima A.
[1982] On M-ideals and the best approximation, *Indiana Univ. Math. J.* 31 pp.27-36

Linde W.
[1983] *Probability in Banach spaces – Stable and infinitely divisible distributions*, B.G. Teubner Verlagsgesellschaft, Leipzig

Lindenstrauss J.
[1964] *Extension of compact operators*, Memoirs Amer. Math. Soc. 48

Lindenstrauss J. and Pełczyński A.
[1968] Absolutely summing operators in $\mathcal{L}_p$-spaces and their applications, *Studia Math.* 29 pp.275-326
[1971] Contributions to the theory of the classical Banach spaces, *J. Funct. Anal.* 8 $N°2$ pp.225-249

Lindenstrauss J. and Rosenthal H. P.
[1969] The $\mathcal{L}_p$-spaces, *Israel Math. J.* 7 pp.325-349

Lindenstrauss J. and Tzafriri L.
[1977] *Classical Banach spaces I*, Springer Verlag
[1979] *Classical Banach spaces II*, Springer Verlag

Lopez J. M. and Ross K. A.
[1975] *Sidon sets*, Lecture Notes in Pure and Applied Math., Marcel Dekker N.Y.

Lorentz G. G and Tomczak-Jeagermann N.
[1984] Projections of minimal norms, *Longhorn Notes, The University of Texas at Austin Functional Analysis Seminar* 1983-84 pp.167-176

Løw E.
[1982] A construction of inner function in the unit ball in C^p, *Invent. Math.* 67 pp.223-229

Luecking D.
[1980] The compact Hankel operators form an M-ideal in the space of Hankel operators, *Proc. Amer. Math. Soc.* 79 pp.222-224

Lukacs E.
[1970] *Characteristic functions*, second edition, Griffin, London

Lusternik L. A.
[1936] Basic concepts of functional analysis, *Uspekhi Mat. Nauk* 1 pp.77-140 (in Russian)

Lyons R.
[1985] Fourier-Stieltjes coefficients and asymptotic distribution modulo 1, *Annals of Math.* 122 pp.155-170
[1988] A new type of sets of uniqueness, *Duke Math. J.* 57 N^o2 pp.431-458

Maharam D.
[1942] On homogenous measure algebras, *Proc. Nat. Acad. USA* 28 pp.108-111

Marcinkiewicz J.
[1937] Quelques théorèmes sur les séries orthogonales, *Ann. Soc. Polon. Math.* 16 pp.84-96
[1937a] Quelques remarques sur l'interpolation, *Acta Scientiarum Math. (Szeged)* 8 pp.127-130

Matelievič M. and Pavlovič M.
[1984] L_p-behaviour of the integral means of analytic functions, *Studia Math.* 77 pp.219-227

Maurey B.
[1972] Espaces de cotype (p,q) et théorèmes de revement, *Comptes Rendues Acad. Sci. Paris série A-B* 275 pp.A785-788
[1974] Théoremes de factorisation pour les operatèurs linéaires à valeurs dans les espaces L_p, *Astérisque* 11, Société Math. de France

Maurey B. and Pisier G.
[1976] Séries de variables aléatoires vectorielles indépendantes et propriétés géometriqués des espaces de Banach, *Studia Math.* 58 pp.45-90

Mazur S.
[1933] Über convexe Mengen in linearen normierte Räumen, *Studia Math.* 4 pp.70-84

McCarthy Ch. A.
[1967] C_p, *Israel J. Math.* 5 pp.249-271

Menchoff D.
[1916] Sur l'unicité du développement trigonométrique, *Comptes Rendues Acad. Sci. Paris* 163 pp.433-436
[1923] Sur les séries de fonctions orthogonales I, *Fund. Math.* 4 pp.82-105

Michael E. and Pełczyński A.
[1967] A linear extension theorem, *Illinois Math. J.* 11 $N°4$ pp.563-579

Milman V. D.
[1971] The geometric theory of Banach spaces II, *Uspekhi Mat. Nauk* 26 pp.73-149 (in Russian)
[1987] The concentration phenomenon and linear structure of finite dimensional normed spaces, in *Proc. Intern. Congress of Math. Berkeley 1986* pp.961-975, Amer. Math. Soc. Providence R.I.

Milman V. D. and Schechtman G.
[1986] *Asymptotic theory of finite dimensional normed spaces*,
Lecture Notes in Math. 1200, Springer Verlag

Milne H.
[1972] Banach space properties of uniform algebras, *Bull. London
Math. Soc.* 4 pp.323-326

Milutin A. A.
[1966] Isomorphism of spaces of continuous functions on compacta of
cardinality continuum, *Teoria Funktsii, Funkts. Anal. i
Priloż.* 2 pp.150-156, Kharkov (in Russian)

Mitiagin B. S.
[1964] Absolute convergence of the series of Fourier coefficients,
Doklady AN SSSR 157 pp.1047-1050 (in Russian)

Mitiagin B. S. and Pełczyński A.
[1975] On the non-existence of linear isomorphisms between Banach
spaces of analytic functions of one and several complex
variables, *Studia Math.* 56 pp.85-96

Nachbin L.
[1950] A theorem of the Hahn-Banach type for linear
transformations, *Trans. Amer. Math. Soc.* 68 pp.28-46

Nagy B. and Foias C.
[1967] *Analyse harmonique des opérateurs de l'espace de Hilbert*,
Academiai Kiado

Nahoum A.
[1973] Applications radonifiantes dans l'espace des séries
convergentes I, le théorème de Menchov, *Séminaire
Maurey-Schwartz 1972-73*, Ecole Polytechnique, exposè 24

Natanson I. P.
[1949] *Constructive theory of functions*, Gosud. Izdat. Tekn.-Teoret.
Lit. Moscow-Leningrad (in Russian)

von Neumann J.
[1930] Zur Algebra der Funktionaloperationen und Theorie der
normalen Operatoren, *Math. Annalen* 102 pp.370-427

[1951] Eine Spectraltheorie für allgemeine Operatoren eines unitären Raumes, *Math. Nachrichten* 4 pp.258-281

Newman D. J.
[1963] Pseudo-uniform convexity in H^1, *Proc. Amer. Math. Soc.* 14 pp.676-679

Nikishin E. M.
[1970] Resonance theorems and superlinear operators, *Uspekhi Mat. Nauk* 25 $N°6$ pp.129-191 (in Russian)

Nikolskiĭ N. K.
[1980] *Lectures on the shift operator*, Nauka, Moscow (in Russian)

Olevskiĭ A. M.
[1975] *Fourier series with respect to general orthogonal series*, Springer Verlag

Orlicz W.
[1929] Beiträge zur Theorie der Orthogonalentwicklungen II, *Studia Math.* 1 pp.241-255
[1933] Über unbedingte Konvergenz in Funktionen Raumen I, *Studia Math.* 4 pp.33-37

Ørno P.
[1976] A note on unconditionally convergent series in L_p, *Proc. Amer. Math. Soc.* 59 pp.252-254

Ovčinnikov V. I.
[1976] Interpolation theorems that arise from the Grothendieck inequality, *Funct. Anal. Priloż.* 10 $N°4$ pp.45-57 (in Russian)
[1985] Applications of absolutely summing operators in interpolation theorems on reiteration, *Reports to the extended session of a seminar of I. N. Vekua*, Institut Appl. Mat. Tbilisi vol I $N°2$ (in Russian)

Paley R. E. A. C.
[1932] A remarkable series of orthogonal functions I, *Proc. London Math. Soc.* 34 pp.241-264
[1933] On the lacunary coefficients of power series, *Annals of Math.* 34 pp.615-616

Pełczyński A.

[1958] On the isomorphism of spaces m and M, *Bull. Acad. Polon. Sci. ser. math. astr. phys.* 6 pp.695-696

[1960] Projections in certain Banach spaces, *Studia Math.* 19 pp.209-228

[1961] On the impossibility of embedding of the space L in certain Banach spaces, *Colloquium Math.* 8 pp.199-203

[1962] Banach spaces on which every unconditionally converging operator is weakly compact, *Bull. Acad. Pol. Sci. ser. math. astr. phys.* 10 pp.265-270

[1964] A proof of the Eberlein-Šmulian Theorem by an application of basic sequences, *Bull. Acad. Pol. Sci. ser. math. astr. phys.* 12 pp.543-548

[1964a] On simultanous extensions of continuous functions. A generalisation of theorems of Rudin-Carleson and Bishop, *Studia Math.* 24 pp.285-304

[1967] A characterisation of Hilbert-Schmidt operators, *Studia Math.* 28, pp.355-360

[1968] Linear extensions, linear averagings and their applications to linear topological classification of spaces of continuous functions, *Dissertationes Mathematicae* 58

[1977] *Banach spaces of analytic functions and absolutely summing operators*, CBMS Regional Conference Series N° 30, American Mathematical Society, Providence R. I.

[1989] An analogue of the F. and M. Riesz theorem for spaces of differentiable functions, in *Contemporary mathematics vol. 85,* American Mathematical Society, Providence R. I. pp.405-25

Pełczyński A. and Rosenthal H. P.

[1975] Localisation techniques in L_p-spaces, *Studia Math.* 52 pp.263-289

Pełczyński A. and Senator K.

[1986] On isomorphisms of anisotropic Sobolev spaces with 'classical Banach spaces' and a Sobolev type embedding theorem, *Studia Math.* 84 pp.169-215

Pełczyński A. and Wojtaszczyk P.

[1971] Banach spaces with finite dimensional expansions of identity and universal bases of finite dimensional subspaces, *Studia Math.* 40 pp.91-108

Peller V. V.
[1982] Estimates of functions of power bounded operators on Hilbert spaces, *J. Operator Theory* 7 pp.341-372

Persson A.
[1969] On some properties of p-nuclear and p-integral operators, *Studia Math.* 33 pp.213-222

Persson A. and Pietsch A.
[1969] p-nukleare und p-integrale Abbildungen in Banachräumen, *Studia Math.* 33 pp.19-62

Phillips R. S.
[1940] On linear transformations, *Trans. Amer. Math. Soc.* 48 pp.516-541

Pietsch A.
[1963] Zur Fredholmschen Theorie in lokalkonvexen Räumen, *Studia Math.* 22 pp.161-179
[1967] Absolut p-summierende Abbildungen in normierte Räumen, *Studia Math.* 28 pp.333-353
[1978] *Operator ideals*, VEB Deutscher Verlag der Wissenschaften, Berlin
[1986] Eigenvalues of absolutely r-summing operators, in *Aspects of mathematics and its applications*, ed. J. A. Barroso, Elsevier, Amsterdam pp.607-617
[1987] *Eigenvalues and s-numbers*, Akademische Verlagsgesellschaft Geest & Portig K.-G., Leipzig

Pigno L. and Saeki S.
[1973] Measures whose transforms vanish at infinity, *Bull. Amer. Math. Soc.* 79 $N^o 3$ pp.800-801

Pisier G.
[1974] Sur les espaces qui ne contiennent pas de l_n^1 uniformément, *Séminaire Maurey-Schwartz 1973-74*, exposé VII
[1978] Some results on Banach spaces without local unconditional structure, *Compositio Math.* 37 pp.3-19
[1983] Counterexamples to a conjecture of Grothendieck, *Acta Math.* 151, pp.181-208

[1986] *Factorisation of linear operators and geometry of Banach spaces*, CBMS Regional Conference Series N° 60, Amer. Math. Soc., Providence R.I.

[1986a] Factorisation of operators through $L_{p,\infty}$ or $L_{p,1}$ and non-commutative generalisations, *Math. Ann.* 276 pp.105-136

[P] *The volume of convex bodies and Banach space geometry*, Cambridge Tracts in Math., 94, Cambridge University Press, 1989

[P1] Factorisation of operator-valued analytic functions, *Advances in Math.* 93 N^o1 (1992) pp.61-125

Privalov Al. A.

[1987] On increase of degrees of polynomial bases and approximation of trigonometric projections, *Mat. Zametki* 42 N^o2 pp.207-214 (in Russian)

Rademacher H.

[1922] Einige Sätze über Reihen allgemeinen Orthogonalfunktionen, *Math. Ann.* 87 pp.111-138

Rao M. M.

[1982] Harmonisable processes; Structure theory, *L'Enseignement mathématique*, XXVIII (3-4) pp.295-351

Ringrose J. R.

[1959] A note on uniformly convex spaces, *J. London Math. Soc.* 34 p.92

Rosenthal H. P.

[1966] Projections onto translation-invariant subspaces of $L_p(G)$, *Memoirs Amer. Math. Soc.* 63

[1973] On subspaces of L_p, *Annals of Math.* 97 N^o2 pp.344-373

[1976] Some applications of p-summing operators to Banach space theory, *Studia Math.* 58 pp.21-43

Rudin W.

[1956] Boundary values of continuous analytic functions, *Proc. Amer. Math. Soc.* 7 pp.808-811

[1962] Projections onto invariant subspaces, *Proc. Amer. Math. Soc.* 13 pp.429-432

[1962a] *Fourier analysis on groups*, Interscience Publishers, New York

[1969] *Function theory in polydiscs*, Benjamin, New York

[1973] *Functional analysis*, McGraw-Hill Inc.
[1975] Spaces of type $H^\infty + C$, *Annales de l'Institut Fourier* XXV N^o1, pp.99-125
[1980] *Function theory in the unit ball of C^n*, Springer Verlag, Berlin
[1986] *New constructions of functions holomorphic in the unit ball of C^n*, CBMS Regional Conference Series 63, Amer. Math. Soc. Providence R.I.

Rutovitz D.
[1965] Some parameters associated with finite dimensional Banach spaces, *J. London Math. Soc.* 40 pp.241-255

Ryll J. and Wojtaszczyk P.
[1983] On homogenous polynomials on a complex ball, *Trans. Amer. Math. Soc.* 276 N^o1 pp.107-116

Sarason D.
[1979] *Function theory on the unit circle*, Virginia Poly. Inst. and State University, Blacksburg, Virginia

Schatten R.
[1950] *A theory of cross spaces*, Annals of Mathematics Studies N^o26, Princeton N.J.

Schauder M. J.
[1927] Zur Theorie stetige Abblidungen in Funktionalräumen, *Math. Zeit.* 26 pp.47-67
[1928] Einige Eigenschaft der Haarschen Orthogonalsystems, *Math. Zeit.* 28 pp.317-320

Schreier J.
[1930] Ein Gegenbeispiel zur Theorie der schwache Konvergenz, *Studia Math.* 2 pp.58-62

Schur I.
[1921] Über lineare Transformationen in der Theorie der unendlichen Reihen, *J. Reine Angew. Math.* 151 pp.79-111

Schwartz L.
[1970] Probabilités cylindriques et applications radonifiantes, *J. Fac. Sci. Univ. Tokyo Sec. 1A* 18 N^o2 pp.139-286

Semadeni Z.
[1971] *Banach spaces of continuous functions*, Polish Scientific
Publishers, Warszawa

Shields A. L. and Williams D. L.
[1971] Bounded projections, duality and multipliers in spaces of
analytic functions, *Trans. Amer. Math. Soc.* 162 pp.287-302

Simon B.
[1979] *Trace ideals and their applications*, London Math. Soc.
Lecture Notes 35, Cambridge University Press

Singer I.
[1970] *Bases in Banach spaces I*, Springer Verlag, Berlin
[1981] *Bases in Banach spaces II*, Springer Verlag, Berlin

Smith B.
[1983] Two trigonometric designs: one sided Riesz products and
Littlewood products, in "General inequalities 3, Oberwolfach
1981" *Internat. Schriftenreihe Numer.Math. 64 Birkhäuser.
Basel* pp.141-148

Šmulian V.
[1939] Linear topological spaces and their connection with Banach
spaces, *Doklady AN SSSR* 22 N^o8 pp.471-473 (in Russian)
[1940] Über lineare topologische Räume, *Mat. Sbornik* 7(49)
pp.425-448

Sobczyk A.
[1941] Projection of the space m onto its subspace c_o, *Bull. Amer.
Math. Soc.* 47 pp.938-947

Stein E. M.
[1961] On limits of sequences of operators, *Annals of Math.* 74
pp.140-170
[1970] *Singular integrals and differentiability properties of functions*,
Princeton University Press, Princeton N.J.

Stein E. M. and Weiss G.
[1971] *Introduction to Fourier analysis on Euclidean spaces*,
Princeton University Press, Princeton N.J.

Steinhaus H.
   [1919] Additive und stetige Funktionaloperationen, *Math. Zeit.* 5
   pp.186-221

Steinitz E.
   [1913] Bedingt konvergente Reihen und konvexe Systeme, *J. Reine
   Angew. Math.* 143 pp.128-175

Stone M. H.
   [1937] Applications of the theory of Boolean rings to general
   topology, *Trans. Amer. Math. Soc.* 41 pp.375-481

Strömberg J. O.
   [1983] A modified Franklin system and higher order spline systems
   on R^n as unconditional bases for Hardy spaces, in *Conference
   on harmonic analysis in honor of Antoni Zygmund.* ed W.
   Beckner, A. P. Calderon, R. Fefferman, P. W. Jones,
   Wadsworth, Belmont CA, pp.475-494

Szarek S. J.
   [1987] A Banach space without a basis which has the bounded
   approximation property, *Acta Math.* 159 N^o1-2 pp.81-98

Szasz O.
   [1922] Über den Konvergenzexponenten der Fourierischen Reihen
   gewisser Funktionenklassen, *S-B Math. Phys. Kl.* pp.135-150

Szlenk W.
   [1965] Sur les suites faiblement convergentes dans l'espace L, *Studia
   Math.* 25 pp.337-341

Tomczak-Jaegermann N.
   [1974] The moduli of smoothness and convexity and the Rademacher
   averages of trace classes S_p $(1 \le p < \infty)$, *Studia Math.* 50
   pp.163-182
   [1989] Banach-Mazur distances and finite dimensional operator
   ideals, *Pitman Monographs and Surveys in Pure and Appl.
   Math.*

Torchinsky A.
   [1986] *Real-variable methods in harmonic analysis,* Academic Press

Triebel H.

[1978] *Interpolation theory, function spaces, differential operators,*
VEB Deutscher Verlag der Wissenschaften, Berlin &
North-Holland Publishing Company

Ullrich D.C.

[1988] Khinchin's inequality and the zeros of Bloch functions, *Duke
Math. J.* 52 $N°2$ pp.519-535

[1988a] A Bloch function in the ball with no radial limits, *Bull.
London Math. Soc.* 20 pp.337-41

[P] Radial divergence in BMOA *Proc. London Math. Soc.* 68
$N°1$ (1994) pp.145-160

Varopoulos N.

[1966] Sets of multiplicity in locally compact abelian groups,
Annales Institut Fourier, 16 $N°2$ pp.123-158

[1975] A theorem on operator algebras, *Math. Scand.* 37 pp.173-182

[1976] Une remarque sur les ensembles de Helson, *Duke Math. J.* 43
pp.387-390

Veech W. A.

[1971] Short proof of Sobczyk's theorem, *Proc. Amer. Math. Soc.*
28 pp.627-628

Vinogradov S. A.

[1970] Interpolation theorems of Banach-Rudin-Carleson and norms
of embedding operators for some classes of analytic functions,
Zapiski Nauč. Sem. LOMI 19, Issl. po lin. op. i teorii
funktsii 1, pp.6-55 (in Russian)

Voskanian V.V.

[1973] On a projection constant, *Teoria Funk. Funkts. Anal. Priloz.,*
Kharkov 17 pp.183-187 (in Russian)

Whitley R.

[1966] Projecting m onto c_o, *Amer. Math. Monthly* 73 $N°3$
pp.285-286

[1967] An elementary proof of the Eberlein-Šmulian theorem, *Math.
Annalen* 172 pp.116-118

Wojtaszczyk P.
[1979] Decompositions of H_p-spaces, *Duke Math. J.* 46 N^o3 pp.635-644
[1979a] On projections in spaces of bounded analytic functions, with applications, *Studia Math.* 65 pp.147-173
[1982] On functions in the ball algebra, *Proc. Amer. Math. Soc.* 85 N^o2 pp.184-186
[1984] H_p-spaces, $p \leq 1$ and spline systems, *Studia Math.* 77 pp.289-320
[1986] On values of homogenous polynomials in discrete sets of points, *Studia Math.* 84 pp.97-104
[1986a] Two remarks on the Franklin system, *Proc. Edinburgh Math. Soc.* 29 pp.329-333
[1988] Smoothness of functions and Fourier coefficients; a functional analysts approach, *Math. Proc. Cambridge Phil. Soc.* 103 N^o1 pp.117-127
[P] On multipliers into Bergman spaces and Nevanlinna class, *Canadian Math. Bull.* 33 N^o2 (1990) pp.151-161

Wolniewicz T.
[1980] On the Banach-Mazur distance between annulus algebras, *Bull. Acad. Pol. Sci.* 28 N^o3-4 pp.125-130

Wulbert D. E.
[1968] Convergence of operators and Korovkin's theorem, *J. Approx. Theory* 1 pp.381-390

Żelazko W.
[1973] *Banach algebras*, Polish Scientific Publishers, Warszawa

Zygmund A.
[1968] *Trigonometric series vol I and II*, second edition, Cambridge University Press

Index

This index contains concepts and results which appear in the main text of this book. A name appears here only if it is commonly associated with a concept or a result. The Exercises are indexed only when a new concept (or definition) is introduced there.

Printed in the United States
By Bookmasters